KINETICS OF
ENZYME ACTION

KINETICS OF ENZYME ACTION
Essential Principles for Drug Hunters

Ross L. Stein

A JOHN WILEY & SONS, INC., PUBLICATION

Published by John Wiley & Sons, Inc., Hoboken, New Jersey
Published simultaneously in Canada

For general information on our other products and services or for technical support, please contact our Customer Care Department within the United States at (800) 762-2974, outside the United States at (317) 572-3993 or fax (317) 572-4002.

Wiley also publishes its books in a variety of electronic formats. Some content that appears in print may not be available in electronic formats. For more information about Wiley products, visit our web site at www.wiley.com.

Library of Congress Cataloging-in-Publication Data
ISBN 978-0-470-41411-8

Printed in the United States of America

eISBN 9781118084397
oIBSN 9781118084410
ePub ISBN 9781118084403

10 9 8 7 6 5 4 3 2 1

*To my children, Sarah and Zak, and
my three mentors, Mike Matta, Gene Cordes, and Dick Schowen.*

CONTENTS

Preface   xi

1 INTRODUCTION   1
1.1 A Brief History of Enzymology   2
1.2 Goal of Enzymology: The Elucidation of Mechanism   11
1.3 The Emergence of Mechanism from Data   13

2 KINETICS OF SINGLE-SUBSTRATE ENZYMATIC REACTIONS   19
2.1 The Dependence of Initial Velocity on Substrate Concentration
     and the Requirement for an E : S Complex   19
2.2 Derivation of the Rate Equation for a Single-Substrate Reaction
     with the Assumption of Rapid Equilibrium   21
2.3 Derivation of Rate Equations Using the Steady-State Assumption   24
2.4 Methods of Enzyme Assay   26
2.5 Enzyme Kinetics Practicum: Assay Development,
     Experimental Design, Data Collection, and Data Analysis   32

3 KINETICS OF SINGLE-SUBSTRATE ENZYMATIC
REACTIONS: SPECIAL TOPICS   41
3.1 Transition State Theory and Free Energy Diagrams   41
3.2 Kinetic Consequences of an Enzyme : Substrate Complex   46
3.3 Reactions with More Than One Intermediary Complex   51
3.4 Deviations from Michaelis–Menten Kinetics   53
3.5 Kinetics of Enzymatic Action on Substrates with
     Multiple Reactive Centers   65

4 ENZYME INHIBITION: THE PHENOMENON AND
MECHANISM-INDEPENDENT ANALYSIS   73
4.1 Enzyme Inhibition: The Phenomenon   74
4.2 Enzyme Inhibition: The First Quantitative Steps   76

4.3   Enzyme-Inhibitor Systems Misbehaving   77
4.4   Case Studies   83

**5 KINETIC MECHANISM OF INHIBITION OF ONE-SUBSTRATE
ENZYMATIC REACTIONS   89**
5.1   Importance in Drug Discovery   89
5.2   Theoretical Considerations   90
5.3   Analysis of Initial Velocity Data for Enzyme Inhibition   96
5.4   Inhibition of One-Substrate, Two-Intermediate Reactions   108
5.5   Inhibition by Depletion of Substrate   111

**6 TIGHT-BINDING, SLOW-BINDING, AND IRREVERSIBLE
INHIBITION   115**
6.1   Importance in Drug Discovery   115
6.2   Tight-Binding Inhibition   116
6.3   Slow-Binding Inhibition   122
6.4   Irreversible Inhibition   137

7 KINETICS OF TWO-SUBSTRATE ENZYMATIC REACTIONS   141
7.1   Importance in Drug Discovery   141
7.2   Basic Mechanisms   142
7.3   Conceptual Understanding of Sequential
       Mechanisms   143
7.4   Derivation of Rate Equations for Sequential
       Mechanisms   147
7.5   Ping-Pong Mechanisms   153
7.6   Determining the Kinetic Mechanism for Two-Substrate Reactions 156
7.7   A Conceptual Understanding of the Shapes of Secondary Plots   163
7.8   Mistaken Identity: Rapid Equilibrium Random Versus
       Steady-State Ordered   168

**8 KINETIC MECHANISM OF INHIBITION OF TWO-SUBSTRATE
ENZYMATIC REACTIONS   169**
8.1   Importance in Drug Discovery   170
8.2   Mechanisms of Inhibition of Two-Substrate Reactions   170
8.3   Inhibition by Substrate Analogs   185
8.4   Analysis of Sequential Reactions in which Inhibitor
       Binds to Enzyme : Product Complexes   191
8.5   Driving SAR Programs for Two-Substrate Enzymatic Reactions   196

9 ALLOSTERIC MODULATION OF ENZYME ACTIVITY   199

9.1   Mechanisms of Enzyme Modulation   201
9.2   Kinetics of Allosteric Modulation   202
9.3   Meaning of β and γ 208
9.4   Case Studies: Dependence of Allosteric Modulation on
      Structural Features of the Substrate   212

10 KINETICS-BASED PROBES OF MECHANISM   219

10.1 pH Dependence of Enzymatic Reactions   220
10.2 Temperature Dependence of Enzymatic Reactions   229
10.3 Viscosity Dependence of Enzymatic Reactions   235
10.4 Kinetic Isotope Effects on Enzyme-Catalyzed Reactions   239

APPENDIX A BASIC PRINCIPLES OF CHEMICAL KINETICS   251

A.1   One-Step, Irreversible, Unimolecular Reactions   252
A.2   One-Step, Irreversible, Bimolecular Reactions   253
A.3   One-Step, Reversible Reactions   254
A.4   Two-Step, Irreversible Reactions   257
A.5   Two-Step Reaction, with Reversible First Step   259

APPENDIX B TRANSITION STATE THEORY AND ENZYMOLOGY:
ENZYME CATALYTIC POWER AND INHIBITOR DESIGN   263

B.1   Catalytic Power of Enzymes   263
B.2   Transition State Analog Inhibition   268

APPENDIX C SELECTING SUBSTRATE CONCENTRATIONS
FOR HIGH-THROUGHPUT SCREENS   275

C.1   Balancing the Steady State for One-Substrate Reactions   276
C.2   Balancing the Steady State for Two-Substrate, Rapid
      Equilibrium-Ordered Enzymatic Reactions   276
C.3   Balancing the Steady State for Two-Substrate, Rapid
      Equilibrium Random Enzymatic Reactions   279
C.4   Balancing the Steady State for Nonequilibrium Enzymatic
      Reactions Involving a Second Steady-State Intermediate   282
C.5   Balancing the Steady State for Two-Substrate, Ping-Pong
      Enzymatic Reactions   283

Index   287

PREFACE

All too often, textbooks of enzymology present enzyme kinetics simply as a catalogue of standard mechanisms and their rate laws. Little effort is made to give the reader what one might call a "gut-feeling" for the connection between a mechanism and its rate law; that is, how a mechanism is reflected in the form of its rate law, and how a rate law predicts the behavior of its corresponding mechanism. Furthermore, these kinetic treatments seldom provide adequate conceptual resources for the analysis of real-world enzyme kinetic data, which frequently reflect underlying mechanisms that deviate from standard mechanisms. In this book, I aim to develop a distinctly different treatment of enzyme kinetics.

The two overarching goals of this book are to present (1) theoretical aspects of enzyme kinetics within a conceptual framework that underscores the connections between mechanism and rate law and (2) practical aspects of enzyme kinetics with a focus on mechanism-independent methods of data analysis.

An underlying premise of this book is that the conceptual understanding of an enzyme mechanism, independent of any mathematical treatment, allows a more informed use of the mechanism's rate law, for both experimental design and data analysis. Such an understanding can be achieved through the construction and analysis of free energy diagrams for enzymatic mechanisms. This approach is used throughout this book to provide conceptual underpinnings for single-substrate reactions with one or two intermediate species, enzyme inhibition, and mechanisms of two-substrate reactions.

Analysis of enzyme kinetic data is presented in this book as a process based on a mechanism-independent approach, the goal of which is to provide an alternative to the standard analytical approach in which mechanisms are imposed on a data set. Instead, an analytical method is presented that allows the mechanism to emerge from the data. Significantly, this approach reveals departures from non-ideal behavior in enzyme kinetic data that reflect non-standard mechanisms.

Another goal of this book is to provide a resource for practitioners of enzymology who work in early stage drug discovery. So, throughout this book, the theoretical and practical principles just described are presented in the context of drug discovery, using examples and case studies drawn from the literature.

Finally, it is my hope that this book will provide enzymologists, at all stages of their careers, with a new way of thinking about the theory and practice of enzyme kinetics.

ROSS L. STEIN

1

INTRODUCTION

Enzymes are effectors of chemical change. Through their action, enzymes bring about the transformation of one chemical substance into another. Like all change, the chemical change brought about by enzymes involves a temporal aspect that can be expressed as the rate at which the change occurs. The systematic study of these rates defines the biochemical field known as enzyme kinetics.

In this book, I describe the various approaches that are used to study the kinetics of enzyme catalysis and inhibition. The chapters of this book are arranged in order of increasing complexity of the system under study, moving from single substrate enzymatic reactions and their inhibition, to two substrate reactions and their inhibition. Along the way are chapters devoted to special areas, such as the utility and construction of free energy diagrams, kinetics of multi-intermediate reactions, and the kinetic analysis of tight-binding and time-dependent inhibitors. While this book should be useful to any investigator involved in kinetic examinations of enzymatic reactions, it is aimed at those involved in drug discovery research, where kinetic characterization is fundamental to drug discovery programs that have enzymes as their therapeutic target.

In this introductory chapter, I discuss three topics that set the stage for the remainder of the book. I first provide a historical context for our subsequent discussions of

Kinetics of Enzyme Action: Essential Principles for Drug Hunters, First Edition. Ross L. Stein.
© 2011 John Wiley & Sons, Inc. Published 2011 by John Wiley & Sons, Inc.

enzyme kinetics. Here we will discuss the historical connections between enzymology and other branches of chemistry, early developments in the kinetics of enzymatic reactions, and several aspects of contemporary enzymology. In the second section of this chapter, we dissect the concept of "mechanism of action" into component parts, or "sub-mechanisms," and see what is actually involved in elucidating an enzyme's mechanism. Finally, in the third section, we discuss how an accurate description of enzymatic mechanisms can emerge from kinetic data.

1.1   A BRIEF HISTORY OF ENZYMOLOGY

The history I present below is incomplete; space does not permit a full historical account of enzymology. Rather, this summary was written to give the reader some sense of how enzymology developed into the sophisticated, quantitative science it has now become. More often than not, I let the many pioneering figures of enzymology speak for themselves in quotations drawn from their original works.

1.1.1   A History of the Interplay between Organic and Biochemistry

Since its beginning in the closing decades of the nineteenth century, enzymology has developed in close relation with both biochemistry and organic chemistry. Advances in organic chemistry have informed and enriched both biochemistry and enzymology, and there have always been strong synergistic interactions between enzymology and biochemistry. Organic and biochemistry are, of course, subdivisions that grew out of chemistry itself.

Chemistry is the most ancient of the special sciences, with origins that can be traced to the dye and perfume makers of Mesopotamia, Egypt, India, and China of the third millennia B.C., and before that to metallurgists of prehistory. Following them were alchemists, who, in their search for the Philosopher's Stone, flourished from the beginning of the common era on into medieval times. It was not until Robert Boyle published his *The Sceptical Chymist* in 1661 that the distinction between alchemy and chemistry was clearly articulated, the latter relying on the "scientific method" and inductive logic as laid out by Francis Bacon in the late sixteenth century. However, the birth of chemistry is usually dated to Antoine Lavoisier's discovery in 1783 of the law of conservation of mass that refuted and laid to rest the phlogiston theory of combustion.

Our chief concern in this section is with two subdivisions of chemistry, the allied fields of organic and biochemistry. The origin of organic chemistry is usually said to be Friedrich Wöhler's synthesis of urea in 1828. This simple laboratory procedure produced the principal organic component of urine from the inorganic salt ammonium cyanate, and in one stroke dispelled the vitalistic notion that only living organisms have the ability to produce organic substances. Fully understanding the significance of his accomplishment, Wöhler exclaimed to his mentor, the Swedish chemist Jöns Jakob Berzelius: "I can no longer, so to speak, hold my chemical water and must tell you that I can make urea without needing a kidney, whether of man or dog."

Wöhler's landmark 1828 synthesis not only triggered a wave of interest in organic chemistry, especially in Germany, but also motivated scientists throughout Europe to study the chemical basis of biological phenomenon. This new field of study would eventually become known as "biochemistry," a term that would not exist until 1903, when it was coined by German chemist Carl Neuber. The true beginning of biochemistry can be traced to the 1833 studies of French chemist Anselme Payen that resulted in the production of barley extracts that contained heat-labile components with the remarkable ability to convert starch into sugar. Previous to Payen's studies, it was thought that activities such as these could occur only in intact organisms, such as the grain berry itself. The extracts that Payen investigated were called "diastase," which we know now to be a mixture of related amylase enzymes.

These early studies marked the beginnings of organic chemistry and biochemistry and were characterized by investigations of the macroscopic. During the infancy of biochemistry, we hear of "substances" and "factors" with no mention yet of molecules. For example, in the passage below from a publication that appeared in 1827, British physician and chemist William Prout first introduces and then goes on to describe results of his studies that were aimed at understanding the "organized bodies" (i.e., starch, fat, and protein droplets and globules) that initially form during digestion of food stuffs and serve as the "principal alimentary matters" used by animals to extract nourishment.

> The subject of digestion had for a long time occupied my particular attention: and by degrees I had come to the conclusion, that the principal alimentary matters employed by man, and the more perfect animals, might be reduced to three great classes, namely, the *saccharine* [starch], the *oily* [fat], and the *albuminous* [protein]: hence, it was determined to investigate these in the first place, and their exact composition being ascertained, to inquire afterwards into the changes induced in them by the action of the stomach and other organs during the subsequent processes of assimilation. . . . It was known from the very infancy of chemistry, that all organized bodies, besides the elements of which they are essentially composed, contain minute quantities of different foreign bodies, such as the earthy and alkaline salts, iron, etc. These have usually been considered as mere mechanical mixtures accidentally present; but I can by no means subscribe to this opinion. Indeed, much attention to this subject for many years past has satisfied me that they perform the most important functions; in short, that organization cannot take place without them. . . . Thus, starch I consider as *merorganized* sugar, the two substances having the same essential composition, but the starch differing from sugar by containing minute portions of other matters, which we may presume, prevent its constituent particles from arranging themselves into the crystalline form, and thus cause it to assume different sensible properties. (Prout 1827; italics in the original)

We see here not a hint of Prout using molecular theory to describe the macromolecules (i.e., carbohydrates, lipids, and proteins) that concern him.

Similarly, German chemist Moritz Traube explained fermentation and related processes in terms of substances and not molecules: "The putrefaction and decay ferments are definite chemical compounds arising from the reaction of the protein substances with water, arising thus from a chemical process" (Traube 1858a). We see here that while Traub recognized that the underlying bases of fermentation and putrefaction were chemical, he did not describe the results of his experiments in molecular terms as the chemical transformation of molecules.

Like Traub, many chemists of the mid-nineteenth century were reluctant to accept the existence of entities that that they could not see with their own eyes. This reluctance was despite the fact that Amedeo Avogadro first proposed the existence of molecules in 1811 and the enormous inroads that Friedrich Kekulé made from 1850–1870 into the understanding of carbon's multivalency and thus its ability to form complex structure; that is, "molecules."

It would not be until the beginning of the twentieth century that molecular theory, finally embraced by most organic chemists during the closing decades of the nineteenth century, would become part of the interpretational apparatus of biochemical studies. In 1902, Franz Hofmeister reports that "the protein *molecule* is mainly built-up from amino acids" (Hofmeister 1902; italics mine). Three years later, a paper appeared in first volume of the American publication *Journal of Biological Chemistry* by biochemist Phoebus A. T. Levene that extended Hofmeister's observations to try to understand how differences in amino acid composition of various proteins render them more or less susceptible to degradation by the action of trypsin and other digestive enzymes of the gut. Levene remarks that "polypeptides composed of the lower amino-acids are decomposed by trypsin less readily than polypeptides containing in their *molecule* the higher acids" (Levene 1905; italics mine).

Organic chemistry and biochemistry were rapidly becoming molecular sciences. During the twentieth century, organic chemistry would develop in many directions, spawning a host of subspecialties (e.g., synthetic organic chemistry, physical organic chemistry, organo-metallic chemistry) and entire industries (e.g., polymers, pharmaceuticals, petroleum products). At the same time, biochemists would unravel the intricacies of the many metabolic pathways that comprise cellular physiology, and probe the structure and function of the principal macromolecules of all living organisms—DNA, RNA, proteins, carbohydrates, and lipids. The history of the development of organic chemistry and biochemistry in the twentieth century is a fascinating story, in which the two disciplines at times separate to only merge again several years later in their intertwined and symbiotic relationship.

1.1.2 Early Developments in the Quantitative Study Enzyme-Catalyzed Reactions: Kinetics, Catalysis, and Inhibition

Even in its infancy, enzymology possessed a quantitative aspect that organic chemistry and biochemistry largely lacked. A key concept that allowed the development of enzymology as a quantitative science was the idea that enzymes are chemical in nature, "definite chemical compounds" (Traube 1858b). Wilhelm Kuhne, who coined the term *enzyme*, explained that enzymatic reactions are "simple chemical changes" and that enzyme "activity can occur without the presence of the organisms and outside the latter" (Kuhne 1877). Buchner, in his studies of fermentation, insisted that "an apparatus as complicated as the yeast cell is not required to institute the fermenting process" (Buchner

1897). He also had the insight that "the carrier of the fermenting activity of the press juice must be a dissolved substance, undoubtedly a protein." (Buchner 1897).

It is interesting to note that even in the face of these advances toward establishing the chemical nature of enzymes, there persisted the sense that enzymes must still possess, in some manner, the "vital force" of the organism from which they were extracted. For example, in 1901, Joseph Kastle published a paper in *Science* entitled "On the Vital Activity of the Enzymes," where he concluded that "the enzymes are active in the same sense of retaining certain of the vital activities of the living cell" (Kastle 1901).

1.1.2.1 Chemical Kinetics, the Concept of the Active Site, and Enzyme Kinetics.

One of the goals of enzymology, both then and now, is to establish quantitative and predictive relationships between reaction velocities and experimental variables, such as enzyme and substrate concentration. Methods to accurately measure the rates of enzymatic reactions and the sense that these measurements could be made reproducibly and lead to testable hypotheses concerning how reaction rates depend on experimental variables, grew out of the rapidly evolving field of chemical kinetics.

Chemical kinetics was born with Ludwig Wilhelmy's 1850 publication on the kinetics on the acid-catalyzed hydrolysis of sucrose (Wilhelmy 1850). In this publication, he not only provided methodology for accurate measurement of this reaction, but also set forth the important hypothesis that the rates of chemical reaction follow "general laws of nature."

> It is known that the action of acids on cane sugar converts it into fruit sugar, which rotates the plane [of polarized light] to the left. Since readings of how far this change has proceeded can be made with great ease, it seemed to me to offer the possibility of finding the laws of this process. . . . This is certainly only one member of a greater series of phenomena which all follow general laws of nature. (Wilhelmy 1850)

The methods used by Wilhelmy in his kinetic studies, relying on polarimetry, would be used in subsequent years to follow the invertase-catalyzed hydrolysis of sucrose. But even more important than providing new methodologies with which to study enzymatic reactions, studies such as these in chemical kinetics provided the theoretical underpinnings for enzyme kinetics.

A key to the development of kinetic methods to treat enzymatic reactions, and an important advance toward modern concepts of enzyme mechanism, was the recognition that enzymes must form complexes with their substrates prior to chemical transformation of the substrate. This was first articulated by Emil Fischer, with his now famous "lock-and-key" model (Fischer 1894).

Fischer had observed that the enzyme invertase, which he had isolated from brewer's yeast, hydrolyzes α-glucosides but not β-glucosides, while the related enzyme emulsin hydrolyzes β-glucosides but not α-glucosides. Reflecting on these observations, Fischer speculated that "enzyme and glucoside have to fit to each other like lock and key in order to exert a chemical effect on each other" (Fischer 1894). This was a profound insight that accounted not only for the stereoselectivity that these enzymes possess, but also led to a construal of enzyme action in which a snug and intimate fit between enzyme and substrate is a necessary requirement for catalysis.

Studies of lipase led Henry Dakin to similar conclusions (Dakin 1903). Dakin found that esters of racemic mandelic acid were hydrolyzed by pig liver lipase in a stereospecific manner, resulting in the production of D-mandelic acid, with little of the L-isomer being produced. He concluded that lipase is a "powerfully optically active substance" and that "actual combination takes place between the enzyme and with the ester undergoing hydrolysis" (Dakin 1903). In trying to account for the observed differences in hydrolysis rates of D- and L-mandelate esters, Dakin reasoned that when esters of D-mandelic acid combine with an optically active lipase molecule, it forms a diastereomer that is chemically more reactive than the diastereomer formed upon combination of the enzyme with esters of L-mandelic acid. He stated: "Since the additive compounds thus formed in the case of the dextro and levo components of the ester would not be optical opposites, they may be decomposed with unequal velocity, and thus account for the liberation of optically active mandelic acid" (Dakin 1903).

These concepts concerning mandatory formation of enzyme : substrate complexes were tied to the kinetic behavior of enzymes by Adrian Brown. In his 1902 paper, Brown proposes "that one molecule of an enzyme combines with one molecule of a reacting substance, and that the compound molecule exists for a brief interval of time during the further actions which end in disruption and change" (Brown 1902). He went on to describe what we now refer to as "saturation kinetics."

> If a constant amount of enzyme is in the presence of varying quantities of a reacting substance, and in all cases the quantity of reacting substance ensures a greater number of molecular collisions in unit time than the possible number of molecular changes, then a constant amount of substance will be changed in unit time in all the actions. (Brown 1902)

Brown's hypothesis was restated in the landmark paper of Leoner Michaelis and Maud Menten (Michaelis and Menten 1913). Their studies of invertase resulted in the proposal that "the rate of breakdown at any moment is proportional to the concentration of the sucrose-invertase compound; and the concentration of this compound at any moment is determined by the concentration of the ferment and of the sucrose" (Michaelis and Menten 1913). What was shown to be true of invertase, is true for all enzymes, that velocities of enzyme-catalyzed reactions are proportional to the concentration of the enzyme : substrate complex, where the constant of proportionality is k_{cat}, the first-order rate constant for decomposition of the complex into enzyme and products.

From the concept of the catalytically active enzyme : substrate complex emerged the idea that substrates bind at a specific site on the enzyme, and it is the chemistry of this site that leads to transformation of substrate into product. One of the first enzymologist to articulate this view was George Falk who explained that "some definite grouping in the complex enzyme molecule were responsible for a given enzyme action. . . . The problem therefore resolves itself into a study of the chemical nature of this grouping" (Falk 1918).

A decade later, Barnet Woolf would propose the "addition compound theory of enzyme action":

> An enzyme is a definite chemical compound which is able to form an unstable addition compound with all its substrates, each at its own specific combining group in the enzyme molecule. The process of catalysis then consists of a series of tautomeric changes in the enzyme-substrate complex, as a result of which, in a certain proportion of cases, the complex is able to dissociate into free enzyme plus the products of the catalyzed reaction. (Woolf 1931)

Woolf's hypothesis gives us an early view of the contemporary concept of the "active-site," and how it is the seat of chemical change and catalysis.

1.1.2.2 Enzyme Catalysis.
To this point, we have not spoken of the catalytic nature of enzymes. By the opening years of the twentieth century, catalysis was a well-known phenomenon in chemistry that originated in 1836, with the insights of Jacob Berzelius. In describing "a new force for developing chemical activity," he named it "the *catalytic force* of bodies and the breakdown caused by it *catalysis*" (Berzelius 1836). Berzelius explained that the "catalytic force appears to consist intrinsically in this: that bodies through their mere presence may awaken affinities slumbering at this temperature. So that as a result of this the elements in a complex body arrange themselves in altered relations."

Years later, in 1894, Wilhelm Ostwald would provide a more detailed description of catalysis based on chemical energetics:

> Catalysis is the acceleration of a chemical reaction, which proceeds slowly, by the presence of a foreign substance. . . . There are numerous substances or combinations of substances which in themselves are not stable but undergo slow change and only seem stable to us because their changes occur so slowly that during the usual short period of observation they do not strike us. Such substances or systems often attain an increased reaction rate if certain foreign substances, that is, substances which are not in themselves necessary for the reaction, are added. This acceleration occurs without alteration of the general energy relations, since after the end of the reaction the foreign body can again be separated from the field of the reaction. (Ostwald 1894)

In this statement, Oswald explained one of the central principals of chemical catalysis, that the catalyst can only influence rate, not the free energy difference between the reactant and product.

These ideas help answer what had become one of the principal questions in enzymology: How is it that "a very small amount of an enzyme can transform a relatively very large amount of another compound" (Loew 1899)? In trying to answer this question, Oscar Loew likened enzymes to "machines [that] transform heat into chemical action," this transformation involving "two or even more labile [i.e., reactive] groups in one molecule of an enzyme." (Loew 1899)

> As soon as we understand the close connection between [reactivity] and activity, and that enzymes are capable of transforming heat energy into chemical energy, we

can . . . understand that their chemical energy may be transferred to other compounds. And when these other compounds are of such a character that their atoms are easily set in motion, we can further understand that by lessening certain affinities in them another grouping of atoms may result. . . . Such chemical action produced by the mere transmission of chemical energy by a certain substance, which remains chemically unaltered, are called *catalytic*. (Loew 1899)

The ideas we have been discussing, from Traube's concept that enzymes are "definite chemical substances" to Loew's ideas about the catalytic nature of enzymes to Woolf's "addition compound theory of enzyme action," are all concerned with the enzymatic transformation of reactants into products, and set the stage for our modern ventures into this area. In the course of the studies that gave rise to these ideas, investigators noted that certain substances can retard or inhibit the enzymatic reactions they were studying.

1.1.2.3 Enzyme Inhibition and the Prospect of Designing Drugs. In 1904, E. F. Armstrong reported on experiments he had conducted "with the object of ascertaining by direct observation whether and to what extent the action of a given enzyme is affected by one or more of the products formed under its influence" (Armstrong 1904). Armstrong examined the effect of certain hexoses on the rate of hydrolysis catalyzed by a number of sugar-splitting enzymes. He found that in those cases where the hexoses were decomposition products of hydrolysis they inhibited the progress of the enzymatic reaction (Armstrong 1904). Adrian Brown had made similar observations for invertase (Brown 1902).

In their studies of xanthine oxidase, Dixon and Thurlow noted that the purine bases adenine and guanine inhibited reactions catalyzed by this enzyme (Dixon and Thurlow 1924). To explain this, they conjectured that "the effect might be due to adsorption by the enzyme of the inhibitory substances, thus preventing the adsorption of one or both of the reactants." They noted that the effect was "remarkably specific. Caffeine showed no trace of inhibitory effect. The pyrimidine substances uracil, cytosine, and thymine were also tried but show no effect" (Dixon and Thurlow 1924). Herbert Coombs noted the same effect, and extended these studies to include a number of synthetic purine derivatives, developing perhaps the first structure–activity relationship for inhibition: "The complete purine skeleton—the two ring structure—is necessary for adsorption. The introduction of an amino-group strongly favors adsorption; and the introduction of methyl groups, particularly in the iminazole ring, tends to prevent adsorption" (Coombs 1927).

These studies laid the groundwork for the use of enzyme inhibitors as mechanistic probes as wells as for therapeutics (Sizer 1957). D. D. Woods summarized the latter sentiment in his introduction to a 1950 volume of the *Annals of the New York Academy of Sciences* devoted to antimetabolites (Woods 1950). He stated that there is a "general concept that substances of related structure may compete with others having physiological action" and in so doing these substances exert their physiological/pharmacological effects. In this introductory chapter, Woods illustrated this concept with the sulfonamide antibiotics, noting these these drugs compete with the structurally related bacterial metabolite *p*-aminobenzoic for enzymes essential to bacteria.

The use of enzyme inhibitors as therapeutics agents was reviewed in 1955 by F. Edmund Hunter and Oliver Lowry. They concluded their review with the following passage, whose sentiment is still expressed today:

> At the present time, most of the successful drugs are developed empirically or through synthesis of congeners of pre-existing drugs. One wishes for a rational approach which would permit the deliberate development of new drugs specifically to affect particular enzymes. There have been many attempts to develop drugs on a rational basis. Perhaps these attempts will be more successful in the near future. But it must be admitted that the total enzyme matrix of the living organism is more complex and subtle than formerly visualized. Although new information concerning enzymes and their sensitivity to inhibition is being obtained at a constantly accelerated rate, each increment of knowledge also shows the goal to be a little further away than it had appeared to be. (Hunter and Lowry 1955)

1.1.3 Contemporary Enzymology

We now possess an understanding of enzyme action that would have seemed unobtainable and utterly fantastic to the scientists that worked in the intertwined fields of organic chemistry, biochemistry, and enzymology, from the early part of the nineteenth century through the mid-twentieth century. The understanding we now possess is built upon their pioneering work, and arises from a coupling of new concepts and new technologies. In this section, we explore some of the central themes that define contemporary enzymology.

1.1.3.1 Enzyme Kinetic Theory. One of the principal advancements in enzymology was the theory developed by W. W. Cleland for handling the steady-state kinetics of enzymes with two or more substrates (Cleland 1963a; Cleland 1963b; Cleland 1963c). To apply these theories, Cleland went on to insist that enzyme kinetic data should no longer be analyzed by linearization methods (e.g., Lineweaver–Burk plots), but rather by fitting nontransformed kinetic data to rate equations by nonlinear least squares analysis (Cleland 1963d).

It was also recognized that for complex enzyme mechanisms involving multiple intermediates, the information available from steady-state kinetic analysis is limited to estimation of macroscopic constants, such as k_c and K_m, while microscopic constants, which define the interconversion of intermediates, remain largely inaccessible. To solve this problem, the analysis of pre-steady-state kinetics was needed. This was made possible through the development of the appropriate, and often quite complex, rate laws that govern the pre-steady state, and the introduction of rapid-kinetics instrumentation (Gutfreund 1975).

With the recognition that steady-state kinetic parameters k_c and k_c/K_m are often composites of several microscopic rate constants came the appreciation that more than one of these rate constants could contribute to rate limitation. This idea was captured and put to use in the analysis of isotope effects by Dexter Northrop and his "commitment to catalysis" factors (Northrop 1981), and by the elegant concept of the virtual transition state introduced by Richard Schowen (Schowen 1978).

1.1.3.2 Transition State Theory and Its Application to Problems of Enzymology. Transition state rate theory has had a profound effect on chemical thinking since its introduction in 1935 by Henry Eyring. In this theory, Eyring proposed that for the chemical transformation of a molecule to occur, the molecule must pass through a high-energy "transition state" that separates reactants from products (Eyring 1935a,b; Wynne-Jones and Eyring 1935). Eyring called the molecular species that exists in the transition state the "activated complex" and posited that it has the properties of a stable molecule except for translation along the reaction coordinate, which leads to reaction. A decade later, Linus Pauling proposed that enzyme catalysis results not because the substrate is bound tightly by enzyme, but rather because the activated complex of the reaction is bound tightly (Pauling 1946, 1948). A practical outcome of Pauling's idea was Richard Wolfenden's proposal that new types of enzyme inhibitors with extraordinary potencies might be designed based on structural properties of the transition state (Wolfenden 1969). Wolfenden's hypothesis has borne much fruit, being used for the successful design of hundreds of transition state analog inhibitors.

1.1.3.3 Kinetic Isotope Effects. Of all the tools of the physical organic chemist that have been brought to bear on problems of enzyme mechanism, the kinetic isotope effect is by far the most useful (see Chapter 10). Perhaps the first isotope effect measured for an enzyme-catalyzed reaction was the primary deuterium isotope effect for the alcohol dehydrogenase-catalyzed reduction of acetaldehyde by NADH(D) (Mahler and Douglas 1957). Two years later, a heavy atom isotope effect was measured by Westheimer, who reported a ^{13}C isotope effect for oxaloacetate decarboxylase (Seltzer et al. 1959). Since these early studies, kinetic isotope effects have been used by numerous investigators to probe the mechanisms and transition state structures for a host of enzymes (Cook 1991).

1.1.3.4 Structural Investigations of Enzymes. While the principle focus of this introduction is on kinetic and methodological aspects of the history and development of enzymology, I would be remiss if not stating the extraordinary impact that structural studies have had on our understanding of enzyme action. Among the first X-ray structures of enzymes to be published were those of lysozyme in 1965 (Johnson and Phillips 1965) and chymotrypsin in 1967 (Matthews et al. 1967). With the solution of these structures came a physical context for the interpretation of kinetic data. In the case of lysozyme, the structure of the bound glycan substrate suggested a mechanism involving formation of an oxocarbenium ion intermediate and general-based catalysis, which have since been substantiated by α-deuterium isotope effects (Dahlquist et al. 1969) and solvent isotope effects (Banerjee et al. 1975). The solved structure of chymotrypsin suggested that a grouping of active site amino acids, Ser195, His, 57, and Asp102, now known as the "charge relay system," plays a critical role in catalysis by this enzyme. The actual operation of the charge relay system has been demonstrated not only for chymotrypsin but also for a number of other serine proteases (Elrod et al. 1976, 1980; Quinn et al. 1980; Stein et al. 1987). In the years since these structures appeared, hundreds more have been solved for enzymes of all mechanistic classes.

These structures have been a great aid not only in understanding mechanism, but also in the design of inhibitors.

1.1.4 Enzyme Catalytic Power: The Final Step in Our Understanding Enzyme Action

The most intriguing aspect of enzyme action, their tremendous catalytic power, remains an open question today. While a great many hypotheses have been advanced to explain enzymatic catalysis (Jenks 1975; Fersht 1999; Bugg 2001), it seems now that the answer lies in the coupling of protein conformational isomerization to the reaction coordinate during the transformation of the Michaelis complex to product (Case and Stein 2003; Hengge and Stein 2004). While this is not a new idea (Lumry and Biltonen 1969; Careri et al. 1979; Welch et al. 1982; Somogyi et al. 1984; Welch 1986), interest in it has recently been renewed with the observation of the apparent coupling of protein motions to active site hydrogen tunneling in a number of hydrogen transfer reactions (Antoniou et al. 2002; Knapp and Klinman 2002; Hay et al. 2008).

An increased understanding of how protein dynamics is coupled to catalysis will not only boost our understanding of other enzymatic phenomenon, such as allostery, but will also have far reaching practical consequences, such as the design of synthetic enzymes, with catalytic efficiencies approaching those of natural enzymes, and new classes of inhibitors and activators.

1.2 GOAL OF ENZYMOLOGY: THE ELUCIDATION OF MECHANISM

1.2.1 Mechanism of Substrate Turnover

One of the principal goals of enzymology is to understand how enzymes transform their substrates into products. When attained, this understanding allows investigators to describe an enzyme's *mechanism of action*. Accurate descriptions of an enzyme's mechanism of action must include accounts of four underlying mechanisms:

- **Kinetic Mechanism**—the pathway of steady-state intermediates leading from substrates to products.
- **Chemical Mechanism**—the pathway of chemical transformations leading from substrates to products.
- **Dynamics Mechanism**—the protein conformational changes required for substrate turnover.
- **Catalytic Mechanism**—the means by which an enzyme effects rate acceleration.

The division of an enzyme's overall mechanism of action into these "sub"-mechanisms is, in a sense, artificial. The intermediates that populate the steady state of an enzyme-catalyzed reaction include both stable conformational isomers as well as the intermediates that form during the course of a substrate's chemical conversion into products. And the means by which an enzyme effects catalysis is dictated both by the nature of the chemical transformation and the protein dynamic potential of the enzyme. An enzyme's overall mechanism of action is, in fact, the *integration* of the four underlying mechanisms. We see then, for a mechanistic proposal to be accurate, it must incorporate information about all these mechanisms.

While perhaps artificial, this division is operationally useful, in that it allows experiments to be designed and hypotheses to be constructed that are of a limited and manageable scope. For an investigator to elucidate an enzyme's chemical mechanism, it is unnecessary to consider the catalytic power that the enzyme brings to bear on the reaction. By the same token, the protein dynamical changes that attend an enzymatic reaction can be probed and ultimately tied to catalytic power without complete knowledge of the enzyme's chemical mechanism. And, as we will see throughout this book, kinetic mechanisms can be probed and understood in the complete absence of information regarding the other three mechanisms.

The definitions given above for the four underlying mechanisms that comprise an enzyme's overall mechanism of action are qualitative in nature. However, associated with each of these mechanisms are specific parameters whose values provide a quantitative element to the mechanistic description. For example, elucidating the kinetic mechanism of a two substrate reaction does not only include determining the order of the two substrates to the enzyme, but also the equilibrium and rate constants for the addition steps. The chemical and dynamics mechanisms will not be fully elucidated until rate constants are determined for interconversion of the various chemical intermediates and conformational isomers, respectively. Finally, an understanding of the catalytic mechanism will require an estimate of the catalytic enhancement over the uncatalyzed reaction and, if possible, a quantitative breakdown of catalytic enhancement into the various elements that comprise the catalytic mechanism, including utilization of general acid–base chemistry, substrate binding energy, and energy derived from the heat sink of bulk solvent.

The integration of the underlying mechanisms mentioned above will require establishing quantitative relationships among rate constants for the kinetic, chemical, and dynamics mechanisms. More specifically, this integration will require the investigator to determine how the steady-state rate parameters relate to the rate constants for the interconversion of the various chemical intermediates and conformational isomers that attend enzymatic turnover of substrate.

1.2.2 Mechanism of Inhibition

The division of mechanism of action into various underlying mechanisms is also useful when studying enzyme inhibition. Here, the four mechanisms become

- **Kinetic Mechanism**—knowledge of the enzyme forms to which an inhibitor binds, and the pathway of intermediates leading to stable enzyme:inhibitor complexes.
- **Chemical Mechanism**—the pathway of chemical transforms, if any, required to form stable enzyme:inhibitor complexes.

- **Dynamics Mechanism**—the protein conformational changes required for formation of the various stable enzyme : inhibitor complexes.
- **Inhibitory Mechanism**—the binding-site interactions that stabilize enzyme : inhibitor complexes, and the means by which the enzyme establishes these interactions.

As above, this mechanistic "deconstruction" is a useful, albeit somewhat artificial, way to approach the study of how an enzyme inhibitor works. Any mechanistic hypothesis for how an inhibitor works that seeks to be complete must integrate information concerning all four of the sub-mechanisms.

1.2.3 Kinetic Mechanisms of Substrate Turnover and Inhibition

The focus of this book is the kinetic mechanism. For both substrate turnover and inhibition, we will derive rate equations for many standard mechanisms, discuss how to set up steady-state kinetic experiments to probe a mechanism, and then how to proceed from the experimental data to the final elucidation of the kinetic mechanism. We will discuss these topics with relatively little reference to the other "sub"-mechanisms. While a complete description of an enzyme's overall mode of action does, of course, require an understanding of the chemical, dynamics, and catalytic mechanisms, lack of this knowledge in no way hinders the determination of kinetic mechanisms.

The conceptual disengagement of kinetics from other mechanistic components has important consequences. It means that kinetic mechanistic information (e.g., the order of addition of substrates) can be properly used for various purposes (e.g., assay design) without knowledge of the chemical, dynamical, or catalytic mechanistic features.

1.3 THE EMERGENCE OF MECHANISM FROM DATA

This book is based on a short course I periodically taught for the American Chemical Society for about 3 years. Like this book, the course was aimed at scientists in the biotech and pharmaceutical industries whose jobs involved the kinetic analysis of enzymatic reactions. Despite the responsibilities of their jobs, the nearly 200 scientists who participated in this course had only the most basic understanding of enzyme kinetics. What became clear in teaching this course is that the availability of inexpensive curve fitting software designed to fit enzyme kinetic data to the rate laws for standard mechanisms had made scientists in the industry intellectually lazy.

All too frequently, investigators fit kinetic data to rate equations for several standard models, and then pronounce as winner the mechanism that resulted in the best fitting statistics. While easy and convenient, this approach can result in inaccurate descriptions of mechanism since it does not require the investigator to carefully examine primary data for subtle trends that may signal nonstandard mechanisms. And, of course, recognizing the existence of such trends is the crucial first step in the accurate description of mechanism.

What is required is an analytic method that readily allows the identification of these trends. Such a method would not *impose* a mechanism *on* the data, but rather would allow the mechanism to *emerge from* the data. The key lies in the stepwise analysis of kinetic data, where, at each step, the data are examined for departures from ideal behavior.

Data from the simplest kinetic experiments, in which initial velocities are determined as a function of a single independent variable (e.g., inhibitor or substrate concentration), require only a single step for analysis, in which the data are examined for departures from simple Michaelis–Menten kinetics. For kinetic experiments in which initial velocities are determined as a function of two independent variables (e.g., concentrations of two substrates, or concentrations of a substrate and an inhibitor), a two-step, sequential process is used:

1. Analyze the dependence of initial velocity on one independent variable, for all instances of the second independent variable.
2. Analyze the dependence of the parameters derived from the first step of analysis on the second independent variable.

For example, we will see in Chapter 5 that in determining the kinetic mechanism of an inhibitor, initial velocities are determined as a function of both substrate and inhibitor concentration. The first step in the analysis of the data set will be to plot initial velocities versus substrate concentration, at each of the inhibitor concentrations, and examine each plot for departures from simple Michaelis–Menten kinetics. In the second step of the analysis, the observed steady-state rate parameters (i.e., $(V_{max})_{obs}$ and $(V_{max}/K_m)_{obs}$) are plotted as a function of inhibitor concentration. Again, at this second stage, the two plots are examined for departures from simple inhibitor binding behavior. Finally, from these replots, one can construct a mechanism and calculate inhibition constants. At each stage of analysis, departures from ideal behavior are recognized as important mechanistic features.

While subtle departures from nonideal behavior can readily be identified using the method just described (i.e., the method of replots), they can easily be overlooked if the data are fit globally. The relative merits of global fitting versus the method of replots will be discussed at length in Chapters 5, 7, and 8, for inhibition of one substrate reactions, kinetics of two-substrate reactions, and inhibition of two-substrate reactions, respectively. In these chapters, I advocate the use of the method of replots because at each step of the analysis, departures from nonideal behavior can be identified and then incorporated into the evolving mechanistic picture.

REFERENCES

Antoniou, D., et al. (2002). "Barrier passage and protein dynamics in enzymatically catalyzed reactions." *Eur. J. Biochem.* **269**: 3102–3112.

Armstrong, E. F. (1904). "Studies on enzyme action. III. The influence of the products of change on the rate of change by sucroclastic enzymes." *Proc. Roy. Soc.* **73**: 516.

Banerjee, S. K., et al. (1975). "Reaction of N-acetylglucosamine oligosaccharides with lysozyme. Temperature, pH, and solvent deuterium isotope effects; equilbrium, steady state, and pre-steady state measurements." *J. Biol. Chem.* **250**(11): 4355–4367.

Berzelius, J. J. (1836). "Einige Ideen uber eine bei der Bildung Organischer Verbindungen in der lebenden Natur wirksame, aber bisher nicht bemerkte Kraft [A few ideas about a force active in the formation of organic compounds in living nature but hitherto not oberserved]." *Jahres-Bericht uber die Fortschritte Physischen Wissenschaften* **237**: 237–248.

Brown, A. (1902). "Enzyme action." *J. Chem. Soc.* **81**: 377–388.

Buchner, E. (1897). "Alcoholic fermentation without yeast cells." *Ber. Chem. Ges.* **30**: 117.

Bugg, T. D. H. (2001). "The development of mechanistic enyzmology in the 20th century." *Nat. Prod. Rep.* **18**: 465–493.

Careri, G., et al. (1979). "Enzyme dynamics: The statistical physics approach." *Annu. Rev. Biophys. Bioeng.* **8**: 69–97.

Case, A. and R. L. Stein (2003). "Mechanistic origins of the substrate selectivity of serine proteases." *Biochemistry* **42**: 3335–3348.

Cleland, W.W. (1963a). "The kinetics of enzyme catalyzed reactions with two or more substrates or products. I. Nomenclature and rate equations." *Biochim. Biophys. Acta* **67**: 104–137.

Cleland, W.W. (1963b). "The kinetics of enzyme catalyzed reactions with two or more substrates or products. II. Inhibition: nomenclature and theory." *Biochim. Biophys. Acta* **67**: 173–187.

Cleland, W.W. (1963c). "The kinetics of enzyme catalyzed reactions with two or more substrates or products. III. Prediction of initial velocity and inhibition patters by inspection." *Biochim. Biophys. Acta* **67**: 188–196.

Cleland, W.W. (1963d). "Computer programmes for processing enzyme kinetic data." *Nature* **198**: 463–465.

Cook, P. F. (1991). *Enzyme Mechanism from Isotope Effects.* New York, CRC Press.

Coombs, H. I. (1927). "Studies on xanthine oxidase. The specificity of the system." *Biochem. J.* **21**: 1260–1265.

Dahlquist, F. W., et al. (1969). "Application of secondary alpha-deuterium kinetic isotope effects to studies of enzyme catalysis. Glycoside hydrolysis by lysozyme and beta-glucosidase." *Biochemistry* **8**(10): 4214–4221.

Dakin, H. D. (1903). "The hydrolysis of optically inactive esters by means of enzymes." *J. Physiol.* **30**: 253–263.

Dixon, M. and S. Thurlow (1924). "Studies on xanthine oxidase." *Biochem. J.* **18**(976–988): 976–988.

Elrod, J. P., et al. (1976). "Proton bridges in enzyme catalysis." *Faraday Symp. Chem. Soc.* **10**: 145–153.

Elrod, J. P., et al. (1980). "Protonic reorganization and substrate structure in catalysis by serine proteases." *J. Am. Chem. Soc.* **102**: 3917–3922.

Eyring, H. (1935a). "The activated complex and the absolute rates of chemical reactions." *Chem. Rev.* **17**: 65–77.

Eyring, H. (1935b). "The activated complex in chemical reactions." *J. Phys. Chem.* **3**: 107–115.

Falk, K. G. (1918). "A chemical study of enzyme action." *Science* **47**: 423–429.

Fersht, A. R. (1999). *Structure and Mechanims in Protein Science—A Guide to Enzyme Catalysis and Protein Folding.* New York, W.H. Freeman and Co.

Fischer, E. (1894). "Influence of the configuration on the activity of the enzyme." *Ber. Chem. Ges.* **27**: 2985.

Gutfreund, H. (1975). *Enzymes: Physical Principles*. New York, John Wiley & Sons.

Hay, S., et al. (2008). "Atomistic insight into the origin of the temperature-dependence of kinetic isotope effects and H-tunnelling in enzyme systems is revealed through combined experimental studies and biomolecular simulation." *Biochem. Soc. Trans.* **36**(Pt 1): 16–21.

Hengge, A. C. and R. L. Stein (2004). "Role of protein conformational mobility in enzyme catalysis: Acylation of α-chymotrypsin by specific peptide substrates." *Biochemistry* **43**(3): 742–747.

Hofmeister, F. (1902). "On the structure and grouping of the protein molecule." *Ergeb. Physiol.* **1**: 759.

Hunter, F. E. and O. H. Lowry (1955). "The effects of drugs on enzyme systems." *Pharmacol. Rev.* **7**: 89–135.

Jenks, W. P. (1975). "Binding energy, specificity, and enzymatic catalysis: the Circe effect." *Adv. Enzymol. Relat. Areas Mol. Biol.* **43**: 219–410.

Johnson, L. N. and D. C. Phillips (1965). "Structure of some crystalline lysozyme-inhibitor complexes determined by X-ray analysis at 6 Angstrom resolution." *Nature* **206**: 761–762.

Kastle, J. H. (1901). "On the vital activity of the enzymes." *Science* **13**: 765–771.

Knapp, M. J. and J. P. Klinman (2002). "Environmentally coupled hydrogen tunneling—Linking catalysis to dynamics." *Eur. J. Biochem.* **269**: 3113–3121.

Kuhne, W. (1877). "Ueber das Verhalten Verschiedener Organisirter und Sog Ungeformter Fermente [On the behavior of different organized and so-called unformed ferments]." *Verhandlungen des Naturehistorisch-Medicinischen Vereins* **1**: 190.

Levene, P. A. T. (1905). "The cleavage products of proteoses." *J. Biol. Chem.* **1**: 45–58.

Loew, O. (1899). "On the chemical nature of enzymes." *Science* **10**: 955–961.

Lumry, R. and R. Biltonen (1969). "Thermodynamic and kinetic aspects of protein conformations in relation to physiological function." In *Structure and Stability of Biological Macromolecules*, S. Timasheff and G. Fasman, eds. New York, Marcel Dekker.

Mahler, H. R. and J. Douglas (1957). "Mechanisms of enzyme-catalyzed oxidation-reduction reactions. I. An investigation of the yeast alcohol dehydrogenase reaction by means of the isotope rate effect." *J. Am. Chem. Soc.* **79**: 1159–1166.

Matthews, B. W., et al. (1967). "Three-dimensional structure of tosyl-alpha-chymotrypsin." *Nature* **214**: 652–656.

Michaelis, L. and M. Menten (1913). "The kinetics of invertase action." *Biochem. Z.* **49**: 333.

Northrop, D. B. (1981). "The expression of isotope effects on enzyme-catalyzed reactions." *Annu. Rev. Biochem.* **50**: 103–131.

Ostwald, W. F. (1894). "Abstract on catalysis." *Zeitschrift Physikalische Chemie* **15**: 705–706.

Pauling, L. (1946). "Molecular architecture and biological reactions." *Chem. Eng. News* **24**: 1375–1377.

Pauling, L. (1948). "The nature of forces between large molecules of biological interest." *Proc. R Inst. G B* **34**: 181–187.

Prout, W. (1827). "On the ultimate composition of simple alimentary substances; with some preliminary remarks on the analysis of organized bodies in general." *Philos. Trans. R Soc.* **117**: 355–388.

Quinn, D. M., et al. (1980). "Protonic reorganization in catalysis by serine proteases: Acylation by small substrates." *J. Am. Chem. Soc.* **102**: 5358–5365.

Schowen, R. L. (1978). Catalytic power and transition-state stabilization. In *Transition States of Biochemical Processes*, R. D. Gandour and R. L. Schowen, eds. New York, Plenum Press: 77–114.

Seltzer, S., et al. (1959). "Isotope effects in the enzymatic decarboxylation of oxalacetic acid." *J. Am. Chem. Soc.* **81**: 4018–4024.

Sizer, I. W. (1957). "Chemical aspects of enzyme inhibition." *Science* **125**: 54–59.

Somogyi, B., et al. (1984). "The dynamic basis of energy transduction in enzymes." *Biochim. Biophys. Acta* **768**: 81–112.

Stein, R. L., et al. (1987). "Catalysis by human leukocyte elastase. 7. Proton inventory as a mechanistic probe." *Biochemistry* **26**: 1305–1314.

Traube, M. (1858a). "On the theory of fermentation and decay." *Ann. Physiol.* **103**: 331.

Traube, M. (1858b). "On the theory of fermentation and decay phenomena, also of ferment activity in general." *Annalen Physik Chemie* **103**: 331.

Welch, G. R. (1986). *The Fluctuating Enzyme*. New York, John Wiley & Sons.

Welch, G. R., et al. (1982). "The role of protein fluctuations in enzyme action: A review." *Prog. Biophys. Mol. Biol.* **39**: 109–146.

Wilhelmy, L. (1850). "The law by which the action of acid on cane sugar occurs." *Annalen Physik Chemie* **81**: 413–433.

Wolfenden, R. (1969). "Transition state analogues for enzyme catalysis." *Nature* **223**(5207): 704–705.

Woods, D. D. (1950). "Biochemical significance of the competition between p-Aminobenzoic acid and the sulphonamides." *Ann. N Y Acad. Sci.* **52**: 1199–1211.

Woolf, B. (1931). "The addition compound theory of enzyme action." *Biochem. J.* **25**: 342–348.

Wynne-Jones, W. F. K. and H. Eyring (1935). "The absolute rate of reactions in condensed phases." *J. Chem. Phys.* **3**: 492–402.

2

KINETICS OF SINGLE-SUBSTRATE ENZYMATIC REACTIONS

We begin this book with two chapters devoted to the examination of single-substrate enzymatic reactions. In the present chapter, we learn that to explain enzyme kinetic behavior, one must hypothesize the formation of a reaction intermediate, comprising substrate and enzyme, on the pathway from substrate to product. Moving from theory to practical matters, we discuss methods of enzyme assay, the design of enzyme kinetic experiments, and the collection and analysis of kinetic data. In Chapter 3, we develop a more detailed understanding of the kinetic consequences that attend the existence of the enzyme : substrate $(E : S)$ complex, and then consider enzymatic reactions that have more than this single intermediate species. The concepts of these two chapters provide the needed theoretical underpinnings for all of the topics we will explore throughout this book.[1]

2.1 THE DEPENDENCE OF INITIAL VELOCITY ON SUBSTRATE CONCENTRATION AND THE REQUIREMENT FOR AN E : S COMPLEX

The catalytic activity of an enzyme can be measured either as the time-dependent consumption of substrate or, as illustrated in the simulation of Figure 2.1, the formation

---

[1] As foundational to our study of enzyme kinetics, a review of basic kinetic principles of nonenzymatic kinetics is provided in Appendix A.

Kinetics of Enzyme Action: Essential Principles for Drug Hunters, First Edition. Ross L. Stein.
© 2011 John Wiley & Sons, Inc. Published 2011 by John Wiley & Sons, Inc.

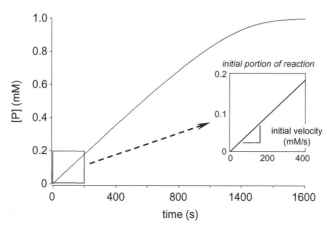

Figure 2.1. Time course for an enzyme-catalyzed reaction following simple Michaelis–Menten kinetics. Inset is an expansion of the initial part of the curve. Simulation conditions: $[E]_o = 0.1\,\mu M$, $[S]_o = 500\,\mu M$, $K_m = 100\,\mu M$, and $k_c = 100\,s^{-1}$.

of product. While full reaction progress curves, which record the formation of product until all substrate has been consumed, can be analyzed under favorable circumstances to yield kinetic parameters, it is more common in enzymology to restrict attention to the initial part of the reaction, prior to consumption of more than 5–10% substrate. In this early stage of the reaction, the dependence of product concentration on time is linear, with slope equal to the initial reaction velocity, v_o. As we will see throughout this book, mechanism emerges from an analysis of how the initial velocity depends on various reaction parameters, including substrate and inhibitor concentration. Our main concern in this chapter is to see what we can learn from the dependence of v_o on the initial substrate concentration, $[S]_o$.

For the enzyme-catalyzed conversion of S into P, v_o depends on $[S]_o$ in a complex multi-phasic manner (see Fig. 2.2). At low $[S]_o$, v_o is first order in both $[S]_o$ and $[E]_o$ and thus is linearly dependent on both $[S]_o$ and $[E]_o$. At these concentrations of substrate, the reaction is governed by the second-order rate constant k_{II}. At high $[S]_o$, v_o is still first order in $[E]_o$, but is now zero order in $[S]_o$, meaning the v_o is independent of $[S]_o$. This reaction is governed by first-order rate constant k_I. And finally, at intermediate concentration, v_o remains first order in $[E]_o$ but is now mixed order in $[S]_o$.

This sort of behavior confounded the earliest attempts of investigators to understand enzyme action in terms of simple theories of reaction kinetics. In the early years of the twentieth century, British chemist Adrian Brown had been studying the invertase-catalyzed hydrolysis of sucrose to produce glucose and fructose. From the pioneering kinetic studies of Ludwig Wilhelmy in 1850, the time course of the acid-catalyzed reaction was known to follow the simple first-order rate law:

$$[S] = [S]_o\, e^{-kt} \qquad (2.1)$$

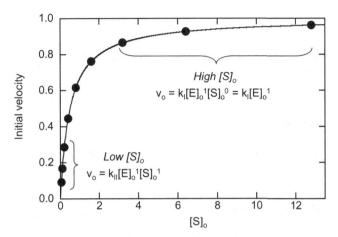

Figure 2.2. Dependence of initial velocity on $[S]_o$. At low $[S]_o$, reaction is first order in both $[E]_o$ and $[S]_o$, while at high $[S]_o$, reaction remains first order in $[E]_o$, but is zero order in $[S]_o$.

which upon differentiation yields

$$-\frac{d[S]}{dt} = k[S] \tag{2.2}$$

From these simple relationships, Brown knew that the dependence of initial reaction velocity on $[S]_o$ would be linear with a slope equation to first-order rate constant, k. This was in stark contrast to what he observed for the invertase-catalyzed reaction, which changes from first order at low sugar concentrations to zero order at high concentrations of sugar (Brown 1902). To explain this behavior, Brown proposed the existence of an enzyme:substrate complex along the reaction pathway from substrate to product: "It will be noted that the author's theory demands not only the formation of a molecular compound of enzyme and reacting substance, but the existence of this molecular compound for an interval of time previous to final disruption and change" (Brown 1902, p. 388). A decade later, working again with invertase, Leonor Michaelis and Maud Menten made the same observation and, like Brown, hypothesized that "the rate of breakdown (of sucrose) at any moment is proportional to the concentration of the sucrose-invertase compound; and that the concentration of this compound at any moment is determined by the concentration of the ferment (i.e., crude invertase preparation) and of the sucrose" (Michaelis and Menten 1913).

2.2 DERIVATION OF THE RATE EQUATION FOR A SINGLE-SUBSTRATE REACTION WITH THE ASSUMPTION OF RAPID EQUILIBRIUM

The existence of an $E:S$ complex leads to the well-known minimal mechanism for enzyme-catalyzed reactions (Fig. 2.3). According to this mechanism, enzyme combines

$$E + S \underset{K_s}{\overset{}{\rightleftharpoons}} E{:}S \xrightarrow{k_c} E + P$$

Figure 2.3. Minimal mechanism for an enzyme-catalyzed reaction in which free enzyme is in equilibrium with the enzyme:substrate complex.

Experimental condition: $[E]_o \ll [S]_o$

Conservation equations: $[E]_o = [E] + [ES]$

$[S]_o = [S] + [ES] = [S]$

Definition of binding constants: $K_s = \dfrac{[E][S]}{[ES]}; \quad \dfrac{[E]}{[ES]} = \dfrac{K_s}{[S]} = \dfrac{K_s}{[S]_o}$

Statement of rate dependence: $v_o = k_c[ES]$

Derivation of rate equation:

$$\frac{v_o}{[E]_o} = \frac{k_c[ES]}{[E] + [ES]} \tag{2.3}$$

$$\frac{v_o}{[E]_o} = \frac{k_c}{\dfrac{[E]}{[ES]} + 1} \tag{2.4}$$

$$\frac{v_o}{[E]_o} = \frac{k_c}{\dfrac{K_s}{[S]_o} + 1} \tag{2.5}$$

$$v_o = \frac{V_{max}[S]_o}{K_s + [S]_o} \tag{2.6}$$

$$V_{max} = k_c[E]_o$$

Figure 2.4. Setup and derivation of the rate equation, under the rapid-equilibrium assumption, for the mechanism of Figure 2.3.

with substrate to form the $E{:}S$ complex, from within which chemistry occurs to transform substrate into product. As written, this mechanism assumes an equilibrium between the two enzyme forms, E and $E{:}S$. This equilibrium is governed by the unitless constant $[S]/K_s$, the ratio of substrate concentration to dissociation constant K_s. For this equilibrium to exist, it must be case that the conversion of $E{:}S$ into enzyme and product is much slower than the rate for the establishment of the equilibrium. Thus, in so-called *rapid equilibrium* mechanisms, the reaction is rate-limited by turnover and not formation of $E{:}S$.

The rapid equilibrium assumption of this mechanism allows for a simple derivation of a rate expression that can account for the complex dependence of v_o on $[S]_o$ that Brown observed. This derivation is outlined in Figure 2.4 and serves as a general recipe for the derivation of the rate expression for any enzyme-catalyzed reaction in which binding of substrate, and other ligands, is more rapid than the turnover of the central complex to product and free enzyme. In other chapters, we will see that the category of "other ligands" includes inhibitors and a second substrate for two substrate reactions.

The recipe-like nature of Figure 2.4 results from the fact that the derivation can be broken down into five stages. First, it is assumed that $[E]_o$ is much less than $[S]_o$, where, in practice, "much less" means that $[E]_o \leq [S]_o/10$. If $[E]_o$ is similar to $[S]_o$, both the derivation and final form of the rate expression will be quite complex (Segel 1975, pp. 72–77).

In the second stage of the derivation, we write conservation of matter equations. We see that enzyme exists either as free enzyme or in complex with substrate, and that substrate exists as free substrate or in complex with enzyme. A key simplifying feature of the conservation equation for substrate is that the concentration of free substrate $[S]$ is equal to $[S]_o$. This is the case because $[E]_o \ll [S]_o$, thus making $[E:S]$ negligently small relative to $[S]$.

The third stage of this recipe is to define the various binding constants for the mechanism. In this case, there is only a single binding constant K_s, which is defined as the ratio of $[E][S]$ to $[ES]$. Once the binding constants have been defined, it is useful to write a rearranged form in which a ratio of enzyme forms is equated to a ratio of dissociation constant and ligand concentration. With these rearranged forms handy, subsequent steps of the derivation become a bit easier.

But beyond this convenience, these rearranged forms emphasize an important concept in derivations of rate expressions for enzyme-catalyzed reactions: *always strive to express unknown in terms of what is known and what can be known*. Here, $[E]$ and $[ES]$ cannot be readily measured, but their ratio can be expressed as the ratio of $[S]_o$, what is known, and K_s, what can be known.

In the fourth stage of the derivation, one writes the statement of rate dependence, $v_o = k_c[ES]$. The rate of an enzyme-catalyzed reaction will always be equal to the product of the catalytically competent complex and the first-order rate constant for reaction of the complex to form product.

Stage five is the actual derivation where we bring together the various expressions from the other stages. The first step in the derivation is to divide the rate dependence by the conservation of enzyme. The left side of the resultant expression is the ratio $v_o/[E]_o$, which has the units of a first-order rate constant, reciprocal time. This expression tells us that $v_o/[E]_o$ is equal to some fraction of the first-order rate constant k_c. If we rearrange this term to isolate k_c in the numerator, we obtain $k_c/([E]/[ES] + 1)$. From the binding constant expression of stage three, we immediately see that $[E]/[ES]$ is equal to and can be replaced by $K_s/[S]_o$. Upon rearrangement, we arrive at the familiar Michaelis–Menten equation, in which the initial velocity is equal to the maximal velocity, $V_{max} = k_c[E]_o$, multiplied by a term, $[S]_o/K_s + [S]_o$, which expresses the fractional saturation of enzyme with substrate.

For mechanisms involving more than a single ligand, the denominator of Equation 2.4 will be expanded to include a term $[L]/[EL]$ for each ligand, which always can be expressed as $K_L/[L]_o$. Here as before, the unknown is expressed in terms of what is known and what can be known.

We can now see how Equation 2.6 can explain the complex, multi-phasic dependence of v_o on $[S]_o$ that perplexed Brown and other early enzymologists. At low $[S]_o$, Equation 2.6 reduces to

$$v_o = \frac{V_{max}}{K_s}[S]_o = \frac{k_c}{K_s}[E]_o[S]_o \tag{2.7}$$

which tells us that the initial velocity is first order in both enzyme and substrate concentration and predicts a linear dependence of v_o on $[S]_o$, with slope equal to (k_c/K_s) $[E]_o$. We see that k_c/K_s is the second-order rate constant k_{II} of Figure 2.2 and has the appropriate units of $M^{-1}s^{-1}$.

At high $[S]_o$, Equation 2.6 becomes

$$v_o = V_{max} = k_c[E]_o \tag{2.8}$$

which tells us that the initial velocity is zero order in substrate and predicts a dependence of v_o on $[S]_o$ with a slope of zero.

Equations 2.7 and 2.8 illustrate a point that will be made throughout this text—when considering enzyme action, attention should be paid to k_c and k_c/K_s, and not to K_s. Enzymology is about acceleration of substrate transformation, not about substrate binding. It is by convention only that enzyme catalysis is generally described today in terms of k_c and K_s, and not k_c and k_c/K_s. If the history of enzyme kinetics could be rewritten, convention would have us speaking of enzyme action in terms of k_E, the second-order rate constant for the reaction enzyme and substrate to form product, and k_{ES}, the first-order rate constant for turnover of $E:S$ to form product.[2] More will be said about these constants, their meaning, and their utility in descriptions of enzyme catalysis.

2.3 DERIVATION OF RATE EQUATIONS USING THE STEADY-STATE ASSUMPTION

For the most complete and general approach to analyzing the kinetics of enzyme-catalyzed reactions, mechanisms should be written with microscopic rate constants rather than equilibrium constants. This is illustrated in Figure 2.5, where K_s has been replaced with k_1, the second-order rate for combination of enzyme and substrate to form

$$E + S \underset{k_{-1}}{\overset{k_1}{\rightleftharpoons}} E:S \xrightarrow{k_2} E + P$$

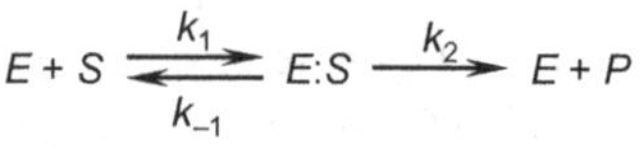

Figure 2.5. Minimal mechanism for an enzyme-catalyzed reaction in which no assumption is made about the equilibrium status of enzyme and enzyme:substrate complex.

[2] This nomenclature was first introduced by physical-organic chemist Richard Schowen in 1978. R. L. Schowen, "Catalytic Power and Transition-State Stabilization," in *Transition States of Biochemical Processes*, R. D. Gandour and R. L. Schowen, eds. (New York: Plenum Press, 1978), 77–114.

$E:S$, and k_{-1}, the first-order rate constant for dissociation of this complex. For consistency, k_c is replaced with k_2.

The derivation of this general expression is based on the "steady-state assumption," a concept originating from the observations and insights of German physical chemist Max Bodenstein (Bodenstein 1913) and later applied to enzyme kinetics by J. B. S. Haldane (Briggs and Haldane 1925). The steady-state assumption is that in a multistep reaction, the rate of formation of an intermediate equals its rate of decay, so that each intermediate exists at a constant concentration. For the simple enzymatic reaction of Figure 2.5, it would be assumed that the concentration of $E:S$ is constant throughout the course of the reaction. In reality, the steady state of an enzymatic reaction exists for only a brief period. The steady state is preceded by a pre-steady-state phase in which enzyme and substrate bind, and is followed by a much longer phase of substrate depletion, in which the concentration of ES changes with time. This is discussed in more detail below.

The derivation of the rate expression for the mechanism of Figure 2.5 using the steady-state assumption is given in Figure 2.6 and is similar to the derivation using the rapid equilibrium assumption. The key difference lies in how the concentration of ES is expressed. In the context of the rapid equilibrium assumption, $[ES] = [E][S]/K_m$,

Experimental condition:	$[S]_o \gg [E]_o$
Conservation equations:	$[S]_o = [S] + [ES] = [S]$
	$[E]_o = [E] + [ES]$
Statement of rate dependence:	$v_o = k_2[ES]$
Steady-state condition:	$d[ES]/dt = 0$

$$k_1[E][S] = (k_{-1} + k_2)[ES]$$

$$[ES] = \frac{k_1}{k_{-1} + k_2}[E][S]$$

$$\frac{v_o}{[E]_o} = \frac{k_2[ES]}{[E] + [ES]} \tag{2.9}$$

$$v_o = \frac{k_2[E]_o\left\{\dfrac{k_1}{k_{-1} + k_2}[E][S]\right\}}{[E] + \dfrac{k_1}{k_{-1} + k_2}[E][S]} \tag{2.10}$$

$$v_o = \frac{k_2[E]_o[S]_o}{\left\{\dfrac{k_{-1} + k_2}{k_1}\right\} + [S]_o} \tag{2.11}$$

$$v_o = \frac{V_{max}[S]_o}{K_m + [S]_o} \tag{2.12}$$

Derivation of rate equation:

Figure 2.6. Setup and derivation of rate equation, under the steady-state assumption, for the mechanism of Figure 2.5.

while using the steady-state assumption, $[ES] = [E][S](k_1/[k_{-1} + k_2])$. With this defini-tion, the familiar Equation 2.9 becomes Equation 2.10. The rearrangement of Equation 2.10 to Equation 2.11 involves the isolation of k_2 in the numerator. When this is done, we arrive at the Michaelis–Menten equation of 2.12 where,

$$V_{max} = k_2[E]_o \tag{2.13}$$

$$K_m = \frac{k_{-1} + k_2}{k_1} \tag{2.14}$$

and

$$\frac{V_{max}}{K_m} = \frac{k_2}{\dfrac{k_{-1} + k_2}{k_1}}[E]_o = \frac{k_1 k_2}{k_{-1} + k_2}[E]_o \tag{2.15}$$

Note that when $k_2 < k_{-1}$, V_{max}/K_m and K_m simplify to k_2/K_s and $k_{-1}/k_1 = K_s$, respec-tively, which are identical to V_{max}/K_m and K_m derived using the rapid equilibrium assumption.

To illustrate some of the principles that we have been discussing, simulations of reaction progress curves for the four species of the mechanism in Figure 2.5 are shown in Figure 2.7. The rate constants for the enzyme of these simulations are similar to those of a "real-world" enzyme, and $[E]_o$ and $[S]_o$ are similar to those of a typical experiment one might conduct.

In the upper panel, we see that the concentration of the E:S complex increases exponentially to a steady-state concentration of 8.2 nM, a value that can be calculated from the steady-state condition of Figure 2.6 by using the rate constants of the simula-tion and substituting $[E]_o - [ES]$ for $[E]$. This is the steady-state concentration of ES and is reached after the reaction has proceeded for 0.8 ms. The steady state is maintained for another 20 s, after which time $[ES]$ falls to zero due to substrate depletion.

During the 20 s of steady state, product formation and substrate consumption are linear with time, as illustrated in the lower panel of Figure 2.7. Pre-steady-state phases of these curves are not observed because of the rapidity of these phases and the small signals that are generated relative to those of substrate and product measurements. Thus, the value of v_o that can be calculated from the slope of the curves between 0 s and 20 s is an accurate estimate of v_{ss}, the reaction velocity in the steady-state phases of the curves. Note that in cases of a pre-steady-state phase that extends over seconds or minutes rather than milliseconds, v_o will not be equal to v_{ss}, and the velocity measure-ment will need to be taken after the pre-steady-state has resolved into the steady state.

2.4 METHODS OF ENZYME ASSAY

In previous sections, we explored kinetic theories of how to interpret the dependence of initial velocity on substrate concentrate. Our concerns in this and the following

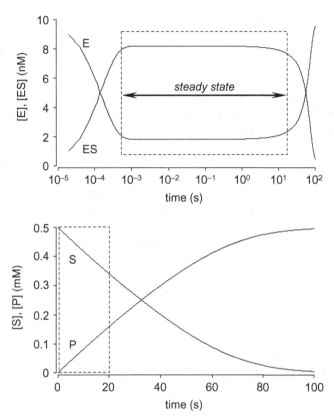

Figure 2.7. Simulation of time courses for progress of an enzyme catalyzed reaction in which the *E* and *ES* complex do not attain equilibrium. *Upper Panel:* Time courses for *E* and *ES*. A steady state exists from about 0.8 ms to 20 s. Note the logarithmic time scale. *Lower Panel:* Time courses for *S* and *P*. Linear region of the progress curves extend to about 20 s, the duration of the steady state. It is in this region where initial velocity measurements can be made. Simulation conditions: $k_1 = 10^7\,M^{-1}s^{-1}$, $k_{-1} = 10^2\,s^{-1}$, $k_2 = 10^3\,s^{-1}$, $[E]_o = 10\,nM$, and $[S]_o = 0.5\,mM$.

section will be of more practical matters. Here, we will discuss the types of methods available to measure an enzyme's catalytic activity, and in the next section, experimental design and data analysis.

The methods available to measure an enzyme's catalytic activity fall into one of two broad categories: continuous assays and discontinuous assays. Each of these can be divided into methods involving the measurement of either substrate consumption or production formation.

2.4.1 Continuous Assays

Continuous assays are by far the most convenient way to measure an enzyme's catalytic activity. These assays involve recording over time a change in some physical property of the reaction solution. The most common physical property observed in continuous assays are changes in UV-visible absorbance or fluorescence. These changes result from either loss of a signal as substrate is consumed or increase in signal as product is formed.

Loss of a substrate-related signal occurs, for example, in NADH-dependent, reduction reactions. NADH has a UV-vis absorbance at 340 nm ($\varepsilon_{340} = 6200$) which disappears as it is oxidized to NAD^+, in the course of reduction of the second substrate. The consumption of NADH is also accompanied by a decrease in fluorescence signal 465 nm ($\lambda_{ex} = 340$ nm). Monitoring the fluorescence loss of NADH affords assays that are about 10-fold more sensitive than monitoring loss of UV-vis signal at 340 nm.

Gain of signal with product formation is observed with proteases in which chromogenic or fluorogenic leaving groups are used. For example, the activity of many serine proteases is monitored as the increase in *p*-nitroaniline ($\varepsilon_{410} = 8800$) that accompanies the hydrolysis of peptide-*p*-nitroanilides. Much more sensitive assays result if one monitors the increase in fluorescence that accompanies the hydrolysis of peptide-7-amido-4-methylcoumarins. While the intact substrate has no fluorescence, the product 7-amino-4-methylcoumarin fluoresces strongly at 460 nm ($\lambda_{ex} = 360$ nm). Other protease assays are based on peptides that have a fluorescent moiety on one side of the scissile bond and a fluorescence quenching moiety on the other. The observation that the metalloproteinase stromelysin can cleave substance *P* and derivatives led to the design of the substrate N-(2,4-dinitrophenyl)-Arg-Pro-Lys-Leu-Nva-TrpNH$_2$ (Niedzwiecki et al. 1992). In the intact peptide, the fluorescence of Trp ($\lambda_{ex} = 280$ nm, $\lambda_{ex} = 420$ nm) is quenched by the 2,4-nitropheny moiety. This quenching is released upon hydrolysis of the peptide at Leu-Nva bond and the two products diffuse apart in solution.

2.4.2   Discontinuous Assays

It is often the case that an enzymatic reaction is not accompanied by a change in a physical property that can be continuously monitored over time. In such case, aliquots are taken at predetermined times from a reaction solution, and the concentration of product or unreacted substrate are measured. There are two general protocols here.

2.4.2.1   *Direct Measurement.*   For certain reactions, quantization can be made directly on the aliquot by the addition of some reagent(s) that converts unreacted substrate or product to a substance whose concentration can be measured. For example, the activity of protein phosphates and the ATPase activity of certain kinases can be measured by assays in which aliquots of the reaction solution are treated with a solution of malachite green molybdate. The latter reacts with the liberated free phosphate to form a complex which absorbs strongly at 620–640 nm.

In all cases in which one wishes to make a direct measurement of the reaction solution, the investigator must be certain that there is no interference by nonspecific reaction of reaction components with the reagent. For example, when quantifying the activity of a phosphatase using malachite green molybdate, one of course cannot use a phosphate buffer.

2.4.2.2 Measurement after Separation. For many reactions, there is simply too much background interference to allow measurements to be made directly on the reaction solution. Here, it is necessary to separate product from unreacted substrate. The method of choice for such separations is high-performance liquid chromatography (HPLC) with UV-vis or fluorescence detection of the eluate. After the chromatographic separation, the peak area or height for the substrate and/or product is quantified with the use of standard curves.

A particularly useful variant on standard HPLC methods of measuring enzyme activity is the so-called "semi-continuous" method of HPLC analysis (Harrison et al. 1989). According to this method, the reaction solution of enzyme and substrate is placed in the sample tray of an HPLC autosampler. Aliquots are then withdrawn and injected onto the column at preset times. In a fully automated manner under control of an interfaced computer, areas of chromatographic peaks corresponding to substrate and product are calculated and converted into units of molarity using a calibration curve stored in the computer's memory. To take full advantage of this method, one first must develop an isocratic separation of product from substrate, in which the two are separated cleanly in no more than about 10 min. Isocratic elution avoids the time for column reequilibration that would be necessary if a gradient were to be used, and the short separation time allows for aliquots of reaction solution to be taken at time intervals whose length are consistent with typical enzyme assays.

This method is illustrated in Figure 2.8 for the stromelysin-catalyzed hydrolysis of substance P (SP, Arg-Pro-Lys-Pro-Gln-Gln-Phe-Phe-Gly-Leu-MetNH2; cleavage at Gln^6-Phe^7 bond) (Harrison et al. 1989). To implement a semi-continuous HPLC assay, the investigators first developed an elution method that allows injection of the reaction solution aliquots at 3-min intervals. While this elution does not allow quantization of unreacted SP or product SP^{1-6}, both of which are buried in the solvent front along with the enzyme, it is possible to cleanly isolate product SP^{7-11} (upper panel, Fig. 2.8). The areas of these peaks were calculated, converted to micromolar units, and plotted versus reaction time in the lower panel. It is clear from this progress curve that this method affords precise data allowing subtle features of mechanism, such as the pre-steady-state phase, to be studied.

2.4.3 Coupled Enzyme Assays

In the above paragraphs, we discussed several solutions to the general problem of measuring rates of enzymatic reactions in which the product does not produce a readily measurable signal. There is another solution to this problem—the coupled enzyme assay.

The basic principle underlying coupled assays is illustrated in Figure 2.9. Here, we see the enzyme of interest E_1 catalyzing the conversion of S into P_1, a substance whose concentration cannot be conveniently measured. The solution to this dilemma is found in the use of enzyme E_2, which can catalyze the conversion of P_1 into P_2, a substance whose concentration *can* be readily measured.

To reduce theory to practice, there are a number of requirements the investigator must adhere to:

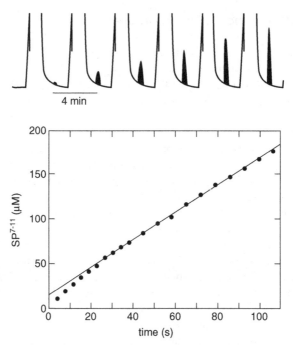

Figure 2.8. Semi-continuous HPLC assay for the stromelysin-catalyzed hydrolysis of substance *P*. *Upper Panel:* Chromatogram for automated injection of a reaction solution of stromelysin and substance *P* onto a column with isocratic elution. Darkened peaks correspond to hydrolysis product, substance P[7-11]. *Lower Panel:* Peak areas from the chromatogram of the upper panel were converted to micromolar units by use of a calibration curve and plotted versus reaction time.

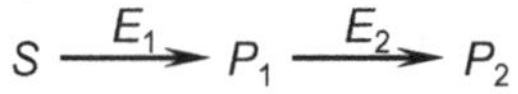

$$S \xrightarrow{\;E_1\;} P_1 \xrightarrow{\;E_2\;} P_2$$

Figure 2.9. Coupled enzyme reaction. Catalytic activity of E_1 can be measured by converting the product of its reaction, P_1, into the a readily measured product P_2.

- Reaction conditions, for the efficient operation of both enzymes, must be compatible. For example, if the activities of the primary and coupling enzyme have different pH optima, a coupled assay system using that particular coupling enzyme will not be able to be developed.
- The primary enzyme must operate in the steady state throughout the assay. As we saw above, this means that one should use only that portion of the reaction progress curve representing <10% substrate turnover.
- The conversion of P_1 into P_2 must be functionally irreversible.

- A sufficiently high concentration of coupling enzyme must be used to ensure that conversion of P_1 into P_2 is much faster than the conversion of S into P_1. In this way, the rate P_2 formation will be a true reflection of the rate of P_1 production. As a consequence of this, P_1 will always be at a concentration far below its K_m for E_2 and thus, the E_2-catalyzed conversion of P_1 into P_2 will follow first-order kinetics with rate constant $(k_{c,2}/K_{m,2})[E]_{2,o} = k_{E,2}[E]_{2,o}$.

Given the conditions above, one can predict the shape of progress curves for coupled enzyme assays. Like any two-step, irreversible chemical process in which the rate of the second step is larger than that of the first, the progress curve for formation of final product will be characterized by a pre-steady-state lag that resolves into the steady-state reaction rate. For coupled enzymatic systems of the type shown in Figure 2.9, the length of the lag will decrease with higher concentrations of E_2. For this coupled system, the rate of P_2 formation is given by

$$\frac{d[P_2]}{dt} = v_{ss}\left(1-e^{-kt}\right) \tag{2.16}$$

in which v_{ss} is the steady-state velocity for the reaction catalyzed by E_1, and k is the pseudo-first order rate constant for the reaction catalyzed by E_2, equal to $k_{E,2}[E_2]_o$.

Integration of this equation yields an expression for the dependence of $[P]$ on time and proceeds in the following way:

$$d[P]=\left[v_{ss}\left(1-e^{-kt}\right)\right]dt \tag{2.17}$$

$$d[P]= v_{ss}dt -\left(v_{ss}e^{-kt}\right)dt \tag{2.18}$$

$$[P]= v_{ss}t -\frac{v_{ss}}{k}e^{-kt} -\frac{v_{ss}}{k} \tag{2.19}$$

$$[P]= v_{ss}t -\frac{v_{ss}}{k}\left(1-e^{-kt}\right) \tag{2.20}$$

Using Equation 2.20, the simulations of Figure 2.10 were drawn. In these simulations, kinetic constants $k_{c,1} = 10\,s^{-1}$ and $K_{m,1} = 100\,\mu M$ were assumed for the E_1-catalyzed conversion of S_1 to P_2. Also assumed here were $[E_1]_o = 0.01\,\mu M$ and $[S_1]_o = 100\,\mu M$. These values yield $v_{ss} = 5 \times 10^{-3}\,\mu M\,s^{-1}$. For the coupling reaction, $k_{E,2}$ was set equal to $0.1\,\mu M^{-1}\,s^{-1}$ and $[E_2]_o$ to the values indicated in Figure 2.10. Also shown in this figure is a linear progress curve representing v_{ss}.

As predicted, all three reaction progress curves for the coupled reactions begin with a lag phase whose duration is shorter with larger concentrations of the coupling enzyme. Only the progress curve with $[E_2]_o = 0.05\,\mu M$ attains linearity, with a slope equal to v_{ss}, before 10% of $[S_1]_o$ is consumed. For the other two curves, pre-steady-state consumption of S_1 continues beyond 10%. Note that while the curve with $[E_2]_o = 0.05\,\mu M$ appears to have resolved into the steady state, the final velocity of the curve is about 85% of the v_{ss}.

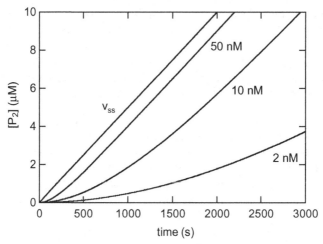

Figure 2.10. Simulated progress curves for a coupled enzyme assay. These curves were drawn according to Equation 2.2 with $v_{ss} = 5 \times 10^{-3} \mu M s^{-1}$ and k, which is the product of $k_{E,2}$, equal to $0.1 \mu M^{-1} s^{-1}$, and the value of $[E_2]_o$ indicated in the figure. Also shown in this figure is a linear progress curve representing v_{ss}.

2.5 ENZYME KINETICS PRACTICUM: ASSAY DEVELOPMENT, EXPERIMENTAL DESIGN, DATA COLLECTION, AND DATA ANALYSIS

In the previous sections, I outlined the theoretical background necessary to conduct kinetic studies of single-substrate reactions and the sorts of assay methods available to conduct these studies. In the final section of this chapter we will examine practical aspects of enzyme kinetics: assay development, designing kinetic studies, data collection, and data analysis.

2.5.1 Assay Development

Under this broad heading, "Assay Development," I include four activities: choice of substrate, establishing the optimal assay buffer, choice of assay method, and finally, the actual implementation of the method using the chosen substrate and optimal buffer solution. I discuss these topics in the order listed with the idea that this order represents a suitable pathway for the development of a new assay.

2.5.1.1 Selection of Substrate. Selection of substrate is an issue only for those enzymes that have some breadth of substrate specificity. For enzymes with a very strict specificity, such as enzymes of intermediary metabolism, selection is unnecessary; one simply uses the substrate that the enzyme has evolved to accept. With enzymes of broader substrate specificity, however, there is more leeway in the choice of substrates. For these enzymes, the selection of substrate will be based on the purpose of the assay. For example, if an investigator is developing an assay for a protease with the intention of using this assay to characterize active-site-directed inhibitors, any substrate will do since all that is required in such an assay is occupancy of the active site and competition with the inhibitor. However, if one wishes to mount a high-throughput screen of that protease with the intention of screening broadly to identify molecules that can bind to an exosite on the enzyme's surface, a full-length, natural protein substrate is required, since only a substrate such as this will occupy the exosite. In general, it is advisable to use full-length natural substrates when screening for inhibitors of protease, kinases, and other enzymes that work on polymeric substrates.

2.5.1.2 Assay Buffer Solution.

Often when we are studying an enzyme, enough is known that we have a good idea of what the optimal assay solution is for that enzyme. Numerous citations might exist that provide guides to the pH, buffer salt, ionic strength, and various additives that are needed for optimal activity. But, let us consider the case where the investigator is studying a newly discovered enzyme. What should this investigator do to establish optimal reaction conditions?

First, the simplest possible assay solution should be used. A guiding principle here is that an assay solution should contain nothing that is not necessary to support the enzymatic reaction under study. A corollary to this is that when using an assay solution prescribed from the literature, the investigator should understand the reason for the inclusion of each of its components.

There are certain guidelines that are useful in constructing the assay solution for the initial studies of the enzyme. The buffer salt of choice is one of the 12 zwitterionic buffer salts developed by Norman Good in the 1960s (Good et al. 1966). These include the familiar MES, HEPES, and CHES buffer salts. Generally, phosphate and Tris should be avoided—phosphate because it is easy to be careless in buffer preparation and have a final ionic strength different from that wanted, and Tris because of amine nucleophilic impurities and its steep dependence of pK_a on temperature. There is seldom a reason for a buffer salt concentration to lie outside the range of 10–50 mM. In assay solutions for these initial experiments, the final ionic strength should be relatively low and around 100 mM. NaCl or KCl should be used to adjust the ionic strength to this value. At this early stage, there is no reason to include additives (e.g., metals, metal chelators, DTT, or other redox "regulators," protein stabilizing agents, or detergents) unless the investigator has an insight based on the chemistry the enzyme catalyzes. For example, if a protease is being characterized that likely has an active site cysteine, one might add DTT and EDTA to prevent oxidation of the thiol group.

If this simple buffer supports the activity of the enzyme under investigation, it is then important to find the optimal pH, buffer salt, and ionic strength. A reasonable protocol is to measure enzyme activity at constant ionic strength at pH values (using the indicated buffer salts): 5 (acetate; $pK_a = 4.8$), 6 (MES; 6.15), 7 (HEPES; 7.5), 8 (HEPES; 8.8), and 9 (CHES; 9.3). When the optimal pH is found, it is then important to examine other buffer salts at that pH to determine if there is buffer-salt-dependent inhibition or activation. For example, suppose the highest activity is found at pH 7 using a HEPES buffer. One should then determine enzyme activity at that using MES and ACES ($pK_a = 6.9$). If one were now to find that enzyme activity in the HEPES-buffered solution is lower than rates determined in the MES- and ACES-buffered solution, then the investigator would conclude that HEPES is an inhibitor of the enzyme. MES or ACES would then be used for further studies.

When the buffer salt has been chosen, assay solutions differing only in ionic strength should be prepared and used in assays of the enzyme. At this point, there is no reason to choose any salt other than NaCl to adjust the ionic strength. So, continuing with the above example, one would prepare solutions all containing 50 mM ACES, but having a range of NaCl concentrations, such as: 0, 0.1, 0.2, 0.4, 0.6, 0.8, and 1.0 M. It is of course important to adjust each ACES/NaCl solution to pH 7, rather than simply adding NaCl to a pH 7 solution of ACES, since the pK_a of buffer salts are ion strength-dependent.

Note in the above example that ACES is being used at a pH value essentially equal to its pK_a. When pH = pK_a for a zwitterionic buffer salt, the buffer salt makes no contribution to the ionic strength of the solution. So the ionic strength for the above solution is identical to the NaCl concentration.

With the completion of these studies, a workable assay solution for initial experiments will be in hand. In the course of these experiments, anomalies or inconsistencies may appear that indicate that the assay solution is overly simple and requires the addition of other components for optimal performance of the assay. If enzyme is unstable, albumen or glycerol may need to be added. Or, if one suspects that the enzyme has an important thiol group, DTT and/or EDTA may need to be added. All of these additions will initially be empirically based and perhaps guided by the literature on similar enzymes.

2.5.1.3 Choice of Assay Method.
Examination of the reaction catalyzed by an enzyme, coupled with a little imagination, generally reveals a number of ways in which the activity of that enzyme can be measured. Given multiple assay methods, the investigator can then choose a particular one based on the specific reason for developing the assay. He would choose one sort of method for high-throughput screening and another for detailed kinetic studies. In the course of an enzyme-based drug discovery program, it is not uncommon for a number of assays to be developed, each for a specific purpose.

2.5.1.4 Implementation of Assay Method.
At this point, the investigator has settled on an assay method and has the appropriate substrate and assay solution. Implementation should require nothing more than assembling the reaction mixture, comprising assay buffer, enzyme, and substrate, and then monitoring the progress of the reaction in accordance with the assay method. However, as we all know, it is indeed rare that the first reaction mix we assemble leads to a meaningful outcome. Assay implementation is actually assay troubleshooting.

In troubleshooting a misbehaved assay, there are a number of questions that should be asked:

• Is the enzyme concentration correct for the length of the assay?
• Is the buffer, which was optimized for the enzyme and some general assay method, appropriate for this specific assay method?

- Are the reaction vessels (e.g., cuvettes or microtiter plates) composed of material appropriate for the detection method?
- For a coupled enzyme assay, is the coupling enzyme(s) in excess over the primary enzyme?
- Do the assays' results change with differences in the order of addition of assay components?

Unfortunately, there is no recipe for troubleshooting an assay. One must be able to identify all the possible assay features that can go wrong, and then rule them out one-by-one.

2.5.2 Experimental Design: Range of $[S]_o$

The design of a kinetic experiment will always depend on the question being asked. In subsequent chapters we will be concerned with the design of experiments to determine such things as K_i values for enzyme inhibitors or the order of substrate addition in a two-substrate reaction. But here we are concerned with the design experiments to determine kinetic constants for a one-substrate reaction.

To arrive at accurate estimates of V_{max} and K_m, the most important consideration is the range of substrate concentration to be used. In general, the largest possible range of $[S]_o$ should be used, with no less than a 10-fold spread, defined by $K_m/3 \leq [S]_o \leq 3K_m$, where, in the earliest kinetic experiments, the K_m value would only be a tentative estimate, generated from preliminary experiments. Ideally, both lower and higher concentrations would be included, to rule the possibility of cooperativity and substrate inhibition, respectively. However, this large range of substrate concentration is sometimes not possible because of assay insensitivity at low $[S]_o$ and substrate insolubility or assay signal interference at high $[S]_o$.

2.5.3 Data Collection

For most enzymatic reactions, the initial velocity is equal to the steady-state velocity. The exceptions are coupled enzyme assays, which I discuss above and comment on again below, and enzyme hysteresis, which will not be covered in this book, but has been the topic of two excellent reviews (Frieden 1979; Neet and Ainslie 1980).

It is critical in initial velocity experiments that the velocity one measures truly reflects the initial velocity. Figure 2.11 helps to illustrate this point and contains three simulated progress curves for an enzymatic reaction obeying Michaelis–Menten kinetics with $k_c = 100\,s^{-1}$, $[S]_o = 10\,mM$, and K_m values of 1, 10, and 100 mM. With K_m equal to 1 mM, the ratio $[S]_o/K_m$ is 10, in which case the steady state exists for 50–60% of substrate consumption. Here there would be no problem at all in measuring the initial velocity. When K_m equals 10 mM, $[S]_o/K_m$ decreases to 1 and the steady state exists only for about 10% of substrate consumption. With care the true initial velocity can be measured, but caution has to be taken not to include data beyond about 10% substrate consumption. If data points in this region are included, the measured velocity will be

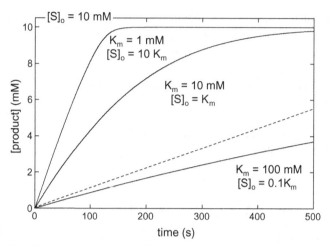

Figure 2.11. Simulated progress curves for enzymatic reactions recorded at identical $[S]_o$ values but three different K_m values.

less than the initial velocity because of the falloff in velocity that accompanies substrate depletion. When $[S]_o/K_m$ is 0.1 or lower, the steady state exists for no appreciable length of time. Under these circumstances, initial velocities are difficult to accurately measure. This is illustrated in Figure 2.11 where the dashed line has a slope equal to the initial velocity of $91\,mM\,s^{-1}$ under the condition of $[S]_o/K_m = 0.1$. Clearly, in cases such as this, extreme care must be exercised if the initial velocity is not to be underestimated.

For coupled enzyme assays, the chief problem that an investigator will encounter is impatience. It is a common mistake not to wait long enough for the lag phase to fully resolve into the steady state. The result of this mistake will be to measure a velocity that underestimates the steady-state velocity. So how long does one need to wait? The answer to this question will come out of an analysis of the simulated progress of Figure 2.12.

The curve of Figure 2.12 was drawn using Equation 2.20 with parameters set at $v_{ss} = 5 \times 10^{-3}\,\mu M\,s^{-1}$ and $k = 5 \times 10^{-3}\,s^{-1}$ (this corresponds to curve of Figure 2.10 with $[E_2]_o = 50\,nM$). Now, at long times when the lag phase has fully resolved into the steady state, Equation 2.20 simplifies to Equation 2.11 which describes a straight line with slope and y-intercept

$$[P] = v_{ss}t - \frac{v_{ss}}{k} \tag{2.21}$$

equal to v_{ss} and $-v_{ss}/k$, respectively. This line crosses the time axis at a point equal to the reciprocal of the rate constant k for the approach to steady state. This time, referred to as the reaction's relaxation time τ, should not be confused with the reaction's half-life $t_{1/2}$, which is equal to $\ln 2/k = 0.69/k$. The lag phase can be considered over when

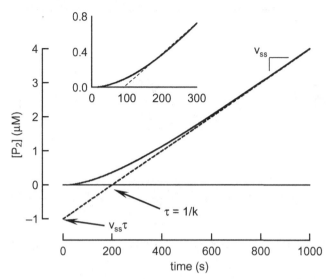

Figure 2.12. Simulated progress for a coupled enzymatic reaction. This simulation was drawn from Equation 2.20 with parameters set at: $v_{ss} = 5 \times 10^{-3} \mu M s^{-1}$ and $k = k_{E,2}[E_2]_o = (0.1 \mu M^{-1} s^{-1})$ $(0.05 \mu M) = 5 \times 10^{-3} s^{-1}$. The inset illustrates velocity determination for a section of the progress curve where the steady state has not yet been reached.

enough time has passed that the exponential term of Equation 2.20 is much less than 1. For the lag phase to have progressed 95% toward its resolution, the exponential term of Equation 2.20 has to equal 0.05. From this we can calculate a time lapse of $(-\ln 0.05)\tau = 3\tau$ between time zero and the beginning of the steady-state velocity.

For the simulated reaction of Figure 2.12, $\tau = 200$ s. Given the above consideration, this means that the coupled system of this example must be allowed to progress for no less than 600 s for the pre-steady-state lag phase to be finished. One can see from the inset of Figure 2.12 how easy it would be to terminate data collection prematurely, leading to erroneous estimates of k and v_{ss} equal to $k = 1.1 \times 10^{-2} s^{-1}$ and $2.5 \times 10^{-3} \mu M s^{-1}$, respectively.

The key to successful estimates of v_s for coupled enzyme assays is to collect data at least three times longer than the time required for the lag phase to have the appearance of "nearly over." In general, one should always err on the side of collecting too much data.

2.5.4 Data Analysis

It is now common practice that enzyme kinetic experiments are conducted with the aid of computers. For continuous assays, progress curves are collected, stored, and analyzed on computers that are interfaced to spectrophotometers or fluorometers. For discontinuous assays, while raw data sets of [P] versus time are collected "by hand" as discrete data points, the data will still be loaded into a computer, where reaction

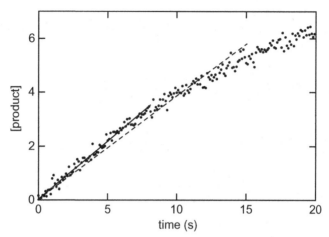

Figure 2.13. Initial part of a progress curve illustrating the region to measure for accurate calculation of an initial velocity.

progress curves are constructed and analyzed. For both types of enzyme assay, dependencies of v_{ss} on $[S]_o$ (aka, Michaelis–Menten plots) will be analyzed on computers to arrive at estimates of V_{max} and K_m.

To calculate the initial velocity from a reaction progress curve, the investigator first needs to examine the curve to ascertain the portion of it that represents the steady-state region. One needs to be sure that in the selected region, pre-steady-state phenomena are fully resolved, and data representing greater than 10% substrate turnover are excluded. Once the steady-state region of the data set is isolated, linear regression is performed to calculate the slope. At this stage, it is important to examine how the theoretical line fits the data. Ideally, one should observe a random distribution of data points above and below the straight line. If the distribution is systematic rather than random, the range of data points included in the calculation was too large and the range narrowed and the slope recalculated. This is illustrated in the simulated progress curve of Figure 2.13. The curve shows clears signs of substrate depletion at about 15 s; thus we know that our initial velocity must not include data beyond this point. The dashed line is a first attempt at calculating the initial velocity using data between $t = 0$ s and 13 s. What is evident here is the systematic distribution of data points, first rising above the line and then dipping below it. When the slope was recalculated with a narrower range of 0–8 s, the distribution of data points about the line is random, as must be the case for accurate estimates of steady-state velocities.

In the analysis of the dependence of v_{ss} on $[S]_o$, one usually starts by fitting the data set to the simple Michaelis–Menten equation. The fitting algorithm will not only generate best-fit values for V_{max} and K_m, but also error limits for the two parameters. What is important to understand is that these error limits do *not* tell us the accuracy with which we know those parameters, but only how well the Michaelis–Menten equation can account for that particular data set. The only way to arrive at accurate estimates

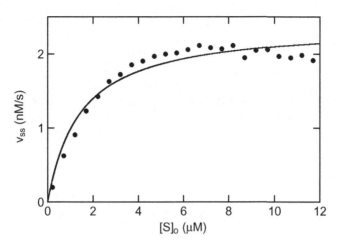

Figure 2.14. Use of the Michaelis–Menten equation to fit a data set generated using Equation 2.22 for substrate inhibition. Data set simulation with the following parameters: $V_{max} = 5\,nM\,s^{-1}$, $K_m = 5\,\mu M$, and $K_{i,s} = 10\,\mu M$. Best-fit values according to the Michaelis–Menten equation are: $V_{max} = 2.4 \pm 0.1\,nM\,s^{-1}$ and $K_m = 1.4 \pm 0.2\,\mu M$.

of steady-state parameters is to conduct the experiment multiple times and calculate the mean and standard deviation of the resultant set of parameter values.

While we may start our analysis using the Michaelis–Menten equation, it is sometimes necessary to fit the data set to a model that reflects a more complex mechanism. Consider the simulated data set in Figure 2.14. When fit to the Michaelis–Menten, the following parameter estimates were obtained: $V_{max} = 2.4 \pm 0.1\,nM\,s^{-1}$ and $K_m = 1.4 \pm 0.2\,\mu M$. What should immediately be clear when the theoretical line through the data set is examined is the systematic distribution of data points about the theoretical line. Despite the small errors associated with these parameters, the simple physical model they reflect cannot account for the data—a more complex model is required by the data. The sort of distribution of data points that we observe in Figure 2.14 is consistent with a phenomenon known as substrate inhibition (see Chapter 3). Indeed, the data set of Figure 2.14 was generated assuming substrate inhibition and using Equation 2.22, where V_{max}, K_m, $K_{i,s}$ were assigned values of $5\,nM\,s^{-1}$, $5\,\mu M$, and $10\,\mu M$, respectively.

$$v_{ss} = \frac{V_{max}[S]_o}{K_m(1+[S]_o / K_{i,s})+[S]_o} \tag{2.22}$$

The consequences of fitting data to an inadequate model is apparent. The values of V_{max} and K_m estimated by fitting the data to the Michaelis–Menten are very different from the actual values for these parameters.

A general method for data analysis emerges here. Fit a new data set to the rate equation for the simplest model that might adequately account the data. If the scatter of data about the theoretical line drawn using the results of this fit is not random, the model must be abandoned, and more complex alternatives must be tried. The formulation of these alternative mechanistic models is a process that draws on the investigator's knowledge of the system, as well as his experience and imagination. After a new model is tried, its adequacy is again judged by the investigator, and, if this model is also found lacking, another round of model development, data fitting, and assessment will be conducted. This iterative process will ultimately lead to the simplest model that can account for the data. This is of course the goal of all mechanistic studies.

REFERENCES

Bodenstein, M. (1913). "Zur Kinetik des Chlorknallgases." *Z. Phys. Chem.* **85**: 329.

Briggs, G. E. and J. B. S. Haldane (1925). "A note on the kinetics of enzyme action." *Biochem. J.* **19**: 338–339.

Brown, A. (1902). "Enzyme action." *J. Chem. Soc.* **81**: 377–388.

Frieden, C. (1979). "Slow transactions and hysteretic behavior in enzymes." *Annu. Rev. Biochem.* **48**: 471–489.

Good, N. E., et al. (1966). "Hydrghen ion buffers for biological research." *Biochemistry* **5**: 467–477.

Harrison, R. K., et al. (1989). "A semi-continuous, HPLC-based assay for stromelysin." *Anal. Biochem.* **80**: 110–113.

Michaelis, L. and M. Menten (1913). "The kinetics of invertase action." *Biochem. Z.* **49**: 333.

Neet, K. E. and G. R. Ainslie (1980). "Hysteretic enzymes." *Methods Enzymol.* **64**: 192–226.

Niedzwiecki, L., et al. (1992). "Substrate specificity of the human matrix metalloproteinase stromelysin and the development of continuous fluorometric assays." *Biochemistry* **31**: 12618–12623.

Schowen, R. L. (1978). Catalytic power and transition-state stabilization. In *Transition States of Biochemical Processes*, R. D. Gandour and R. L. Schowen, eds. New York, Plenum Press: 77–114.

Segel, I. H. (1975). *Enzyme Kinetics*. New York, John Wiley & Sons.

Wilhelmy, L. (1850). "The law by which the action of acid on cane sugar occurs." *Annalen Physik Chemie* **81**: 413–433.

3

KINETICS OF SINGLE-SUBSTRATE ENZYMATIC REACTIONS: SPECIAL TOPICS

In the previous chapter, we discussed basic enzyme kinetic theory as it relates to single-substrate enzymes, as well as a number of practical issues concerning the design of kinetic experiments and the analysis of data. In this chapter, we move our discussion into somewhat more advanced territory, considering how the Michaelis complex determines the meaning of K_m, the meaning of steady-state rate parameters for enzymatic reactions that have more than a single intermediate, and kinetic situations that deviate from simple Michaelis–Menten kinetics, such as that observed with substrate inhibition and cooperativity. Our exploration of these topics will be aided by transition state theory and free energy diagrams, which will be introduced in this chapter's first section. The final special topic we will cover in this chapter is the kinetics of reactions involving substrates that have multiple sites for enzymatic reaction.

3.1 TRANSITION STATE THEORY AND FREE ENERGY DIAGRAMS

3.1.1 Transition State Theory

A goal of chemical kinetics is to move from the empirical determination of rate constants to an understanding of why a rate constant has the value it does and not another.

Kinetics of Enzyme Action: Essential Principles for Drug Hunters, First Edition. Ross L. Stein.
© 2011 John Wiley & Sons, Inc. Published 2011 by John Wiley & Sons, Inc.

Kinetic theorists who seek to answer this question have a lineage originating with Svante Arrhenius, a Swedish physical chemist who worked at the turn of the twentieth century. Arrhenius made the seminal discovery that the routinely observed increase in reaction rate with temperature conforms to a linear dependence of the logarithm of the first-order rate constant on reciprocal temperature Kelvin (Arrhenius 1889):

$$\ln(k) = -\text{slope}\left(\frac{1}{T}\right) + \text{intercept.} \tag{3.1}$$

His genius was to understand that in the rapidly evolving field of chemical thermodynamics laid the physical meaning of the two empirically determined parameters of his equation. Arrhenius recognized that the intercept and slope of his expression must be a probability and energy term, respectively. With this, Arrhenius was able to express Equation 3.2 as

$$k = Ae^{-E_a/RT} \tag{3.2}$$

which posits a first-order rate constant equal to the product of reactant collision frequency, A, and the exponential of the heat of activation, E_a, normalized by universal gas constant R. We see from this expression that even after collision there is an energy barrier that must be surmounted for reactants to be chemically transformed into products.

During the first decades of the twentieth century, Arrhenius' theory served as the starting point for investigators who sought to understand the physical underpinnings of chemical kinetics (reviewed in Eyring 1935a,b). The watershed moment in these investigations was the introduction of transition state rate theory by physical chemist Henry Eyring.

In 1935, Eyring proposed that for the chemical transformation of a molecule to occur, the molecule must pass through a high energy "transition state" that separates reactants from products (Eyring 1935a,b; Wynne-Jones and Eyring 1935). Eyring called the molecular species that exists in the transition state the "activated complex" and posited that it has the properties of a stable molecule except for translation along the reaction coordinate, which leads to reaction.[1] It is instructive to quote from the introduction to Eyring and colleagues' *The Theory of Rate Processes*:

> . . . a chemical reaction is characterized by an initial configuration which passes over by continuous change of the coordinates into the final configuration. There is however always some intermediate configuration which is critical for the process, in the sense

---

[1] One frequently finds the terms "transition state" and "activated complex" used interchangeably in the literature, although from the works of Eyring, "transition state" referred to a specific location (i.e., a saddle point) on a potential energy surface, while "activated complex" referred to the molecular species that occupies the "transition state." But in keeping with the trend in the chemical and biochemical literature, I will use transition state to refer to both the state and the species.

that if this system is attained there is a high probability that the reaction will continue to completion. This critical configuration is called the "activated complex" for the reaction and it is in general situated at the highest point of the most favorable reaction path on the potential energy surface. The activated complex is to be regarded as an ordinary molecule possessing all the usual thermodynamic properties, with the exception that motion in one direction, i.e., along the reaction coordinate, leads to decomposition at a definite rate. (Glasstone et al. 1941, p. 11).

Transition state theory goes on to tell us that the reaction rate for the chemical conversion of S into P is equal to the product of the concentration of S in the transition state (i.e., the concentration of activated complex) and the rate constant for conversion of activated complex into products. This rate constant is frequency, $v^{\neq}$, of the molecular vibration of the activated complex along the reaction coordinate. This vibration has no restoring force and allows only decomposition of the activated complex to products. We see then that the rate of a reaction is expressed as

$$rate = v^{\neq}[S^{\neq}]$$ (3.3)

or

$$rate = \left(\frac{kT}{\hbar}\right)[S^{\neq}]$$ (3.4)

where k^* and h are Boltzmann and Planck constants, respectively.[2]

Now, if $K_{\neq}$, the equilibrium constant for conversion of S into $S^{\neq}$, is defined as $[S]/[S^{\neq}]$, then Equation 3.4 becomes

$$rate = \left(\frac{kT}{\hbar}\right)K_{\neq}[S]$$ (3.5)

or

$$k = \left(\frac{kT}{\hbar}\right)K_{\neq}.$$ (3.6)

Treatment of $K_{\neq}$ as if it were a simple equilibrium constant for conversion of two stable materials allows Equation 3.6 to be rewritten as

$$k = \left(\frac{k^*T}{h}\right)e^{-\Delta G^{\neq}/RT}.$$ (3.7)

[2] The identity of $v^{\neq}$ and $k^*T/\hbar$ is based on statistical mechanical considerations discussed by Eyring (1935a,b).

The pre-exponential term has special meaning in that at room temperature it is equal to 10^{13}/s, or the frequency of a single molecular vibration.[3] This term can be regarded as Nature's chemical "speed limit" since all first-order rate constants are some fraction of it.

Thermodynamic theory allow us to express Equation 3.7 as

$$k = \frac{k*T}{h}\left(e^{\Delta S^{\neq}/R}\right)\left(e^{-\Delta H^{\neq}/RT}\right). \tag{3.8}$$

This expression tells us that the rate constant for a first-order reaction can be considered to comprise factors having to do with the energy required to desolvate and align reactants for reaction, the energy that the molecular system must absorb for it to enter a highly reactive state, and the universal rate constant for decomposition of this activated system to products.

The power of transition state theory[4] is that it allows the chemist to think about the reactions he studies in terms of the principles of chemical thermodynamics, with its emphasis on state-to-state transitions and independence from dynamical or reaction pathway considerations. In transition state theory, a rate constant reflects the free energy difference between two states: the reactant state and the transition state. For complex reactions with multiple intermediates and transition states, the identification of these two states can be aided by free energy diagrams (see below). Now we build on this introduction to transition state theory to construct free energy diagrams.

3.1.2   Free Energy Diagrams

Free energy diagrams plot the energy dependence of the various intermediates of a chemical reaction as the reaction progresses from reactant to product. These diagrams allow immediate visualization of the essentials of a reaction's kinetics and are of great utility in representing and understanding the kinetic features of organic and biochemical reactions. Free energy diagrams are widespread in the literature of enzyme mechanistic studies, and are the forerunner of free energy landscapes that portray the complex dynamics of protein folding.

In Figure 3.1, I have drawn the free energy diagram for the reversible, endothermic chemical transformation of S into P via intermediate X. A number of points common to all free energy diagrams are illustrated in this figure.

3.1.2.1   Axes and Construction of Free Energy Diagrams.   The first thing
to note are the axes. The x-axis is reaction progress, which is simply the sequence of

---

[3] From infrared spectroscopy, we know that organic molecules have absorption bands ranging from about $500\,cm^{-1}$ to $3800\,cm^{-1}$. When multiplied by the speed of light ($3 \times 10^{10}\,cm\,s^{-1}$), we calculate vibrational frequencies of 2×10^{13} to $5 \times 10^{14}\,s^{-1}$.

[4] Transition state theory has played an important role in the development of enzymology. In Appendix B of this book, I discuss two such roles: the development of theories of enzyme catalytic power and the design of enzyme inhibitors.

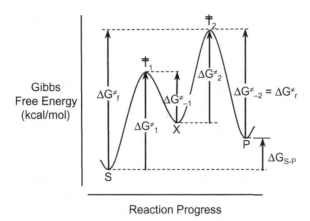

Reaction Progress

Figure 3.1. Free energy diagram for a reversible chemical reaction with a single intermediate.

reaction species as they occur in the course of the reaction. It is a common mistake to think of the x-axis of these diagrams as the reaction coordinate. However, the reaction coordinate exists only in the transition state, and nowhere else along the axis of reaction progression, and corresponds to that specific vibrational mode of the activated complex that lacks a restoring force. The y-axis of these diagrams is Gibbs free energy rather than some other expression of energy, such as potential energy. Of course, energy has no absolute metric to which all species can be related, so the plotted free energy values are all relative to one of the several reaction species. Frequently this species is the substrate in its reactant state, although it need not be.

Free energy diagrams are constructed using $\Delta G^{\neq}$ values for each of the chemical steps that comprise the overall reaction. For a reaction with n individual steps, there are $2n$ $\Delta G^{\neq}$ values that need to be determined to construct an accurate free energy diagram. In this case, there are four values, $\Delta G_1^{\neq}$, $\Delta G_{-1}^{\neq}$, $\Delta G_2^{\neq}$, and $\Delta G_{-2}^{\neq}$.

Recall that $\Delta G^{\neq}$ values can only be calculated from first-order rate constants (see Eq. 3.7). This means that second-order rate constants must be converted to pseudo-first-order rate constants before free energies of activation can be calculated. For the bimolecular reaction $A + B \rightarrow C$ governed by second-order rate constant k_{AB}, pseudo-first-order rate constant k_I is equal to $k_{AB}[X]$, where X is either A or B. Choice of which substrate to use in the calculation is governed by the process one wishes to depict in the diagram. For example, in carbonyl addition reactions, one might choose the nucleophile of the reaction to include in the calculation of the k_I. In a free energy diagram based on k_I, the conversion of the carbonyl moiety into a tetrahedral intermediate would be depicted. Finally, note that the magnitude of k_I, on which $\Delta G^{\neq}$ is dependent, is directly proportion to $[X]$. In the construction of free energy diagrams, one should choose a value of $[X]$ that is relevant to the experimental conditions used in the kinetic characterization of the biomolecular reaction being considered.

3.1.2.2 $\Delta G\neq$ for the Overall Reaction. The rate constant for the "forward" progression of any reversible reaction with multiple sequential steps is based on the free energy difference between the highest energy transition state and species of lowest energy preceding it. The latter is called the reactant state of the reaction, while the former is the rate-limiting transition state of the reaction. Visual inspection of Figure 3.1 tells us immediately that the energy of activation for the forward reaction, $\Delta G_f^{\neq}$, is the free energy difference between S and the transition state for the second reaction step. The same considerations apply for the rate of the reverse reaction. We will see in other sections of this book that reactions containing irreversible segments require special considerations in determining the energy of activation for the overall chemical process.

3.1.2.3 Equilibrium Constant for the Overall Reaction. For a reversible sequential reaction, ΔG_{eq} can be calculated in two ways: as the difference between the activation energies for forward and reverse reactions, $\Delta G_{eq} = \Delta G_f^{\neq} - \Delta G_r^{\neq}$, or as the free energy difference between the two most stable species, $\Delta G_{eq} = \Delta G_{S-P}$. These two calculations must of course be equivalent since, by the principle of microscopic reversibility, $\Delta G_f^{\neq}$ and $\Delta G_r^{\neq}$ reflect the same transition state, thus $\Delta G_f^{\neq} - \Delta G_r^{\neq} = \Delta G_{S-P} = \Delta G_{eq}$.

3.2 KINETIC CONSEQUENCES OF AN ENZYME:SUBSTRATE COMPLEX

We saw in Chapter 2 that in the opening years of the twentieth century, enzymologists posited that enzyme-catalyzed reactions proceed with the initial formation of a binary complex of enzyme and substrate. In this section, we examine the thermodynamics of E:S complex formation and the kinetic partitioning of the complex once it is formed.

3.2.1 Thermodynamics of Saturation Kinetics

A kinetic feature of all enzyme-catalyzed reactions is the observation of so-called "saturation kinetics," in which initial velocities increase with increasing substrate concentration until a limiting velocity is reached, at which point v_o becomes $[S]_o$ independent. As we learned in Chapter 2, this behavior is a consequence of the intermediacy of the E:S complex. Here, we examine the underlying thermodynamics of the formation E:S from E and S.

During the steady state of an enzyme-catalyzed reaction for which rapid equilibrium conditions obtain (see Fig. 2.3), the concentrations of E and E:S are equilibrium concentrations and are thus determined by the ratio $[S]_o/K_S$, which of course equals $[ES]/[E]$. When $[S]_o/K_S \ll 1$, enzyme exists almost entirely as free enzyme, whereas when $[S]_o/K_S \gg 1$, enzyme is captured by substrate and exists predominantly as the E:S complex. Figure 3.2 illustrates these concepts as free energy diagrams.

In the diagrams of Figure 3.2 we assume a K_s of $1\,\mu M$ and a k_c value of $0.1\,s^{-1}$. The diagram on the left illustrates the kinetic situation when $[S]_o/K_S \ll 1$. From this diagram we see that the reversible formation of E : S from E is an endothermic reaction, requiring an input of energy. In contrast, the diagram on the right shows the exothermic formation of E : S that results under the condition $[S]_o/K_S \gg 1$. To under-

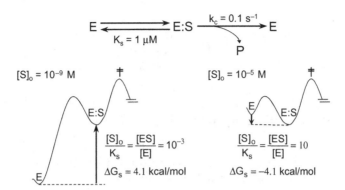

Figure 3.2. Thermodynamics of the formation of the Michaelis complex. The free energy diagrams illustrate how substrate concentration determines ΔG for dissociation of E : S.

stand how it is that substrate concentration determines the energy requirements of this equilibrium process, we take a chemical thermodynamics approach.

We start by noting that $[S]_o/K_S$ is the unitless equilibrium constant governing the reversible formation of E : S from E and, further, that $[S]_o/K_S$ is in the form of an association constant, not a dissociation constant, the norm in enzymology. Expressing this ratio as an association constant allows us to calculate ΔG_s directly from the familiar thermodynamic expression $K_{eq} = \exp(-\Delta G_{eq}/RT)$:

$$\frac{[S]_o}{K_s} = e^{-\Delta G_s / RT} \tag{3.9}$$

$$\Delta G_s = -\ln\left(\frac{[S]_o}{K_s}\right) RT \tag{3.10}$$

Thus, for the diagram on the left side of Figure 3.2 where $[S]_o/K_S = 10^{-3}$, we can calculate a ΔG_s of 4.1 kcal mol^{-1}, while for the right-hand diagram where $[S]_o/K_S = 10$, we calculate a ΔG_s of -1.4 kcal mol^{-1}.

These free energy changes can be broken down into enthalpy and entropy components according to the familiar expression

$$\Delta G_s = \Delta H_s - T\Delta S_s. \tag{3.11}$$

ΔH_s derives its value from changes in chemical bonding that occur when E is transformed into E : S and thus do not vary with substrate concentration. On the other hand, ΔS_s takes its value from two components, as expressed in Equation 3.12,

$$\Delta S_s = \Delta S_{dilution} + \Delta S_{chemistry} \tag{3.12}$$

and while $\Delta S_{chemistry}$ reflects entropy changes associated with the chemistry of E : S formation from E, $\Delta S_{dilution}$ reflects the requirement of bringing enzyme and substrate together from dilute solution. Thus, it is this term that is dependent on substrate concentration, becoming larger with substrate dilution. A larger value of $\Delta S_{dilution}$ results in an overall larger ΔS_s and thus larger ΔG_s. We now see that with lower and lower substrate concentrations, the formation of E:S becomes ever more endothermic.

3.2.2 Kinetic Partitioning of the E:S Complex

The E:S complex has one of two fates: it can turn over to form product or decompose to regenerate free substrate. In this section, we will see how partitioning of E:S sets the meaning of measured steady-state kinetic parameters.

Consider again the mechanism of Figure 2.5 and its rate law in Equation 2.11. This expression can be rearranged to Equation 3.13, to illustrate the impact that the E:S partitioning

$$v_o = \frac{k_2[E]_o[S]_o}{K_s\left(1+\dfrac{k_2}{k_{-1}}\right)+[S]_o} \tag{3.13}$$

ratio k_2/k_{-1} has on the meaning of the measured steady-state rate parameters K_m and k_c/K_m, whose meanings are expressed in Equations 3.14 and 3.15:

$$K_m = K_s\left(1+\frac{k_2}{k_{-1}}\right) \tag{3.14}$$

$$\frac{k_c}{K_m} = \frac{k_2}{K_s\left(1+\dfrac{k_2}{k_{-1}}\right)}. \tag{3.15}$$

When $k_2/k_{-1} \ll 1$, E : S partitions predominantly to enzyme and substrate, and the turnover of E : S limits the reaction rate. Under these circumstances, E and E : S are in equilibrium, and the measured parameters K_m and k_c/K_m equal K_s and k_2/K_s, respectively, and thus are true reflections of substrate binding affinity and catalytic efficacy.

On the other hand, when $k_2/k_{-1} \gg 1$, E:S partitions predominantly to enzyme and product, and substrate dissociation from E:S is rate-limiting. Under these circumstances, Equations 3.14 and 3.15 reduce to $K_m = k_2/k_1$ and $k_c/K_m = k_1$, respectively. We see that under this condition, K_m does not reflect the affinity with which enzyme binds substrate, and k_c/K_m has nothing to do with catalysis; rather, it is equal to the second-order rate constant for the binding of substrate to enzyme.

Examples of how differences in partitioning of E:S result in differences in the meaning of steady-state kinetic parameters come from the serine protease literature. These enzymes hydrolyze amide bonds of proteins and peptides as well as synthetic substrates, including the esters and amides of short peptides and N-blocked amino acids. Serine proteases follow the mechanism of Figure 3.3, in which formation of E:S is followed by acylation of the active site serine with expulsion of the first product P_1,

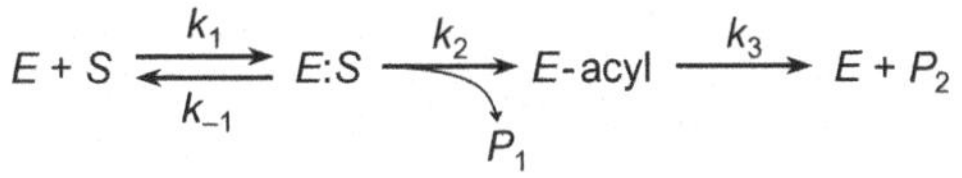

$$E + S \underset{k_{-1}}{\overset{k_1}{\rightleftharpoons}} E{:}S \overset{k_2}{\longrightarrow} E\text{-acyl} \overset{k_3}{\longrightarrow} E + P_2$$
$$\downarrow$$
$$P_1$$

Figure 3.3. Minimal mechanism for reactions catalyzed by serine proteases.

which is an amine in the case of amide substrates or alcohols for ester substrates. Finally, the acyl–enzyme hydrolyzes, producing free enzyme and P_2, a carboxylic acid. In Chapter 5 we will discuss this mechanism in more detail, but here we will restrict our attention to the $E{:}S$ complex and how it partitions between enzyme and acyl-enzyme.

In large part, this partitioning depends on the hydrolytic reactivity of the substrate. Amides are quite unreactive and consequently their acylation rate constants k_2 are small, typically less than $1\,\mathrm{s}^{-1}$. This value is much less than k_{-1}, which can range from 10^2 to $10^3\,\mathrm{s}^{-1}$. Thus, for amide substrates, K_m equals K_s and reflects the affinity of enzyme toward substrate, and k_c/K_m equals k_2/K_s and reflects the catalytic efficiency of the enzyme toward the amide substrate.

Relative to amides, esters, and especially phenyl esters, are very reactive toward hydrolysis, rendering them extremely reactive toward acylation. For example, it has been estimated that acylation of α-chymotrypsin by Ac-Trp p-nitrophenyl ester occurs with a k_2 value that likely exceeds $10^4\,\mathrm{s}^{-1}$; a value much larger than k_{-1}. As a consequence, $k_c/K_m = k_1$ and $K_m = k_3/k_1$ (see Chapter 7). We see then for this class of substrate, neither k_c/K_m nor K_m are what we might think they are: reflections of catalytic efficiency and binding affinity.

3.2.3 Summary

In the brief discussion that follows, I will summarize important concepts of this section.

3.2.3.1 *Relationship between Initial Velocity and Steady-State Kinetic Parameters.* The concentration of substrate sets the kinetic parameter that is reflected in initial velocities. The two limiting conditions and the corresponding initial velocity expression are given below:

$$[S]_o \ll K_m; \quad v_o = \frac{k_c}{K_s}[E]_o[S]_o \tag{3.16}$$

$$[S]_o \gg K_m; \quad v_o = k_c[E]_o. \tag{3.17}$$

3.2.3.2 *Meaning of Steady-State Kinetic Parameters.* For irreversible enzymatic reactions, or a reaction segment preceding the first irreversible step, such as the simple reaction of Figure 3.1 and the acylation portion of the mechanism of Figure 3.3,

- k_c/K_m reflects the free energy difference between the reactant state of free enzyme and the transition state of highest energy.

- k_c reflects the free energy difference between the most stable intermediate and the transition state of highest energy following it.

Analysis of reactions with irreversible segments, such as shown in Figure 3.3, will be discussed in Chapter 7.

$$E + S \underset{K_s}{\rightleftharpoons} E{:}S \underset{k_{-2}}{\overset{k_2}{\rightleftharpoons}} X \xrightarrow{k_3} E + P$$

Figure 3.4. Mechanism for an enzyme proceeding through two reversibly formed intermediates.

Experimental condition:	$[S]_o \gg [E]_o$
Conservation equations:	$[S]_o = [S] + [ES] + [X] = [S]$
	$[E]_o = [E] + [ES] + [X]$
Definition of binding constant:	$K_s = \dfrac{[E][S]}{[ES]}; \quad \dfrac{[E]}{[ES]} = \dfrac{K_s}{[S]} = \dfrac{K_s}{[S]_o}$
Statement of rate dependence:	$v_o = k_3[X]$
Steady-state condition:	$d[X]/dt = 0$

$$k_2[ES] = (k_{-2} + k_3)[X]$$

$$[X] = \frac{k_2}{k_{-2} + k_3}[ES]$$

Derivation of rate equation:

$$\frac{v_o}{[E]_o} = \frac{k_3[X]}{[E] + [ES] + [X]} \tag{3.18}$$

$$\frac{v_o}{[E]_o} = \frac{k_3}{\dfrac{[E]}{[X]} + \dfrac{[ES]}{[X]} + 1} \tag{3.19}$$

$$\frac{v_o}{[E]_o} = \frac{k_3}{\dfrac{[E]}{[ES]}\dfrac{[ES]}{[X]} + \dfrac{[ES]}{[X]} + 1} \tag{3.20}$$

$$\frac{v_o}{[E]_o} = \frac{k_3}{\dfrac{K_s}{[S]_o}\dfrac{k_{-2} + k_3}{k_2} + \dfrac{k_{-2} + k_3}{k_2} + 1} \tag{3.21}$$

$$\frac{v_o}{[E]_o} = \frac{k_3[S]_o}{K_s\dfrac{k_{-2} + k_3}{k_2} + [S]_o\dfrac{k_2 + k_{-2} + k_3}{k_2}} \tag{3.22}$$

$$\frac{v_o}{[E]_o} = \frac{\dfrac{k_2 k_3}{k_2 + k_{-2} + k_3}[S]_o}{K_s\dfrac{k_{-2} + k_3}{k_2 + k_{-2} + k_3} + [S]_o} \tag{3.23}$$

Figure 3.5. Derivation of the rate law for the mechanism of Figure 3.3.

3.3 REACTIONS WITH MORE THAN ONE INTERMEDIARY COMPLEX

To this point, we have primarily been concerned with enzymatic reactions that have a single intermediate, the Michaelis complex. It is far more common in enzymology to encounter enzymes with multiple intermediates where, in addition to the Michaelis complex, there are intermediate species reflecting stepwise chemical events, conformational changes of the enzyme, or accumulation of an enzyme: product complex. In this subsection, we will turn our attention to such mechanisms, using the model of Figure 3.4 as our focus.

The mechanism of Figure 3.4 proceeds through both a Michaelis complex E : S and intermediate X. While the Michaelis complex is in equilibrium with free enzyme, intermediate X may or may not be in equilibrium with the Michaelis complex. For such a mechanism, the rate will be determined by either the formation or the decomposition of X. The rate law for this mechanism is derived by the methods described in Chapter 2 in Figure 3.5. We see that the mechanism follows Michaelis–Menten kinetics with

$$k_c = \frac{k_2 k_3}{k_2 + k_{-2} + k_3} \tag{3.24}$$

$$K_m = K_s \frac{k_{-2} + k_3}{k_2 + k_{-2} + k_3} \tag{3.25}$$

$$\frac{k_c}{K_m} = \frac{k_2 k_3}{K_s (k_{-2} + k_3)}. \tag{3.26}$$

Inspection of these three equations reveals that the identities of k_c and k_c/K_m are determined by relative magnitudes of the microscopic rate constants k_2, k_{-2}, and k_3. As usual we will be concerned with limiting cases.

For k_c/K_m, there are two limiting cases, based on the relative magnitudes of k_{-2} and k_3:

$$k_{-2} > k_3 \rightarrow k_c / K_m = k_3 / K_s K_2 \tag{3.27}$$

$$k_{-2} < k_3 \rightarrow k_c / K_m = k_2 / K_s. \tag{3.28}$$

Now, when $k_{-2} < k_3$, the form that k_c takes is determined by the relative magnitudes of k_{-2} and k_2:

$$k_{-2} > k_2 \rightarrow k_c = k_3 / K_2 \tag{3.29}$$

$$k_{-2} < k_2 \rightarrow k_c = k_3. \tag{3.30}$$

Similarly, when $k_{-2} > k_3$,

$$k_2 < k_3 \rightarrow k_c = k_2 \tag{3.31}$$

$$k_2 > k_3 \rightarrow k_c = k_3. \tag{3.32}$$

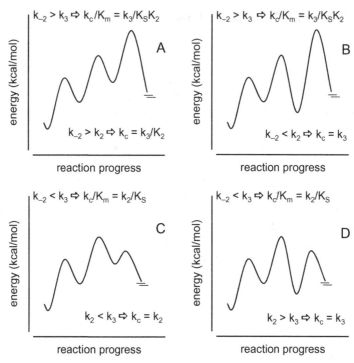

Figure 3.6. Free energy diagrams representing the four limiting kinetic situations for an enzyme-catalyzed reaction with two reversibly formed intermediates.

The four possible kinetic situations that result from these limiting cases are illustrated in Figure 3.6 as free energy diagrams. In these diagrams, the features to which we should be are aware the relative stabilities of E:S and X, and the relative stabilities of $\neq_2$ and $\neq_3$.

Panel A $k_{-2} > k_3$; $k_{-2} > k_2$. In this diagram, E:S is the more stable intermediate and $\neq_3$ is the highest energy transition state. This means that E:S and X are in equilibrium, $\neq_3$ is the rate-limiting transition state for both k_c/K_m and k_c, and that E:S is the reactant state for k_c. Since E:S is more stable than X, k_c reflects the energy difference between E:S and $\neq_3$. This means that k_c must contain the equilibrium constant K_2 (i.e., k_{-2}/k_2). K_m reflects the equilibrium between E and E:S and thus reflects enzyme affinity for substrate.

Panel B $k_{-2} > k_3$; $k_{-2} < k_2$. As in panel A, $\neq_3$ is the highest energy transition state, so k_c/K_m still reflects $\neq_3$. However, X is now the more stable intermediate and as such is the reactant state for k_c. Also in contrast to the energetics of Panel A, K_m reflects the equilibrium between E and X and thus does *not* reflect the affinity of enzyme for substrate.

Panel C $k_{-2} < k_3$; $k_2 < k_3$. In this situation, E:S is more stable than X, and therefore is the reactant state for k_c. This, together with the fact that the highest energy transition state is $\neq_2$, means that $\neq_2$ is reflected in both k_c/K_m and k_c. Note that with these two combined facts, k_3 is kinetically "invisible," meaning that the use of mechanistic probes based on standard steady-state kinetics cannot reveal features of the process governed by k_3.

Panel D $k_{-2} < k_3$; $k_2 > k_3$. Like Panel C, the highest energy transition state is $\neq_2$ and is reflected in k_c/K_m. But here X is the more stable intermediate, meaning that k_c reflects the energy between X and $\neq_3$. Note that k_c/K_m and k_c reflect different transition states and that K_m equals $K_s(k_3/k_2)$. Kinetic terms always appear in K_m when k_c/K_m and k_c reflect different transition states.

This analysis highlights the fact that for most of enzyme reactions we study, which have more than a single intermediate, great care has to be taken in the interpretation of steady-state kinetic parameters. K_m may not be a simple reflection of binding affinity and k_c/K_m may not be a reflection of the transition state for the chemical step of the reaction.

3.4 DEVIATIONS FROM MICHAELIS–MENTEN KINETICS

The dependence of v_o on $[S]_o$ does not always follow simple Michaelis–Menten kinetics. As of 1977, the number of enzymes whose kinetic could not accurately be described by the Michaelis–Menten equation numbered more than 800 (Hill et al. 1977). To account for complex kinetic behavior, William Bardsley and his colleagues at the University of Manchester proposed using the expression of Equation 3.33 in which each coefficient

$$\frac{v_o}{[E]_o} = \frac{\alpha_1[S]_o + \alpha_2[S]_o^2 + ... + \alpha_n[S]_o^n}{\beta_0 + \beta_1[S]_o + \beta_2[S]_o^2 + ... + \beta_m[S]_o^m} \tag{3.33}$$

represents combinations of kinetic and dissociation constants (Bardsley et al. 1980). The greater the complexity of an enzyme's kinetic behavior, the greater the order of the expression that must be used to account for the behavior. The simplest kinetic behavior is Michaelis–Menten kinetics, which is observed when $n = m = 1$. In this case, Equation 3.33 reduces to

$$\frac{v_o}{[E]_o} = \frac{\alpha_1[S]_o}{\beta_0 + \beta_1[S]_o} \tag{3.34}$$

where α_1 is k_{cat}, β_o is K_m, and β_1 equals 1.

The most commonly observed dependencies of v_o on $[S]_o$ that differ from the hyperbolic dependence of Michaelis–Menten kinetics are shown in Figure 3.7 and correspond to Equation 3.33 with $n = m = 2$, leading to Equation 3.35:

$$\frac{v_o}{[E]_o} = \frac{\alpha_1[S]_o + \alpha_2[S]_o^2}{\beta_0 + \beta_1[S]_o + \beta_2[S]_o^2}. \tag{3.35}$$

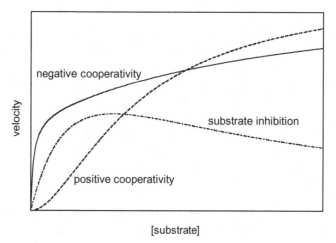

Figure 3.7. Deviations from Michaelis–Menten kinetics.

In the context of rapid-equilibrium mechanisms, these dependencies can be accounted for by substrate inhibition, positive cooperativity, and negative cooperativity. We will discuss each of these in turn, with an emphasis on developing equilibrium mechanisms to account for this kinetic behavior. In the final part of this section, we will see how these dependencies can arise for nonequilibrium mechanisms (Richard et al. 1974; Segel 1975; Whitehead 1976; Neet and Ainslie 1980).

3.4.1 Substrate Inhibition

Substrate inhibition is the reduction of reaction velocities at high substrate concentrations relative to rates that would be predicted from Michaelis–Menten kinetics. In the context of Equation 3.35, substrate inhibition occurs when $\alpha_2 = 0$ and $\beta_1 = 1$, in which case Equation 3.35 becomes

$$\frac{v_o}{[E]_o} = \frac{\alpha_1[S]_o}{\beta_0 + [S]_o + \beta_2[S]_o^2} = \frac{\alpha_1[S]_o}{\beta_0 + [S]_o\left(1 + \beta_2[S]_o\right)}. \tag{3.36}$$

Diminished velocities at high substrate concentrations arise from the $[S]_o^2$ term in the denominator of Equation 3.36. Assuming rapid equilibrium conditions, a squared term in substrate concentration means that an enzyme form accumulates in the steady state comprising enzyme and two molecules of substrate. The simplest mechanism to account for this is shown in Figure 3.8 in which substrate binds to the E:S complex to form an inactive complex S:E:S.

The rate law for substrate inhibition is expressed in Equation 3.37 and was used to draw the

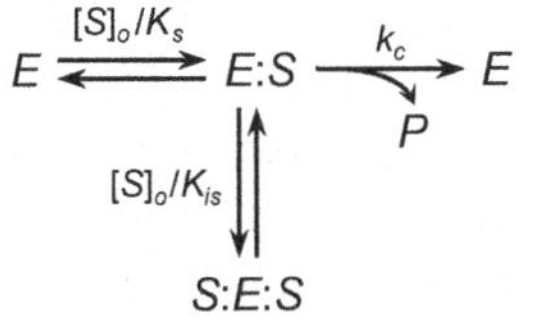

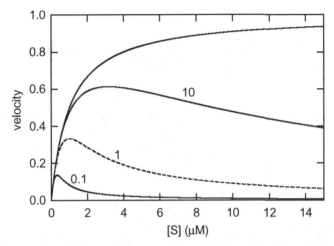

Figure 3.8. Mechanism for substrate inhibition.

Figure 3.9. Dependence of v_o on $[S]_o$ for substrate inhibition at the three indicated ratios of K_{is}/K_s.

$$\frac{v_o}{[E]_o} = \frac{k_c[S]_o}{K_s + [S]_o\left(1 + \dfrac{[S]_o}{K_{is}}\right)} \tag{3.37}$$

simulations of Figure 3.9 at three values of K_{is}/K_s. We see in these simulations that as substrate concentration increases, initial velocities fall off, relative to the case of no substrate inhibition. In principle, values for the three parameters of Equation 3.37 can be solved for by nonlinear least squares fit of a data set to this equation. I say "in principle" because of the difficulty in arriving at accurate initial estimates for these parameters.

Nonlinear least squares analysis always requires initial estimates for the equation's parameters. For many analyses, these estimates can be made by a simple inspection of the data set, mindful of the model to which the data set is to be fit. For the fit of the $[S]_o$ dependence of v_o to the Michaelis–Menten equation, the initial estimate of V_{max} might be the v_o at the highest substrate concentration, and an estimate of K_m would be the substrate concentration corresponding to half of the estimate of V_{max}. Likewise, initial parameter values for the dependence of v_o on $[I]_o$ would be by simple inspection as outlined in the beginning of Chapter 4. We see that for these two examples, Michaelis–Menten kinetics and a simple binding isotherm, there are well-established "landmarks" that allow more-or-less accurate parameter estimates to be made. This is not the case for substrate inhibition. Inspection of the dependence of v_o on $[S]_o$ for a case of substrate inhibition reveals nothing that could be used to guide an investigator in the choice of initial parameter estimates. But there are some readily made calculations that provide parameter estimates.[5]

We start by taking the partial derivative of both sides of Equation 3.37 with respect to $[S]_o$:

$$\frac{\partial v_o}{\partial [S]_o} = \frac{V_{max} - v_o - \dfrac{2v_o[S]_o}{K_{si}}}{K_s + [S]_o \left(1 + \dfrac{[S]_o}{K_{si}}\right)}. \tag{3.38}$$

At the peak of the curve, the partial derivative is equal to zero (see Figure 3.10A). Substituting this into Equation 3.38 and rearranging, we find

$$[S]_{o,max} = \frac{K_{si}(V_{max} - v_{o,max})}{2v_{o,max}} \tag{3.39}$$

where $([S]_{o,max}, v_{o,max})$ is the point at the peak of the curve.

Evaluating the slope of the curve at $[S]_o = 0$, we find

$$\left.\frac{\partial v_o}{\partial [S]_o}\right|_{[S]=0} = \frac{V_{max} - v_o}{K_s} = \frac{V_{max}}{K_s}. \tag{3.40}$$

Now, if we examine Equation 3.40 we see that as $[S]_o$ approaches its two extreme values of zero and infinity, we find

$$(v_o)_{[S]_o \to 0} = \frac{V_{max}[S]_o}{K_s + [S]_o} \tag{3.41}$$

and

$$(v_o)_{[S]_o \to \infty} = \frac{V_{max}}{1 + \dfrac{[S]_o}{K_{si}}}. \tag{3.42}$$

As can be seen in Figure 3.10, these lines intersect at a substrate concentration equal to $[S]_{o,max}$. Setting Equations 3.41 and 3.42 equal at that point, we find

$$[S]_{o,max} = \sqrt{K_s K_{si}}. \tag{3.43}$$

---

[5] I wish to thank my colleague Dr. Kenneth Auerbach for both suggesting the method that appears below and for calculating the partial derivatives.

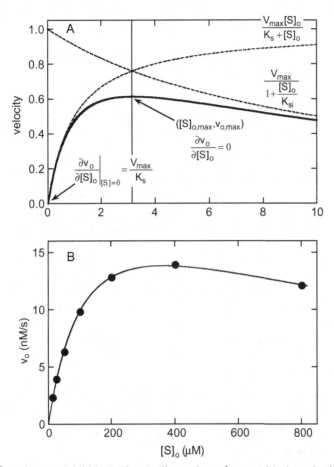

Figure 3.10. Substrate inhibition. *Plot A*: Illustration of a graphical method to estimate parameters for substrate inhibition. *Plot B*: Substrate inhibition for the hydrolysis of Boc-Gly-Arg-Arg-AMC by West Nile protease. Solid points are actual data points, with the solid line through them drawn using Equation 3.37 and best-fit parameters: $V_{max} = 24 \pm 1\,nM\,s^{-1}$, $K_s = 0.13 \pm 0.01\,mM$, and K_{si} of $1.0\,mM \pm 0.1\,mM$. The nonlinear least fit to arrive at these parameters was based on parameter estimates generated by the graphical method of Plot A.

The relationships of Equations 3.41, 3.42, and 3.43 allow us to calculate estimates for K_s, K_{si}, and V_{max}.

Combining Equations 3.40 and 3.43, we find

$$[S]_{o,max} = \sqrt{\frac{V_{max}}{\left. \dfrac{\partial v_o}{\partial [S]_o} \right|_{[S]=0}} K_{si}}, \qquad (3.44)$$

which upon rearranging yields

$$2([S]_{0,max})^2 \left(\frac{\partial v_o}{\partial [S]_o} \bigg|_{[S]=0} \right) = V_{max} K_{si} = \omega. \tag{3.45}$$

This result can be combined with Equation 3.42 to produce

$$K_{si} = \frac{\omega - 2[S]_{o,max} v_{o,max}}{v_{o,max}}. \tag{3.46}$$

Knowledge of K_{si} can be used with Equation 3.43 to calculate an estimate for K_s, and this parameter can be used with Equation 3.41 to calculate V_{max}.

An important feature of this method, is that it allows parameter estimates to be made for situations in which it is impossible to use a sufficiently high range of substrate concentration to fully characterize the substrate inhibition.

To illustrate the utility of this method, I will analyze the substrate inhibition data for the hydrolysis of Boc-Gly-Arg-Arg-AMC by West Nile Virus protease (Tomlinson and Watowich 2008). Figure 3.10B is a plot of the v_o versus $[S]_o$ data that was taken from this paper and clearly indicates substrate inhibition. From this graph we can calculate $\partial v_o/\partial [S]_o|_{[S] = 0}$ as the slope of the line that has an intercept equal to zero and a slope based on the first data point: $(2.3\,nM\,s^{-1})/(12.5\,\mu M) = 0.18\,ms^{-1}$. Our estimate of the point at which $\partial v_o/\partial [S]_o = 0$ must be based on a single data point: $([S]_{o,max}, v_{o,max}) = (400\,\mu M, 13.9\,nM\,s^{-1})$.

From Equation 3.45, we can first calculate, $\omega = \{(400\,\mu M)^2\}(0.18\,ms^{-1}) = 28,800\,\mu M^2\,ms^{-1}$. Next, we use this value together with values of $[S]_{o,max}$ and $v_{o,max}$ with Equation 3.39 to calculate an estimate for K_{si} of 1.3 mM, which allows an estimate of K_s of 0.13 mM to be calculated from Equation 3.43. Finally, we estimate V_{max} as 23 nM s^{-1} using Equation 3.41.

The estimates produced above can be used as initial parameter estimates for a nonlinear least-squares fit of the data of Figure 3.10B to Equation 3.37. Best-fit values are $V_{max} = 24 \pm 1\,nM\,s^{-1}$, $K_s = 0.13 \pm 0.01$ mM, and K_{si} of 1.0 mM $\pm$ 0.1 mM, and were used to draw the solid line through the data in Figure 3.10B.

From this example, one might think that this method not only provides initial estimates for regression analysis, but, in fact, yields bona fide parameter estimates. However, the parameter estimates from this method are often as much as a factor of two different from the actual values. It is best to restrict the use of this method only to parameter estimation for regression analysis.

3.4.2   Cooperativity: General Concepts

Many enzymes function as multimers comprising two or more subunits. Often these subunits are identical, each containing an active site. In these homomultimeric enzymes, the subunits can act independently, each binding substrate with the same affinity, or they can act with *cooperativity*, where binding of one substrate molecule effects binding of subsequent substrate molecules.

The dependence of v_o on $[S]_o$ for a cooperative enzyme can be described by the mechanism-independent expression of Equation 3.47, where K_{sig} is a substrate binding term

$$v_o = \frac{V_{max}}{\left(\dfrac{K_{sig}}{[S]_0}\right)^n + 1} \tag{3.47}$$

with units of M^n, and n determines curve shape as illustrated in Figure 3.11. Sigmoidal curves associated with positive cooperativity are shown in Figure 3.11A, where we see dramatic increases in steepness as n increases from 1, a situation of no cooperativity, to 2 and 4. Figure 3.11B illustrates the curve shapes associated with negative cooperativity with n values of 0.5 and 0.25. In situations of negative cooperativity, the curve is steep at low substrate concentrations but becomes shallow at higher concentrations.

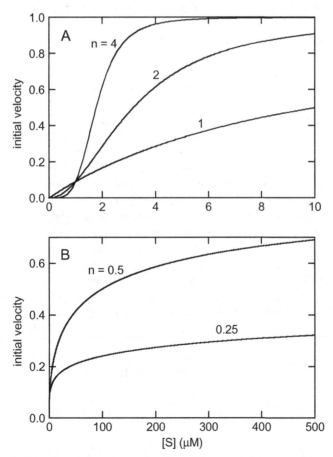

Figure 3.11. Positive (Panel A) and negative (Panel B) cooperativity at various values of slope factor, n.

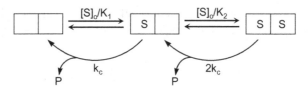

Figure 3.12. Model for cooperativity for a homodimeric enzyme.

In the context of Equation 3.35, a limiting case occurs when $\sqrt{\beta_o} \gg \beta_1$, which allows Equation 3.35 to be simplified to

$$\frac{v_o}{[E]_o} = \frac{\alpha_2}{\dfrac{\beta_0}{[S]_o^2} + \beta_2}. \tag{3.48}$$

Equation 3.48 is of course identical to Equation 3.47, when $\alpha_2 = V_{max}/[E]_o$ and $\beta_2 = 1$.

In structural terms, Equation 3.35 corresponds to the model of Figure 3.12 for a homodimeric enzyme. Binding of substrate to the active site of one of the two subunits produces a complex, $S:E$, that can either turn over to form E and product, or bind another molecule of substrate to form $S:E:S$, which can turn over to form $S:E$ and product. Since $S:E:S$ turns over by two routes, the microscopic rate constant k_c for turnover of one substrate-occupied site must be multiplied by a factor of 2.

Note that the two dissociation constants of this mechanism, K_1 and K_2, reflect substrate/enzyme interactions, not substrate/enzyme-*site* interactions. It is important to understand the differences between these two interactions and how they impact on the interpretation and magnitude of K_1 and K_2. We proceed by considering each dissociative equilibrium as the two rate processes that comprise it.

The rate at which the first molecule of substrate binds to the homodimer to form a singly occupied complex is proportional to the concentration of substrate and the concentration of free sites, which is twice the enzyme concentration. So the rate of $S:E$ formation is equal to $k_1(2[E])[S] = 2k_1[E][S]$, where k_1 is the second-order rate constant for the combination of substrate and a free enzyme site. The rate at which $S:E$ dissociates to substrate and free enzyme is proportional to the concentration of occupied sites. Since there is one occupied site per $S:E$ complex, the rate of dissociation of $S:E$ is equal to $k_{-1}[S:E]$, where k_{-1} is the first-order rate constant for dissociation of an occupied site. At equilibrium,

$$2k_1[E][S] = k_{-1}[S:E] \tag{3.49}$$

and

$$K_1 = \frac{[E][S]}{[S:E:S]} = \frac{k_{-1}}{2k_1} = \frac{K_{S1}}{2} \tag{3.50}$$

where K_{S1} is the dissociation constant for a substrate-occupied site.

We can use the same logic to parse the association of substrate with $S:E$ to form $S:E:S$. The rate of formation of $S:E:S$ from substrate and $S:E$ is proportional to the substrate concentration and the concentration of free sites, or $k_2[S:E][S]$, where k_2 is the second-order rate constant for combination of substrate and the remaining free enzyme site. The rate at which $S:E:S$ dissociates to yield $S:E$ and S is proportional to the concentration of occupied sites, which is twice the concentration of the species $S:E:S$, and equals $k_{-2}(2[S:E:S])$. At equilibrium,

$$k_2[S:E][S] = 2k_{-2}[S:E:S] \tag{3.51}$$

and

$$K_2 = \frac{[S:E][S]}{[S:E:S]} = \frac{2k_{-2}}{k_1} = 2K_{S2} \tag{3.52}$$

where K_{S2} is the dissociation constant for one of the occupied sites of $S:E:S$.

In the rest of this section, we describe cooperativity in terms of K_1 and K_2, so it will be important to bear in mind the two relationships: $K_1 = K_{S1}/2$ and $K_2 = 2K_{S2}$.

The derivation of the rate equation for the mechanism of Figure 3.12 proceeds as follows. We begin with the statement of how the initial velocity relates to rate constants and the two catalytically competent enzyme complexes:

$$\frac{v_o}{[E]_o} = k_c[SE] + 2k_c[SES]. \tag{3.53}$$

Dividing this by the conservation of enzyme produces

$$\frac{v_o}{[E]_o} = \frac{k_c[SE] + 2k_c[SES]}{[E] + [SE] + [SES]}. \tag{3.54}$$

Now, our next move in this derivation is a bit different than what we have done at this stage in previous derivations. Since there is more than a single catalytically competent complex, it is more convenient if we divide through by the concentration of free enzyme rather than the concentration of a catalytically competent complex as we have done previously. Dividing by $[E]$ yields

$$\frac{v_o}{[E]_o} = \frac{k_c\dfrac{[SE]}{[E]} + 2k_c\dfrac{[SES]}{[E]}}{1 + \dfrac{[SE]}{[E]} + \dfrac{[SES]}{[E]}}. \tag{3.55}$$

We now replace each $[X]/[E]$ ratio by the appropriate ratio of total substrate to dissociation constant, to give

$$\frac{v_o}{[E]_o} = \frac{k_c \dfrac{[S]_o}{K_1} + 2k_c \dfrac{[S]_o}{K_1} \dfrac{[S]_o}{K_2}}{1 + \dfrac{[S]_o}{K_1} + \dfrac{[S]_o}{K_1} \dfrac{[S]_o}{K_2}} \tag{3.56}$$

or

$$\frac{v_o}{[E]_o} = \frac{k_c \dfrac{[S]_o}{K_1} + 2k_c \dfrac{[S]_o^2}{K_1 K_2}}{1 + \dfrac{[S]_o}{K_1} + \dfrac{[S]_o^2}{K_1 K_2}}. \tag{3.57}$$

This equation has the form of Equation 3.35, where $\alpha_1 = k_c$, $\alpha_2 = 2k_c$, $\beta_0 = 1$, $\beta_1 = K_1^{-1}$, and $\beta_2 = (K_1 K_2)^{-1}$. Equation 3.57 can account for cases of both positive and negative cooperativity.

3.4.3 Positive Cooperativity

The signature of positive cooperativity is a sigmoidal dependence of v_o on $[S]_o$. When analyzing such a data set, while the investigator will likely start by the use of Equation 3.47 to generate values of n and K_{sig}, he will want to move beyond these mechanism-independent parameters and calculate values of K_1 and K_2 using Equation 3.57. However, if K_1 and K_2 differ by too great an amount, it will be impossible to dissect K_{sig} into the dissociation constants that comprise it.

For the mechanism of Figure 3.12, how large does the ratio K_1/K_2 have to become for failure of Equation 3.57 to yield unique parameter estimates? This problem is addressed in Figure 3.13, where various $\{K_1,K_2\}$ pairs are used together with Equation 3.57, and a constant value of $K_1 K_2$ of $10\,\mu M^2$, to draw dependencies of v_o on $[S]_o$. Note

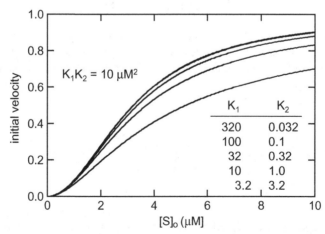

Figure 3.13. Positive cooperativity for values of K_1 and K_2, whose product is $10\,\mu M^2$.

that the three steepest curves, corresponding to K_1/K_2 ratios of 10^4, 10^3, and 10^2, are virtually superimposable. This means that for $K_1/K_2 > 10$, all $\{K_1, K_2\}$ pairs will yield identical dependences of v_o on $[S]_o$, as long as the product $K_1 K_2$ is a constant.

When analyzing a new data set showing a sigmoidal dependence of v_o on $[S]_o$, how can we determine the *possibility* of an accurate dissection of K_{sig}? In fact, the n value calculated from a fit of a data set to Equation 3.47 provides just the indication we seek: as K_1/K_2 increases, n closer approximates 2 and has smaller error limits. This makes intuitive sense, since as K_1/K_2 increases above about 10, Equation 3.51 looks more and more like Equation 3.47. At values K_1/K_2 less than 10, Equation 3.57 no longer can be simplified to Equation 3.47, and attempts at fitting such data sets to Equation 3.47 will be associated with n values that are significantly less than 1 and have large error limits.

To illustrate this concept, consider again the simulated data of Figure 3.13. When these curves are fit to Equation 3.47, the following n values are calculated: $K_1/K_2 = 10^4$, $n = 1.995 \pm 0.001$; $K_1/K_2 = 10^3$, $n = 1.96 \pm 0.01$; $K_1/K_2 = 10^2$, $n = 1.89 \pm 0.02$; $K_1/K_2 = 10$, $n = 1.70 \pm 0.04$; and $K_1/K_2 = 1$, $n = 1.40 \pm 0.06$.

3.4.4 Negative Cooperativity

In the context of the rapid equilibrium mechanism of Figure 3.12, negative cooperativity is observed when $K_1 < K_2$, as shown in Figure 3.14A where data points and the solid line were generated using Equation 3.57 and parameter values: $V_{max} = 0.5$, $K_1 = 0.4\,\mu M$, and $K_2 = 20\,\mu M$. To illustrate the departure from Michaelis–Menten kinetics, the data set was fit to the Michaelis–Menten equation to generate best-fit parameter estimates of $V_{max} = 0.83 \pm 0.04$ and $K_m = 1.4 \pm 0.3\,\mu M$ ($\Sigma(\text{residuals})^2 = 0.040$). These parameter values and the Michaelis–Menten equation were then used to draw the dashed line of this figure.

In contrast to positive cooperativity, estimates of K_1 and K_2 become more accurate as the difference between the two parameters increases. In fact, for K_1 and K_2 values that differ by less than a factor of 10, the data may appear to arise from a noncooperative enzyme following simple Michaelis–Menten kinetics. This is illustrated in Figure 3.14B. The data points and the solid line through them were generated using Equation 3.57 and parameters $V_{max} = 0.5$, $K_1 = 1\,\mu M$, and $K_2 = 10\,\mu M$. When the data set was fit to the Michaelis–Menten equation, the following best-fit parameters were calculated: $V_{max} = 0.93 \pm 0.02$ and $K_m = 2.5 \pm 0.2\,\mu M$ ($\Sigma(\text{residuals})^2 = 0.0043$). The dashed line of the figure was drawn using these parameters and illustrates how well the data set, based on negative cooperativity, can be described by simple Michaelis–Menten kinetics when $K_2/K_1 \leq 10$. This illustration emphasizes the point that it is important to pay attention to small deviations from classical behavior, if these deviations are systematic and repeatable.

3.4.5 Substrate Concentration-Dependent Conformational Transitions

In the discussions above, I have presented departures from Michaelis–Menten kinetics as resulting from cooperativity among subunits of a multimedia enzyme. However, this

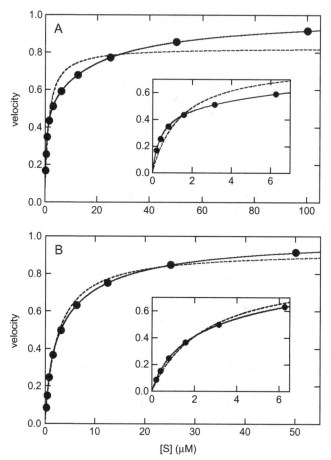

Figure 3.14. Simulations of negative cooperativity, with $K_2/K_1 = 50$ (Panel A) and 10 (Panel B).

sort of kinetic behavior can also be observed for monomeric enzymes if the enzyme exists in two conformations and populations of these forms can be perturbed by substrate binding (Richard et al. 1974; Segel 1975; Whitehead 1976; Neet and Ainslie 1980). Such a mechanism is shown in Figure 3.15, where the enzyme can exist in catalytically active and inactive forms, E and F. If the reversible isomerization between ES and FS is slow relative to the catalytic step governed by k_2, the relative populations of the two enzyme isomers will vary with substrate concentration and influence the overall reaction velocity.

3.4.6 Cooperativity or Conformational Transitions?

The only way to truly demonstrate that deviation from Michaelis–Menten kinetics involves simultaneous binding of more than a single molecule of substrate is by con-

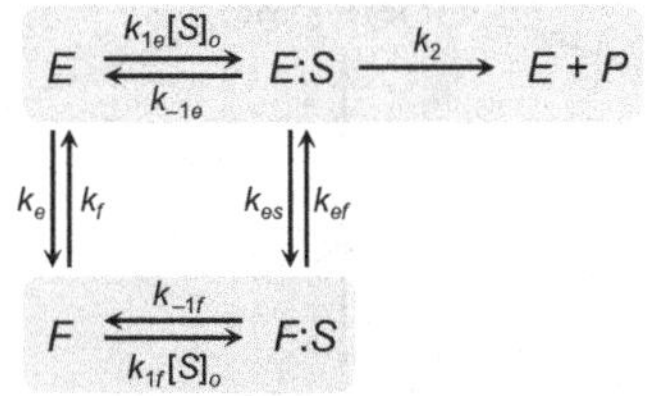

$$E \underset{k_{-1e}}{\overset{k_{1e}[S]_o}{\rightleftharpoons}} E{:}S \xrightarrow{k_2} E + P$$

$$k_e \Big\Updownarrow k_f \qquad k_{es} \Big\Updownarrow k_{ef}$$

$$F \underset{k_{1f}[S]_o}{\overset{k_{-1f}}{\rightleftharpoons}} F{:}S$$

Figure 3.15. Model for non-Michaelis–Menten kinetics involving interconversion of two isomers of enzyme.

ducting a structural or equilibrium experiment. The former includes NMR or X-ray crystallographic studies, while the latter includes equilibrium dialysis. Even if an investigator is dealing with a homodimeric enzyme with data that can be fit to a model involving the binding of two molecules of substrate, he still must conduct studies other than kinetic studies to demonstrate binding of two substrate molecules. In the absence of appropriate structural or equilibrium data, one must interpret the kinetic data cautiously, with the understanding that future studies may prove that models such as shown in Figures 3.8 and 3.12 must be replaced with kinetic models, such as Figure 3.15.

3.5 KINETICS OF ENZYMATIC ACTION ON SUBSTRATES WITH MULTIPLE REACTIVE CENTERS

Our discussion to this point has focused on enzymes that react with simple substrates that have a single reactive center. However, many enzymes react with polymeric substrates at multiple reactive centers. These substrates include proteins, DNA, RNA, polysaccharides, and lignins. The enzymes involved include hydrolases for all these substrates and the various enzymes that perform posttranslational modifications on the monomeric units comprising the polymeric substrates. If we narrow our focus to only those enzymes that work on proteins, these enzymes include proteases, kinases, phosphates, transglutaminases, acetylases, deacetylases, and others. Needless to say, these enzymes all represent targets of drug discovery efforts in the pharmaceutical industry.

These enzymes react with their polymeric substrates by one of two reaction types: processive or nonprocessive. A feature that these two reaction types share is that, in the course of reaction, products are formed which can become substrates. That is, a polymeric substrate with n reactive sites will react with enzyme to produce a product with $n - 1$ reactive sites still remaining. This polymer then becomes a substrate for the enzyme to yield a product with $n - 2$ reactive sites. This cycle of substrate-to-product-to-substrate continues until all reactive sites have been reacted upon by the enzyme.

The feature that distinguishes processive from nonprocessive enzymes is how the two types of enzyme handle these intermediate polymeric species. While nonprocessive enzymes release these polymer intermediates into solution, to be rebound for

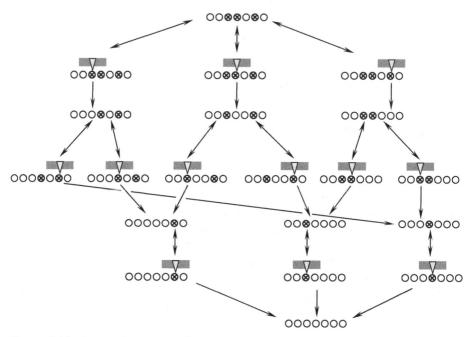

Figure 3.16. Enzymatic turnover of a polymeric substrate with multiple reactive centers by a nonprocessive mechanism.

further reaction, processive enzymes release no intermediate species. The polymer does not dissociate from the enzyme until all the reactive sites have been turned over by enzyme.

Consider the polymer ○○⊗○⊗○ and the enzymes that change ⊗ into ○, to ultimately produce ○○○○○○○. The mechanism by which a nonprocessive enzyme handles this substrate is shown in Figure 3.16. In the first step, the enzyme forms three unique Michaelis complexes with substrate, each centered on one of the three reactive centers. Reaction of these Michaelis complexes yields three unique products, each with two unmodified reactive centers remaining. These three products can now serve as substrates for the enzyme, to form a total of six Michaelis complexes. Reaction of these Michaelis complexes yields three products, each with a single reactive center. These three species react with enzyme to yield the final reaction product ○○○○○○○. In contrast, a processive enzyme would form a single Michaelis complex with the substrate and sequentially transform all ⊗ into ○, with only the product ○○○○○○○ being released into solution.

Examination of the reaction solution of a nonprocessive reaction will reveal unreacted substrate and all possible products. In contrast, the reaction solution of a processive enzyme will contain only unreacted substrate and the final product. This assumes that $[E]_o \ll [S]_o$ and t_{sample}, the time at which the reaction is sampled, is much greater than $1/k_c$. If $[E]_o \sim [S]_o$ and/or $t_{sample} < 1/k_c$, intermediate species will be found even for processive reactions.

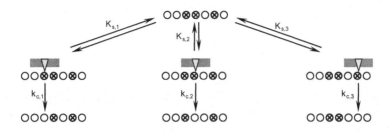

Figure 3.17. Initial velocity conditions for enzymatic polymeric substrate with multiple reactive centers by a nonprocessive (upper scheme) and processive mechanism (lower scheme).

Let us first consider the kinetics of nonprocessive reactions. The mechanism of Figure 3.17 reflects the initial stage of the nonprocessive reaction of Figure 3.16. It is these reaction steps that set the magnitude of the steady-state kinetic parameters determined from initial velocity measurements. According to the mechanism of Figure 3.17, enzyme is seen to bind substrate in three different ways to produce three Michaelis complexes that all turn over to form products with two reactive centers.

Derivation of the rate expression for this mechanism is described below. But before we proceed to this derivation, it is important that we discuss how we will treat the issue of substrate concentrations. If we refer to polymer ○○⊗⊗○⊗○ as S, and the three reactive centers as S_1, S_2, and S_3, we see that

$$[S]_o = [S_1]_o = [S_2]_o = [S_3]_o. \tag{3.58}$$

That is, the initial concentrations of each of the three reactive centers are all identical and equal to the initial concentration of polymer S.

If S_{RC} is a reactive center, then

$$[S_{RC}]_o = [S_1]_o + [S_2]_o + [S_3]_o \tag{3.59}$$

For the most general case,

$$[S_{RC}]_o = \sum_1^{i=n} [S_i]_o, \tag{3.60}$$

but since

$$[S]_o = [S_1]_o = [S_2]_o = ... = [S_n]_o, \tag{3.61}$$

it must be the case that

$$[S_{RC}]_o = n[S_i]_o. \tag{3.62}$$

With these definitions in mind, we can now proceed with the derivation of the initial velocity expression for the reaction of Figure 3.16. If initial velocities are measured for the disappearance of reactive centers or the formation of a common product (e.g., phosphate for phosphatases, or ADP for kinases), we see that

$$v_o = k_{c,1}[(ES)_1] + k_{c,2}[(ES)_2] + k_{c,3}[(ES)_3], \tag{3.63}$$

where $(ES)_i$ is the Michaelis complex for reactive center S_i. Dividing each side by total enzyme gives:

$$\frac{v_o}{[E]_o} = \frac{k_{c,1}[(ES)_1]}{[E]+[(ES)_1]+[(ES)_2]+[(ES)_3]} + \frac{k_{c,2}[(ES)_2]}{[E]+[(ES)_1]+[(ES)_2]+[(ES)_3]}$$
$$+ \frac{k_{c,3}[(ES)_3]}{[E]+[(ES)_1]+[(ES)_2]+[(ES)_3]} \tag{3.64}$$

$$\frac{v_o}{[E]_o} = \frac{k_{c,1}}{\dfrac{[E]}{[(ES)_1]}+1+\dfrac{[(ES)_2]}{[(ES)_1]}+\dfrac{[(ES)_3]}{[(ES)_1]}} + \frac{k_{c,2}}{\dfrac{[E]}{[(ES)_2]}+\dfrac{[(ES)_1]}{[(ES)_2]}+1+\dfrac{[(ES)_3]}{[(ES)_2]}}$$
$$+ \frac{k_{c,3}}{\dfrac{[E]}{[(ES)_3]}+\dfrac{[(ES)_1]}{[(ES)_3]}+\dfrac{[(ES)_2]}{[(ES)_3]}+1}. \tag{3.65}$$

Each of the ratios of enzyme forms in Equation 3.65 can be replaced by the ratio of dissociation constant and $[S_{RC}]_o/3$, the initial concentration of reactive centers divided by n, the number of reactive centers which in this case is 3:

$$\frac{v_o}{[E]_o} = \frac{\dfrac{k_{c,1}}{1+K_{s,1}\left(\dfrac{1}{K_{s,2}}+\dfrac{1}{K_{s,3}}\right)}\dfrac{[S_{RC}]_o}{n}}{\dfrac{K_{s,1}}{1+K_{s,1}\left(\dfrac{1}{K_{s,2}}+\dfrac{1}{K_{s,3}}\right)}+\dfrac{[S_{RC}]_o}{n}} + \frac{\dfrac{k_{c,2}}{1+K_{s,2}\left(\dfrac{1}{K_{s,1}}+\dfrac{1}{K_{s,3}}\right)}\dfrac{[S_{RC}]_o}{n}}{\dfrac{K_{s,2}}{1+K_{s,2}\left(\dfrac{1}{K_{s,1}}+\dfrac{1}{K_{s,3}}\right)}+\dfrac{[S_{RC}]_o}{n}}$$
$$+ \frac{\dfrac{k_{c,3}}{1+K_{s,3}\left(\dfrac{1}{K_{s,1}}+\dfrac{1}{K_{s,2}}\right)}\dfrac{[S_{RC}]_o}{n}}{\dfrac{K_{s,3}}{1+K_{s,3}\left(\dfrac{1}{K_{s,1}}+\dfrac{1}{K_{s,2}}\right)}+\dfrac{[S_{RC}]_o}{n}}. \tag{3.66}$$

Equation 3.61 can be rearranged to produce

$$
\frac{v_o}{[E]_o} = \frac{\dfrac{k_{c,1}/K_{S,1}}{\dfrac{1}{K_{s,1}}+\dfrac{1}{K_{s,2}}+\dfrac{1}{K_{s,3}}}\dfrac{[S_{RC}]_o}{n}}{\dfrac{1}{\dfrac{1}{K_{s,1}}+\dfrac{1}{K_{s,2}}+\dfrac{1}{K_{s,3}}}+\dfrac{[S_{RC}]_o}{n}} + \frac{\dfrac{k_{c,2}/K_{S,2}}{\dfrac{1}{K_{s,1}}+\dfrac{1}{K_{s,2}}+\dfrac{1}{K_{s,3}}}\dfrac{[S_{RC}]_o}{n}}{\dfrac{1}{\dfrac{1}{K_{s,1}}+\dfrac{1}{K_{s,2}}+\dfrac{1}{K_{s,3}}}+\dfrac{[S_{RC}]_o}{n}}
$$

$$
+ \frac{\dfrac{k_{c,3}/K_{S,3}}{\dfrac{1}{K_{s,1}}+\dfrac{1}{K_{s,2}}+\dfrac{1}{K_{s,3}}}\dfrac{[S_{RC}]_o}{n}}{\dfrac{1}{\dfrac{1}{K_{s,1}}+\dfrac{1}{K_{s,2}}+\dfrac{1}{K_{s,3}}}+\dfrac{[S_{RC}]_o}{n}}
\tag{3.67}
$$

and then

$$
\frac{v_o}{[E]_o} = \frac{\dfrac{k_{c,1}/K_{S,1}}{\dfrac{1}{K_{s,1}}+\dfrac{1}{K_{s,2}}+\dfrac{1}{K_{s,3}}}[S_{RC}]_o}{\dfrac{n}{\dfrac{1}{K_{s,1}}+\dfrac{1}{K_{s,2}}+\dfrac{1}{K_{s,3}}}+[S_{RC}]_o} + \frac{\dfrac{k_{c,2}/K_{S,2}}{\dfrac{1}{K_{s,1}}+\dfrac{1}{K_{s,2}}+\dfrac{1}{K_{s,3}}}[S_{RC}]_o}{\dfrac{n}{\dfrac{1}{K_{s,1}}+\dfrac{1}{K_{s,2}}+\dfrac{1}{K_{s,3}}}+[S_{RC}]_o}
$$

$$
+ \frac{\dfrac{k_{c,3}/K_{S,3}}{\dfrac{1}{K_{s,1}}+\dfrac{1}{K_{s,2}}+\dfrac{1}{K_{s,3}}}[S_{RC}]_o}{\dfrac{n}{\dfrac{1}{K_{s,1}}+\dfrac{1}{K_{s,2}}+\dfrac{1}{K_{s,3}}}+[S_{RC}]_o}
\tag{3.68}
$$

The denominator of each of the three terms in the rate expression of Equation 3.68 is the harmonic mean of individual dissociation constants, which is the K_m for the overall reaction:

$$
K_m = \frac{n}{\dfrac{1}{K_{s,1}}+\dfrac{1}{K_{s,2}}+\dfrac{1}{K_{s,3}}}
\tag{3.69}
$$

Equation 3.68 now reduces to

$$
\frac{v_o}{[E]_o} = \frac{\left(\dfrac{K_m/n}{K_{S,1}}k_{c,1}\right)[S_{RC}]_o}{K_m+[S_{RC}]_o} + \frac{\left(\dfrac{K_m/n}{K_{S,2}}k_{c,2}\right)[S_{RC}]_o}{K_m+[S_{RC}]_o} + \frac{\left(\dfrac{K_m/n}{K_{S,3}}k_{c,3}\right)[S_{RC}]_o}{K_m+[S_{RC}]_o}
\tag{3.70}
$$

$$\frac{v_o}{[E]_o} = \frac{\left(\dfrac{\dfrac{K_m}{K_{S,1}}k_{c,1} + \dfrac{K_m}{K_{S,2}}k_{c,2} + \dfrac{K_m}{K_{S,3}}k_{c,3}}{n}\right)[S_{RC}]_o}{K_m + [S_{RC}]_o}.$$ (3.71)

From the above equation, we see that k_c for this reaction is the weighted average of catalytic constants, where the weighting factor is the ratio of the overall K_m value to the individual dissociation constants:

$$k_c = \frac{\dfrac{K_m}{K_{S,1}}k_{c,1} + \dfrac{K_m}{K_{S,2}}k_{c,2} + \dfrac{K_m}{K_{S,3}}k_{c,3}}{n}.$$ (3.72)

This derivation for a nonprocessive reaction of a substrate with three reactive centers can of course be generalized. The dependence of initial velocity on substrate concentration for a substrate with n reactive centers will show a simple hyperbolic dependence on substrate concentration:

$$v_o = \frac{k_c[E]_o[S]_o}{K_m + [S]_o},$$ (3.73)

where K_m is the harmonic mean of all dissociation constants,

$$K_m = \frac{n}{\displaystyle\sum_{i=1}^{n} \frac{1}{K_{s,i}}},$$ (3.74)

k_c is the weighted average,

$$k_c = \frac{\displaystyle\sum_{i=1}^{n} k_{c,i}\frac{K_m}{K_{s,i}}}{n},$$ (3.75)

and k_c/Km is calculated by dividing Equation 3.75 by K_m,

$$\frac{k_c}{K_m} = \frac{\displaystyle\sum_{i=1}^{n} \frac{k_{c,i}}{K_{s,i}}}{n}.$$

We see from Equation 3.74, that K_m predominantly reflects the dissociation of the Michaelis complex of greatest stability. From Equation 3.75, k_c can be seen to be proportional to the sum over all individual $k_{c,i}$, where each is multiplied by a factor $K_m/K_{s,i}$ that gives weight to turnover of the Michaelis complex of greatest stability. Finally, the observed value of k_c/K_m is the simple average of all values of $(k_c/K_m)_i$.

We now turn our attention to processive reaction of Figure 3.17. If we first consider k_c/K_m, we recall that for any enzymatic reaction, k_c/K_m will always reflect the energy barrier between enzyme and substrate free in solution, and the first irreversible step. So, for this reaction,

$$\frac{k_c}{K_m} = \frac{k_1}{K_s}. \tag{3.76}$$

For the series of irreversible steps that comprise substrate turnover for this mechanism, k_c is equal to the reciprocal of the sum of reciprocals of the individual rate constants:

$$k_c = \left(\frac{1}{k_1} + \frac{1}{k_2} + \frac{1}{k_3} \right)^{-1}. \tag{3.77}$$

Thus, k_c will be dominated by the step with the largest energy barrier. From the expression of Equation 3.77, it can be seen that if one step is much slower than the rest, k_c will be numerically equal to the rate constant for that step.

K_m is the ratio of k_c to k_c/K_m, and equals the expression of Equation 3.78 and reflects the most

$$K_m = K_s \frac{\left(\dfrac{1}{k_1} + \dfrac{1}{k_2} + \dfrac{1}{k_3} \right)^{-1}}{k_1} \tag{3.78}$$

stable E:S complex that accumulates in the steady state.

These expressions can be generalized for a processive reaction of a substrate for any number of reactive centers. While the expression for k_c/K_m of course remains the same as in Equation 3.76, k_c and K_m are now expressed as in Equations 3.79 and 3.80, respectively:

$$k_c = \left\{ \sum_{i=1}^{n} (k_i)^{-1} \right\}^{-1} \tag{3.79}$$

$$K_m = K_s \frac{\left\{ \sum_{i=1}^{n} (k_i)^{-1} \right\}^{-1}}{k_1}. \tag{3.80}$$

REFERENCES

Arrhenius, S. (1889). "On the reaction velocity of the inversion of cane sugar by acids." *J. Phys. Chem.* **4**: 226–238.

Bardsley, W. G., P. Leff, J. Kavanagh, and R. D. Waight (1980). "Deviations from Michaelis-Menten kinetics." *Biochem. J.* **187**: 739–765.

Eyring, H. (1935a). "The activated complex and the absolute rates of chemical reactions." *Chem. Rev.* **17**: 65–77.

Eyring, H. (1935b). "The activated complex in chemical reactions." *J. Phys. Chem.* **3**: 107–115.

Glasstone, S., K. J. Laidler, and H. Eyring (1941). *The Theory of Rate Processes.* New York, McGraw-Hill.

Hill, C. M., R. D. Waight, and W. G. Bardsley (1977). "Does any enzyme follow the Michaelis-Menten equation?" *Mol. Cell. Biochem.* **15**: 173–178.

Neet, K. E. and G. R. Ainslie (1980). "Hysteretic enzymes." *Methods Enzymol.* **64**: 192–226.

Richard, J., J. C. Meunier, and J. Buc (1974). "Regulatory behavior of monomeric enzymes 1. The mnemonical enzyme concept." *Eur. J. Biochem.* **49**: 195–208.

Segel, I. H. (1975). *Enzyme Kinetics.* New York, John Wiley & Sons.

Tomlinson, S. M. and S. J. Watowich (2008). "Substrate inhibition kinetic model for West Nile virus NS2B-NS3 protease." *Biochemistry* **47**: 11763–11770.

Whitehead, E. P. (1976). "Simplifications of the derivation and forms of steady-state equations for non-equilibrium random, substrate-modifier, and allosteric enzyme mechanisms." *Biochem. J.* **159**: 449–456.

Wynne-Jones, W. F. K. and H. Eyring (1935). "The absolute rate of reactions in condensed phases." *J. Chem. Phys.* **3**: 492–402.

4

ENZYME INHIBITION: THE PHENOMENON AND MECHANISM-INDEPENDENT ANALYSIS

The notion that an enzyme's catalytic activity can be suppressed by certain chemical compounds is as old as the first systematic studies of enzymes. Recall from Chapter 1 the seminal publication by Adrian Brown in 1902 in which he hypothesized an enzyme-substrate complex as a necessary intermediate during enzyme catalysis (Brown 1902). In this same paper, he also gave what might be the first description of product inhibition, noting that "the action of invertase is influenced prejudicially by the accumulation of its own products of inversion" and that this "arresting influence increases as the amount of the invert sugar increases" (Brown 1902). Some years later, the Dutch scientist Barendrecht, taking up where Brown left off, made the important observation that not only can the products of inversion inhibit invertase, but also structurally related sugars (Barendrecht 1913). These and other observations that were made in the first decade of the twentieth century led to the first structural proposal of how inhibition might come about.

Dixon and Thurlow (1924) and later Coombs (1927) studied the inhibition of xanthine oxidase by purines of varying structure. These systematic studies, the first SAR studies of enzyme inhibition, led to the hypothesis that inhibition "might be due to adsorption by the enzyme of the inhibitory substance, thus preventing the adsorption of one or both of the substrates" (Dixon and Thurlow 1924). Along similar lines,

Kinetics of Enzyme Action: Essential Principles for Drug Hunters, First Edition. Ross L. Stein.
© 2011 John Wiley & Sons, Inc. Published 2011 by John Wiley & Sons, Inc.

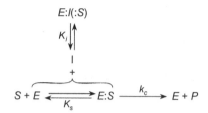

Figure 4.1. Enzyme inhibition—the short story.

Quastel and Wooldridge proposed the "active center theory" to explain inhibition of succinic dehydrogenase (Quastel and Wooldridge 1928). They observed that this enzyme is subject to inhibition by malonate and that this inhibition can be reversed by the addition of succinate. They reasoned that since "enzymes drawn from different sources can activate a particular substrate they should all have the power of adsorbing a particular type of compound of which the active substrate in question is an example." These and other studies conducted during the first quarter of the twentieth century led to the understanding that enzyme inhibition involves the binding of a chemical compound to an enzyme to form a complex that is catalytically inactive.

Like these early studies, this chapter describes enzyme inhibition from a phenomenological perspective, relying on only the most modest mechanistic assumption that inhibition can occur when a molecule binds to free enzyme, the Michaelis complex, or both (see Fig. 4.1 for this minimalist picture).

In this chapter, I develop a general analytical methodology for the dependence of initial velocity on inhibitor concentration, at a fixed concentration of substrate. This method advances in stages from an initial close examination of the data, to quantitative, yet mechanism-independent, determination of inhibition parameters, and finally to mechanistic insights that can be gleaned from kinetic studies conducted at a single substrate concentration. While this chapter considers only one-substrate enzymatic reactions, the methods developed here usually apply well to multisubstrate reactions, in which all substrates are held at a single concentration.

The basic experiment that is the focus of this chapter is the dependence of velocity on inhibitor concentration, and is a necessary prerequisite to more advanced studies that seek to determine the kinetic mechanism of inhibition (see Chapter 5). We will see in this chapter that a great deal of information is available from the simple plot of v_o versus $[I]_o$.

4.1   ENZYME INHIBITION: THE PHENOMENON

Figure 4.2 illustrates the kinds of information that can be garnered from visual inspection of the plotted results of an inhibitor titration experiment. Note that in this simulation, inhibitor concentration spans four orders of magnitude. As we see below, such a broad range is necessary to learn what we need to learn from a first appraisal of the

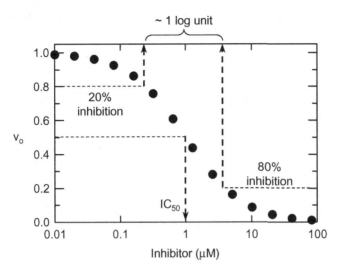

Figure 4.2. Qualitative analysis of the dependence of initial velocity on inhibitor concentration.

phenomena of enzyme inhibition. And, as always, when the independent variable's range covers more than about two orders of magnitude, a logarithmic scale is used to allow visualization of all data points.

Three features of inhibition are revealed from an inspection of Figure 4.1:

- **Completeness of Inhibition.** The first thing to note in this figure is that the velocity is driven to zero by high concentrations of inhibitor. In contrast to this are inhibitor titration curves characterized by nonzero velocities persisting at high concentration of inhibitors, the hallmark of so-called partial inhibitors.

- **Potency of Inhibition.** This is reflected in the inhibitor concentration range that is needed to yield a certain extent of inhibition. An inhibitor is considered potent if its effect is seen at low concentrations. A qualitative measure of inhibitory potency is the IC_{50} value, which is the concentration of inhibitor that reduces the control velocity by a factor of 2. As illustrated in Figure 4.2, the IC_{50} value can be read directly off the plot of v_o versus $[I]_o$. With this, we get our first sense of the affinity with which enzyme binds the inhibitor. However, it is important to understand that an IC_{50} value is but a rough estimate of inhibitor potency, since its magnitude is dependent on a great many variables, including incubation time of enzyme and inhibitor, kinetic mechanism of substrate turnover, kinetic mechanism of inhibition, and substrate concentration. An IC_{50} value can be very different from the true K_i value.

- **Simplicity of Inhibition.** Finally, we see in Figure 4.1 that approximately a 10-fold increase in inhibitor concentration is needed to move from 20% to 80% inhibition, indicating that the inhibition process can be described by a standard binding isotherm (see below).

4.2   ENZYME INHIBITION: THE FIRST QUANTITATIVE STEPS

To move beyond the qualitative assessment described in the previous section to something more quantitative, we must formalize the basic features of an inhibitor titration curve. That is, we need to be able to express the dependence of reaction velocity on inhibitor concentration in terms of the four parameters that define the inhibition of an enzymatic reaction: control velocity, inhibitor potency, completeness of inhibition, and simplicity of inhibition.

The four-parameter function that captures these features is expressed in Equation 4.1. This is

$$v_o = \frac{v_{control} - v_{bkg}}{1 + \left(\dfrac{[I]_o}{K_{i,app}}\right)^n} + v_{bkg} \tag{4.1}$$

the most general of functions that can be used to quantitatively analyze enzyme inhibition data. The only mechanistic assumption of this function is that inhibition occurs through a physical interaction between enzyme and inhibitor that results in the formation of an inactive complex. This assumption provides the general form of the equation, which is a Langmuir binding isotherm.

Other than this, the four-parameter function of Equation 4.1 is mechanism independent, meaning that no mechanism, of either inhibition or substrate turnover, need to be assumed to arrive at a quantitative assessment of the key features of inhibition, whose qualitative assessment we discussed above.

Fitting a v_o versus $[I]_o$ data set to the expression of Equation 4.1 will provide best-fit estimates of the four parameters that describe a binding isotherm:

- v_{contol}, velocity in the absence of inhibitor.
- v_{bkg}, velocity at infinite concentration of inhibitor. This is the measure of the completeness of inhibition, values greater than zero reflecting partial inhibition.
- $K_{i,app}$, apparent inhibition constant.
- n, slope factor or Hill coefficient. This is the measure of binding simplicity. Nonunity values of n reflect complex forms of inhibition that cannot be described by a simple binding isotherm.

For a well-behaved inhibitor, v_{contol} will equal the velocity determined at $[I]_o = 0$, v_{bkg} will equal zero, and n will equal one. Under these conditions, Equation 4.1 becomes the simple expression below:

$$v_o = \frac{v_{control}}{1 + \dfrac{[I]_o}{K_{i,app}}}. \tag{4.2}$$

In Equations 4.1 and 4.2, inhibitor potency is reflected in the term $K_{i,app}$. It is instructive to consider how $K_{i,app}$ differs from IC_{50}, and why it is a more accurate estimate of inhibitory potency. IC_{50} is somewhat of a subjective estimate, $K_{i,app}$ is not; rather it is obtained by curve fitting via a computer-driven algorithm that performs nonlinear regression analysis. Not only do $K_{i,app}$ values eliminate the subjectivity that is inherent in estimates of IC_{50} values, they also provide more accurate potency estimates in cases where v_{bkg} is greater than zero. In these cases, IC_{50} values will always be greater than the $K_{i,app}$. But like IC_{50}, $K_{i,app}$ is still dependent on a number of factors that can that can cause it to differ significantly from the true K_i, whose value we will learn to estimate in Chapter 5.

4.3 ENZYME-INHIBITOR SYSTEMS MISBEHAVING

Not all inhibitors are ideally behaved. This is especially true of compounds identified in high-throughput screens. Often it is seen that $v_{control}$ of Equation 4.1 does not equal the observed velocity at zero inhibitor concentration, the slope factor is significantly different from one, or v_{bkg} is greater than zero. Nonideal behavior of this sort can have origins relating to features of inhibitor mechanism or can be due to assay artifacts. Of course, not every case of nonideal behavior will be readily accounted for. In the discussion that follows, the artifacts that I discuss clearly do not exhaust the list of artifacts that can confound our attempts to discern mechanism.

4.3.1 $v_{control}$ Is Not Equal to the Velocity Determined at Zero Inhibitor Concentration

These days, inhibitor titrations will more likely than not be run in microtiter plates. While convenient, this format can sometimes lead to artifacts caused by well position within the plate. In inhibitor titration studies, it is often the case that the investigator will place all the $[I]_o = 0$ controls in one location, such as the outermost row or column. However, this can lead to invalid controls if these "control wells" are subject to conditions not experienced by wells containing inhibitor, such as might be caused by defects in plate structure or nonuniform temperature across the plate. Under such a circumstance, the "control wells" will exhibit reaction rates that cannot serve as control velocities for the inhibitor titration.

The situation of v_{contol} not being equal to the velocity at zero inhibitor concentration can also occur if velocities are not measured at sufficiently low concentrations of inhibitor. When such a data set is fit to Equation 4.1, the estimated value of v_{contol} can differ significantly from $v_{c,[I] = 0}$. This is illustrated in Figure 4.3. In this simulation, the data points, both filled and unfilled circles, were generated using Equation 4.1, assuming the following parameter values: $v_{control} = 1$, $v_{bkg} = 0$, $K_{i,app} = 1\,\mu M$, $n = 1$, and 3% random error. The data points represented by the black circles are for an inhibitor titration that did not include sufficiently low inhibitor concentrations. When these data are fit to Equation 4.1, the following best fit estimates are obtained: $v_{control} = 0.87 \pm 0.02$, $v_{bkg} = 0.00 \pm 0.01$, $K_{i,app} = 1.2 \pm 0.1\,\mu M$, and $n = 1.3 \pm 0.1$ (solid line). In contrast,

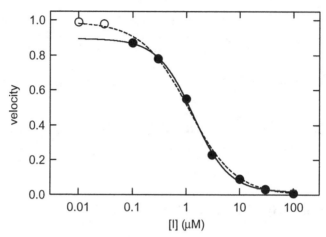

Figure 4.3. Consequence of initial velocities not being measured at sufficiently low concentrations of inhibitor: $v_{contol} \neq v_{c,[I]=0}$. Data points were generated using Equation 4.1, with: $v_{control} = 1$, $v_{bkg} = 0$, $K_{i,app} = 1\,\mu M$, $n = 1$, and 3% random error. Black circles are for an experiment which did not include sufficiently low $[I]_o$. Fitting to Equation 4.1 yields $v_{control} = 0.87 \pm 0.02$, $v_{bkg} = 0.00 \pm 0.01$, $K_{i,app} = 1.2 \pm 0.1\,\mu M$, and $n = 1.3 \pm 0.1$ (solid line). When all the data points were fit, $v_{control} = 0.99 \pm 0.02$, $v_{bkg} = 0.00 \pm 0.01$, $K_{i,app} = 1.1 \pm 0.1\,\mu M$, and $n = 1.0 \pm 0.1$ (dashed line).

when all the data points were fit, $v_{control} = 0.99 \pm 0.02$, $v_{bkg} = 0.00 \pm 0.01$, $K_{i,app} = 1.1 \pm 0.1\,\mu M$, and $n = 1.0 \pm 0.1$ (dashed line). This simulation teaches that accurate parameter estimates will only be obtained when a sufficiently broad range of inhibitor concentration is used.

4.3.2 Slope Factor Is Not Equal to One

To illustrate how the steepness of titration curves increases with increasing values of n, I ran the simulations shown in Figure 4.4 using Equation 4.1 and $v_{control} = 1$, $v_{bkg} = 0$, $K_{i,app} = 10\,\mu M$, and n equal to values of 0.5, 1, and 2. Recall that when $n = 1$, an increase of one order of magnitude in inhibitor concentration is needed to drive inhibition from 20% to 80%. In contrast, for $n = 0.5$, nearly a thousand-fold increase in $[I]_o$ is required, and for $n = 2$, only a threefold increase in $[I]_o$ is required. Below we consider the possible causes of slope factors greater or less than 1.

Slope factors greater than unity have, at least, the following four potential causes:

- *Poor solubility of inhibitor.* If with increasing concentration, the inhibitor forms aggregates that can bind to and denature the enzyme, a slope factor greater than one will be observed. The magnitude of n will be related to the steepness of the dependence of $[I]_{aggregate}$ on $[I]_o$. In these situations, it is not uncommon to see slope factors greater than two or three.

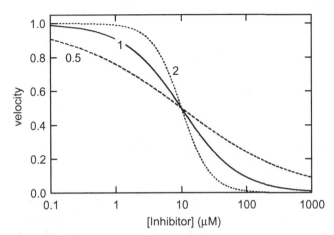

Figure 4.4. Inhibitor titrations with varying slope factors. Curves were generated using Equation 4.1 with $v_{control} = 1$, $v_{bkg} = 0$, $K_{i,app} = 10\,\mu M$, and the indicated n values.

- *Inhibition is time dependent.* Titrations for so-called "time-dependent" inhibitors can exhibit slope factors greater than one. While a detailed account of this phenomenon will be given in Chapter 6, a few words of explanation are in order now.

   The attainment of equilibrium between enzyme and the enzyme : inhibitor complex (E : I) is dependent on both time and inhibitor concentration. Most inhibitors are "classical inhibitors," in that their inhibitory effects seem instantaneous and to depend only on inhibitor concentration. In reality, the attainment of the E/E : I equilibrium for such inhibitors occurs on the time scale of milliseconds or seconds, and we simply do not observe the time-dependent aspect. However, there are inhibitors whose full inhibitory effects can take minutes or even hours to attain. For these inhibitors, the attainment of E to E : I equilibrium can more readily be recognized to have a time-dependent component.

   During the initial characterization of an inhibitor, the potential for time-dependent inhibition is often not taken into account. Often, titrations are based on single time-point assays that do not include a preincubation period that would allow for the attainment of the E to E : I equilibrium. Titrations using such a protocol will have a steep slope (i.e., $n > 1$) and an $K_{i,app}$ value that underestimates the true potency.

- $K_{i,app} \sim [E]_o$. If $K_{i,app}$ for an inhibitor is similar in magnitude to $[E]_o$, the range of inhibitor concentrations that will be needed to estimate the $K_{i,app}$ value will include inhibitor concentrations less than $[E]_o$. Such inhibitors are referred to as tight-binding inhibitors (see Chapter 6). The shape of a titration curve for such an inhibitor will be characterized by a linear dependence of v_o on $[I]_o$ at low concentrations $[I]_o$ and when fit to the four-parameter function will generate slope factors greater than one.

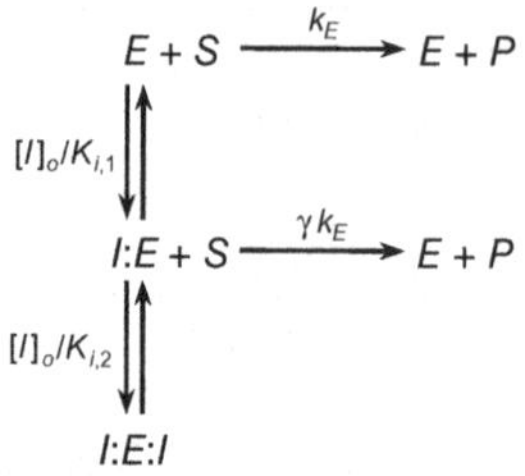

Figure 4.5. General mechanistic scheme for enzyme inhibition requiring binding of two molecules of inhibitor to achieve complete inhibition.

- *I/E stoichiometry is greater than one.* If the inhibitor is neither slow- or tight-binding, and there are no solubility problems or other artifacts, slope factors greater than one suggest that more than a single molecule of inhibitor must be bound to the enzyme to achieve inhibition. Such a mechanism is shown in Figure 4.5, the rate law for which is given in Equation 4.3:

$$v_o = \frac{k_E[S]_o}{1 + \dfrac{[I]_o}{K_{i,1}} + \dfrac{[I]_o^2}{K_{i,1}K_{i,2}}} + \frac{\gamma k_E[S]_o}{\dfrac{K_{i,1}}{[I]_o} + 1 + \dfrac{[I]_o}{K_{i,2}}} = \frac{v_o}{1 + \dfrac{[I]_o}{K_{i,1}} + \dfrac{[I]_o^2}{K_{i,1}K_{i,2}}} + \frac{\gamma v_o}{\dfrac{K_{i,1}}{[I]_o} + 1 + \dfrac{[I]_o}{K_{i,2}}}. \quad (4.3)$$

If $\gamma = 0$ and $K_{i,1} > K_{i,2}$, the dependence of v_o on $[I]_o$ will have a slope factor greater than 1 if this dependence is analyzed using the four-parameter function. This is illustrated in Figure 4.6A, where the titration "data points" were generated with Equation 4.3 and parameter values: $v_0 = 1$, $\gamma = 0$, $K_{i,1} = 50\,\mu M$, and $K_{i,2} = 2\,\mu M$. The solid line through the data was generated with the best-fit values according to Equation 4.1: $v_{control} = 1.00 \pm 0.01$, $v_{bkg} = 0.00 \pm 0.01$, $K_{i,app} = 10.1 \pm 0.1$, and $n = 1.80 \pm 0.01$.

We now consider the causes of slope factors that are less than one.

- *Poor solubility of inhibitor.* If the aggregates that form at high concentration of inhibitor are non-denaturing, they will serve to remove monomeric inhibitor from solution. Since this reduces $[I]_{free}$ below $[I]_o$, enzyme activity will be higher at these concentrations of inhibitor than it would if aggregates were not forming. Under these circumstances, a shallow titration curve will be observed with a slope factor less than one.

- *I/E stoichiometry is greater than one.* Inhibitor titration curves with $n < 1$ are much less common than those with $n > 1$. In the absence of solubility issues, the principal identifiable reason for slope values less than one is mechanistic in origin. If the complex that forms between enzyme and inhibitor is still partially active, and can bind a second molecule of inhibitor to form a catalytically inactive ternary complex, the titration curve for such an inhibitor will have $n < 1$ when fit to the four-parameter function. This is illustrated in Figure 4.6B, where the "data points" were generated with Equation 4.3 and parameter values: $v_0 = 1$, $\gamma = 0.5$, $K_{i,1} = 2\,\mu M$, and $K_{i,2} = 5\,\mu M$. The solid line through the data was gener-

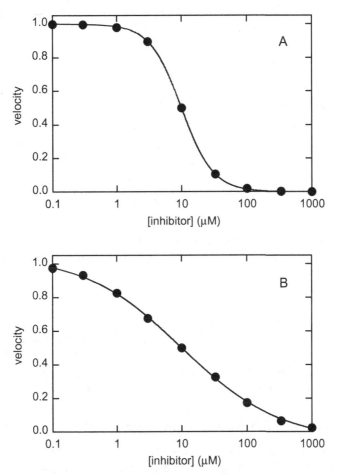

Figure 4.6. Enzyme inhibitory mechanisms involving the binding of two inhibitors can manifest slope factors that are either greater or less than one. *Plot A:* "Data points" generated with Equation 4.3 and $v_0 = 1$, $\gamma = 0$, $K_{i,1} = 50\,\mu M$, and $K_{i,2} = 2\,\mu M$. The line generated with best-fit values according to Equation 4.1: $v_{control} = 1.00 \pm 0.01$, $v_{bkg} = 0.00 \pm 0.01$, $K_{i,app} = 10.1 \pm 0.1\,\mu M$, and $n = 1.80 \pm 0.01$. *Plot B:* "Data points" generated with Equation 4.3 and $v_0 = 1$, $\gamma = 0.5$, $K_{i,1} = 2\,\mu M$, and $K_{i,2} = 5\,\mu M$. The line was generated with the best-fit values according to Equation 4.1: $v_{control} = 1.05 \pm 0.02$, $v_{bkg} = -0.05 \pm 0.02$, $K_{i,app} = 10 \pm 1\,\mu M$, and $n = 0.58 \pm 0.03$.

ated with the best-fit values according to Equation 4.1: $v_{control} = 1.05 \pm 0.02$, $v_{bkg} = -0.05 \pm 0.02$, $K_{i,app} = 10 \pm 1$, and $n = 0.58 \pm 0.03$.

4.3.3 v_{bkg} Greater Than Zero

Figure 4.7 illustrates the difference in appearance among titration curves with $v_{bkg} = 0$, 0.1, and 0.2. Perhaps the most common cause of greater than zero v_{bkg} values has to do with an assay method that exhibits a reaction velocity in the absence of enzyme. If

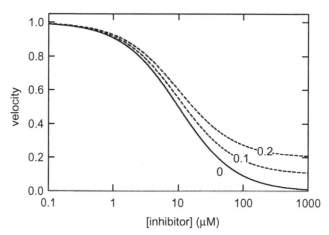

Figure 4.7. Inhibitor titration curves with varying background velocities. Lines were drawn using Equation 4.1 with $v_{control} = 1$, $K_{i,app} = 10$, $n = 1$, and the indicated value of v_{bkg}.

control experiments confirm this for a particular system, this background rate can simply be subtracted from the velocities of an inhibitor titration experiment. If there is a v_{bkg} even after the assay background is taken into account, the cause is likely mechanistic in origin. In this case, the complex that forms between enzyme and inhibitor retains some catalytic activity. Such an inhibitor is called a partial inhibitor.

4.3.4   Kinetic and Mechanistic Ambiguity

Throughout this book, we will from time to time be faced with a situation in which a data set can be fit equally well by more than a single rate law. Even at this early stage of the analysis of an inhibitor's mechanism, there can be kinetic ambiguity. Consider the simulations of Figure 4.8. The six "data points" were generated using the four-parameter function of Equation 4.1 with parameters: $v_{control} = 1$, $v_{bkg} = 0.07$, $n = 0.8$, and $K_{i,app} = 1.1\,\mu M$. Note that two of the four parameters, v_{bkg} and n, have values that give rise to nonideal behavior.

Now, the two lines of this figure were drawn using Equation 4.1 and parameter sets in which only one of the four parameters has a nonideal value, the solid line with $v_{bkd} = 0.1$ and the dashed line with $n = 0.6$. In the context of Figure 4.5, these two parameter sets represent very different mechanisms. The case where $v_{bkg} = 0.1$ and $n = 1.0$ represents a case of partial inhibition (i.e., $\gamma = 0.1$, $K_{i,2} \geq 10^3 K_{i,1}$). In contrast, the case where $v_{bkg} = 0$ and $n = 0.6$ represents a situation in which formation of I : E : I is required for the observation of inhibition (i.e., $\gamma = 0$, $K_{i,2} < K_{i,1}$).

We see from this illustration that the four-parameter function is mechanistically agnostic, and that departure from "classical" behavior may not reflect an assay artifact, but rather may reflect an important feature of mechanism, requiring additional studies to sort out.

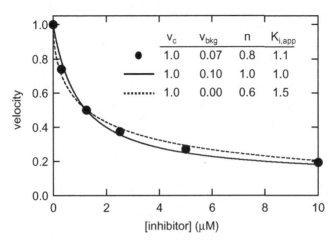

Figure 4.8. Kinetic and mechanistic ambiguity.

4.4 CASE STUDIES

In the final section of this chapter, I illustrate some of the points made in previous sections with two case studies.

4.4.1 Inhibition of Ubiquitin C-Terminal Hydrolase-L1 (UCH-L1) by LDN-91946

UCH-L1 is part of the ubiquitin-proteasome pathway and catalyzes the hydrolysis of low-molecular-weight C-terminal esters and amides of ubiquitin. This enzyme has been genetically linked to Parkinson's disease (PD) (Leroy et al. 1998; Maraganore et al. 2004) and is a component of the insoluble protein deposits (Lewy bodies) that form in the brains of PD patients (Lowe et al. 1990). UCH-L1 is also overexpressed in some cancerous tissues, including primary lung (Hibi et al. 1999), colorectal (Yamazaki et al. 2002), and pancreatic cancers (Tezel et al. 2000).

As part of a program to identify modulators of the activity of UCH-L1, the Harvard drug discovery center LDDN (Laboratory for Drug Discovery in Neurodegeneration (Stein 2003, 2004) executed a screen of its collection of drug-like molecules. In the course of this screen, LDN-91946 was identified as an inhibitor of UCH-L1, giving 80% inhibition at the screening concentration of $10\,\mu M$. As their first step in the mechanistic analysis of LDN-91946, Mermerian et al. (2007) determined the inhibitor concentration dependence of initial velocity for the UCH-L1 catalyzed hydrolysis of the fluorogenic substrate Ub-AMC (see Fig. 4.9).

In this experiment a broad range of inhibitor concentration was used, spanning over three orders of magnitude. This allowed useful insights to be garnered simply from examination of the raw data. Three points are noteworthy: v_o goes to zero at high concentrations of LDN-91946, the IC_{50} of $2\text{--}3\,\mu M$ can be estimated, and a 10-fold increase in inhibitor concentration is required to drive the inhibition from 20% to 80%. These

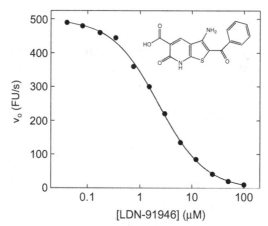

Figure 4.9. Titration curve for the inhibition of UCH-L1-catalyzed hydrolysis of Ub-AMC by LDN-91946. The line was generated with the best-fit values according to Equation 4.1: $v_{control} = 504 \pm 23\,\mathrm{FU\,min^{-1}}$, $v_{bkg} = 2 \pm 4\,\mathrm{FU\,min^{-1}}$, $n = 0.97 \pm 0.02$, and $K_{i,app} = 2.3 \pm 0.1\,\mu\mathrm{M}$.

qualitative observations are borne out when the data are fit to Equation 4.1. We see that v_{bkg} is zero, the slope factor is one, and the $K_{i,app}$ is 2.3 μM. The fact that LDN-91946 is a classical inhibitor allowed the Harvard group to proceed to the next stage of mechanistic analysis, in which experiments were conducted to determine the kinetic mechanism of inhibition (see Chapter 5).

4.4.2 Inhibition of Tissue Transglutaminase (TGase) by GTP

TGase is a Ca^{++}-dependent enzyme that catalyzes cross-linking of intracellular proteins through a mechanism that involves isopeptide bond formation between Gln and Lys residues and is allosterically regulated by GTP. TGase is thought to play a pathogenic role in neurodegenerative diseases by promoting aggregation of disease-specific proteins that accumulate in these disorders (Gentile et al. 1998; de Cristofaro et al. 1999; Junn et al. 2003). Given the role that TGase plays in neurodegenerative disorders, the LDDN initiated a research program to discover inhibitors of this enzyme that might ultimately be developed into therapeutic agents.

The mechanism of TGase-catalyzed transamidation involves acyl-transfer to and from an active site Cys residue (Folk 1983; Folk and Chung 1985; Case and Stein 2003). According to this mechanism, combination of TGase with the Gln-donating substrate to form a Michaelis complex is followed by nucleophilic attack of the active site Cys to generate a covalent acyl-enzyme intermediate and an equivalent of ammonia. In the presence of a suitable primary amine, the acyl-enzyme will undergo aminolysis to regenerate free enzyme and the isopeptide product. In the absence of nucleophile, the acyl-enzyme hydrolyzes. Significantly, transamidation is inhibited by GTP and other nucleotides through an allosteric mechanism in which nucleotide binds to a site on the enzyme that is distinct from the active site where acyl-transfer chemistry occurs

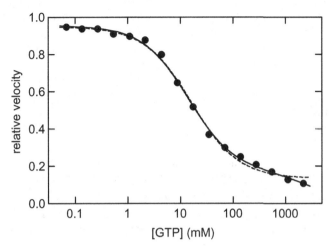

Figure 4.10. Titration curve for the inhibition of the TGase-catalyzed hydrolysis of Z-Glu(γ-AMC)-Gly by GTP. The dashed line was generated with the best-fit values according to Equation 4.1: $v_{control} = 0.95 \pm 0.01$, $v_{bkg} = 0.14 \pm 0.01$, $K_{i,app} = 16 \pm 1\,\mu M$, and $n = 1.0 \pm 0.1$. The solid line was generated with the best-fit values according to Equation 4.3: $v_{o,relative} = 0.96 \pm 0.01$, $\gamma = 0.18 \pm 0.02$, $K_{i,1} = 14 \pm 1\,\mu M$, $K_{i,2} = (3.2 \pm 1.5) \times 10^3\,\mu M$.

(Bergamini 1988; Iismaa et al. 2000; Venere et al. 2000). Binding of GTP is thought to induce a conformational change of TGase that weakens the enzyme's interaction with catalytically essential Ca^{++} cations. Conversely, Ca^{++} can antagonize the inhibition of TGase by GTP.

One of the goals of the screen was to have Ca^{++} cations at a concentration that was high enough to promote reasonable transamidation activity but low enough to allow compounds to bind to the GTP site, should such compounds exist in the library (Case et al. 2005).

In the course of finding just the right Ca^{++} concentration, the effect of GTP on activity was investigated. Figure 4.10 is a plot of the $[GTP]_o$ dependence of reaction velocity for the TGase-catalyzed hydrolysis of Z-Glu(γ-AMC)-Gly. Due to the solubility of GTP, it was possible to conduct the titration over a range of five orders of magnitude of GTP concentration. The requirement for such a broad range of GTP concentration was established in preliminary experiments.

Examination of the data reveals that to drive the inhibition from 20% to 80% requires an increase in $[GTP]_o$ of over two orders of magnitude. This was an immediate alert that inhibtion of TGase by GTP does not follow a simple mechanism. Fitting the data to the four-parameter function provided the following best-fit parameters: $v_{control} = 0.95 \pm 0.01$, $v_{bkg} = 0.14 \pm 0.01$, $K_{i,app} = 16 \pm 1\,\mu M$, and $n = 1.0 \pm 0.1$, suggesting that GTP is a partial inhibitor (Fig. 4.5, $K_{i,2} \geq 10^3 K_{i,1} = 16,000\,\mu M$ and $\gamma = 0.14$). The dashed line in Figure 4.10 was drawn using these parameters, and although it describes the data reasonably well, there is systematic departure from the line at high $[GTP]_o$. A model that can account for the data is the model of Figure 4.5. When the data set was fit to Equation 4.3 the following parameters were obtained: $v_{o,relative} = 0.96 \pm 0.01$, $\gamma = 0.18 \pm 0.02$, $K_{i,1} = 14 \pm 1\,\mu M$, $K_{i,2} = (3.2 \pm 1.5) \times 10^3\,\mu M$. The solid drawn with these parameters is a good fit to the data, suggesting that the mechanism of Figure 4.5 might be at work here.

REFERENCES

Barendrecht, H. F. (1913). "Enzyme-action, facts and theory." *Biochem. J.* **7**: 549–561.

Bergamini, C. M. (1988). "GTP modulates calcium binding and cation-induced conformational changes in erythrocyte transglutaminase." *FEBS Lett.* **239**: 255–258.

Brown, A. (1902). "Enzyme action." *J. Chem. Soc.* **81**: 377–388.

Case, A., et al. (2005). "Development of a mechanism-based assay for tissue transglutaminase—Results of a high-throughput screen and discovery of inhibitors." *Anal. Biochem.* **338**: 237–244.

Case, A. and R. L. Stein (2003). "Kinetic analysis of the action of tissue transglutaminase on peptide and protein substrates." *Biochemistry* **42**: 9466–9481.

Coombs, H. I. (1927). "Studies on xanthine oxidase. The specificity of the system." *Biochem. J.* **21**: 1260–1265.

de Cristofaro, T., et al. (1999). "The length of polyglutamine tact, its level of expression, the rate of degradation, and the transglutaminase activity influence the formation of intracellular aggregates." *Biochem. Biophys. Res. Commun.* **260**: 150–158.

Dixon, M. and S. Thurlow (1924). "Studies on xanthine oxidase." *Biochem. J.* **18**(976–988): 976–988.

Folk, J. E. (1983). "Mechanism and basis for specificity of transglutaminase-catalyzed ε-(γ-glutamyl)lysine bond formation." *Adv. Enzymol. Relat. Areas Mol. Biol.* **54**: 1–56.

Folk, J. E. and S. Chung (1985). "Transglutaminases." *Methods Enzymol.* **113**: 358–375.

Gentile, V., et al. (1998). "Tissue transglutaminase-catalyzed formation of high-molecular-weight aggregates in vitro is favored with long polyglutamine domains." *Arch. Biochem. Biophys.* **352**: 314–321.

Hibi, K., et al. (1999). "PGP9.5 as a candidate tumor marker for non-small-cell lung cancer." *Am. J. Pathol.* **155**(3): 711–715.

Iismaa, S., et al. (2000). "GTP binding and signaling by G_h/transglutaminase II involves distinct residues in a unique GTP-binding pocket." *J. Biol. Chem.* **275**: 18259–18265.

Junn, E., et al. (2003). "Tissue transglutaminase-induced aggregation of α-synuclein: Implications for Lewy body formation in Parkinson's disease and dementia with Lewy bodies." *Proc. Natl. Acad. Sci.* **100**: 2047–2052.

Leroy, E., et al. (1998). "The ubiquitin pathway in Parkinson's disease." *Nature* **395**: 451–452.

Lowe, J., et al. (1990). "Ubiquitin carboxyl-terminal hydrolase (PGP9.5) is selectively present in ubiquitinated inclusion bodies characteristic of human neurodegenerative diseases." *J. Pathol.* **161**: 153–160.

Maraganore, D. M., et al. (2004). "UCHL1 is a Parkinson's disease susceptibility gene." *Ann. Neurol.* **55**(4): 512–521.

Mermerian, A., et al. (2007). "Structure–activity relationship, kinetic mechanism, and selectivity for a new class of ubiquitin C-terminal hydrolase-L1 inhibitors." *Bioorg. Med. Chem. Lett.* **17**: 3729–3732.

Quastel, J. H. and W. R. Wooldridge (1928). "Some properties of the dehydrogenating enzymes of bacteria." *Biochem. J.* **22**: 689–702.

Stein, R. L. (2003). "A new model for drug discovery—Meeting our societal obligations." *Drug Discov. Today* **8**: 245–248.

Stein, R. L. (2004). "High-throughput screening in academia: The Harvard experience." *J. Biomol. Screen.* **8**: 615–619.

Tezel, E., et al. (2000). "PGP9.5 as a prognostic factor in pancreatic cancer." *Clin. Cancer Res.* **6**(12): 4764–4767.

Venere, A. D., et al. (2000). "Opposite effects of Ca2+ and GTP calcium binding on tissue trans-glutaminase tertiary structure." *J. Biochem.* **275**: 3915–3932.

Yamazaki, T., et al. (2002). "PGP9.5 as a marker for invasive colorectal cancer." *Clin. Cancer Res.* **8**(1): 192–195.

5

KINETIC MECHANISM OF INHIBITION OF ONE-SUBSTRATE ENZYMATIC REACTIONS

In the previous chapter, we discussed the first steps in the elucidation of the kinetic mechanism by which an inhibitor has its effect on an enzyme. These studies were all based on the examination of the dependence of initial velocity on inhibitor concentration, at a single concentration of substrate. In this chapter, I build on this information and describe how one determines the enzyme species to which an inhibitor binds and the equilibrium constants for dissociation of inhibitor from these species.

5.1 IMPORTANCE IN DRUG DISCOVERY

Perhaps the most important stage in the drug discovery process where knowledge of the kinetic mechanism of inhibition can have an impact is during medicinal chemical optimization of an inhibitor series. The kinetic mechanism of inhibition is a "kinetic dissection" of $K_{i,app}$ into true dissociation constants for all enzyme : inhibitor (E : I) complexes. This information allows a quantitative assessment of the consequences of inhibitor structural variation on the binding of the inhibitor and its analogs to the various forms of the enzyme. The chemist can then decide which constant to track as structural variations are made to the inhibitor, and then modify his optimization strategy to enhance or

Kinetics of Enzyme Action: Essential Principles for Drug Hunters, First Edition. Ross L. Stein.
© 2011 John Wiley & Sons, Inc. Published 2011 by John Wiley & Sons, Inc.

eliminate binding to a particular enzyme species. For example, in a situation where an inhibitor can bind both to free enzyme and the complex of enzyme and substrate, the decision might be made to optimize binding to the enzyme:substrate (E:S) complex because the enzyme is part of a linear metabolic pathway, and inhibition of this enzyme would ideally be insensitive to the rise and fall of substrate concentration.

5.2   THEORETICAL CONSIDERATIONS

5.2.1   Basic Mechanisms and Rate Expressions

For a one substrate reaction of general form

$$E + S \leftrightarrows ES \rightarrow E + P$$

inhibitor can bind to either E or E:S, or to both, as illustrated in the mechanism of Figure 5.1. The rate rate law for this mechanism, Equation 5.7, is derived in Figure 5.2 using the rapid equilibrium assumption. This rate law tells us that in the presence of an inhibitor, the reaction follows Michaelis–Menten kinetics with

$$(V_{max})_{obs} = \frac{V_{max}}{1 + \dfrac{[I]_o}{\beta K_i}} \tag{5.8}$$

and

$$(V_{max} / K_s)_{obs} = \frac{V_{max} / K_s}{1 + \dfrac{[I]_o}{K_i}}. \tag{5.9}$$

In this mechanism, which is referred to as "mixed inhibition," β is a cooperativity factor. If $\beta = 1$, inhibitor binds to E and E:S with the same affinity. If $\beta > 1$, I binds to E with less affinity than to E:S. And finally, if $\beta < 1$, I binds to E with greater affinity than to E:I. Since the binding events in this mechanism constitute a thermodynamic cycle, the same arguments hold true for the binding of S to E and E:I.

Limiting cases for β allow simplification of the mechanism and lead to the three classical mechanisms of enzyme inhibition:

- $\beta \gg 1$—Competitive inhibition
- $\beta \ll 1$—Uncompetitive inhibition
- $\beta = 1$—Noncompetitive inhibition

These three mechanisms are shown in Figure 5.3 together with their rate laws and how $(V_{max})_{obs}$ and $(V_{max}/K_s)_{obs}$ depend on inhibitor concentration. These dependencies follow directly from the enzyme form(s) to which inhibitor binds, as discussed below.

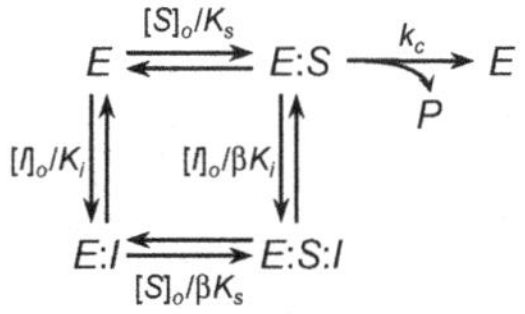

Figure 5.1. Mechanistic scheme for mixed inhibition.

Experimental condition: $[E]_o \ll [S]_o, [I]_o$

Conservation equations:
$$[E]_o = [E] + [ES] + [EI] + [ESI]$$
$$[S]_o = [S] + [ES] + [ESI] = [S]$$
$$[I]_o = [I] + [EI] + [ESI] = [I]$$

Definition of binding constants:

$$K_s = \frac{[E][S]}{[ES]}; \quad \frac{[E]}{[ES]} = \frac{K_s}{[S]} = \frac{K_s}{[S]_o} \qquad \beta K_s = \frac{[EI][S]}{[ESI]}; \quad \frac{[EI]}{[ESI]} = \frac{\beta K_s}{[S]} = \frac{\beta K_s}{[S]_o}$$

$$K_i = \frac{[E][I]}{[EI]}; \quad \frac{[E]}{[EI]} = \frac{K_i}{[I]} = \frac{K_i}{[I]_o} \qquad \beta K_i = \frac{[ES][I]}{[ESI]}; \quad \frac{[ES]}{[ESI]} = \frac{\beta K_i}{[I]} = \frac{\beta K_i}{[I]_o}$$

Statement of rate dependence: $v_o = k_c[ES]$

Derivation of rate equation:

$$\frac{v_o}{[E]_o} = \frac{k_c[ES]}{[E] + [ES] + [EI] + [ESI]} \tag{5.1}$$

$$\frac{v_o}{[E]_o} = \frac{k_c}{\dfrac{[E]}{[ES]} + 1 + \dfrac{[EI]}{[ES]} + \dfrac{[ESI]}{[ES]}} \tag{5.2}$$

$$\frac{v_o}{[E]_o} = \frac{k_c}{\dfrac{K_s}{[S]_o} + 1 + \dfrac{[EI]}{[E]}\dfrac{[E]}{[ES]} + \dfrac{[I]_o}{\beta K_i}} \tag{5.3}$$

$$\frac{v_o}{[E]_o} = \frac{k_c}{\dfrac{K_s}{[S]_o} + 1 + \dfrac{[I]_o}{K_i}\dfrac{K_s}{[S]_o} + \dfrac{[I]}{\beta K_i}} \tag{5.4}$$

$$\frac{v_o}{[E]_o} = \frac{k_c[S]_o}{K_s + [S]_o + K_s\dfrac{[I]_o}{K_i} + [S]_o\dfrac{[I]}{\beta K_i}} \tag{5.5}$$

$$\frac{v_o}{[E]_o} = \frac{k_c[S]_o}{K_s\left(1 + \dfrac{[I]_o}{K_i}\right) + [S]_o\left(1 + \dfrac{[I]}{\beta K_i}\right)} \tag{5.6}$$

$$v_o = \frac{\dfrac{V_{max}}{\left(1 + \dfrac{[I]}{\beta K_i}\right)}[S]_o}{K_s\dfrac{\left(1 + \dfrac{[I]_o}{K_i}\right)}{\left(1 + \dfrac{[I]}{\beta K_i}\right)} + [S]_o} \tag{5.7}$$

Figure 5.2. Derivation of the rate expression for mixed inhibition.

Competitive

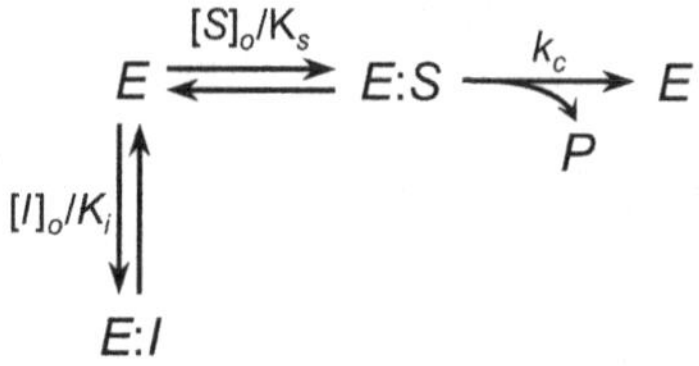

$$v_o = \frac{V_{max}[S]_o}{K_s\left(1+\dfrac{[I]_o}{K_i}\right)+[S]_o} \quad (5.10)$$

$$\left(\frac{V_{max}}{K_s}\right)_{obs} = \frac{V_{max}/K_s}{1+\dfrac{[I]_o}{K_i}} \quad (5.11)$$

$$(V_{max})_{obs} = V_{max} \quad (5.12)$$

Uncompetitive

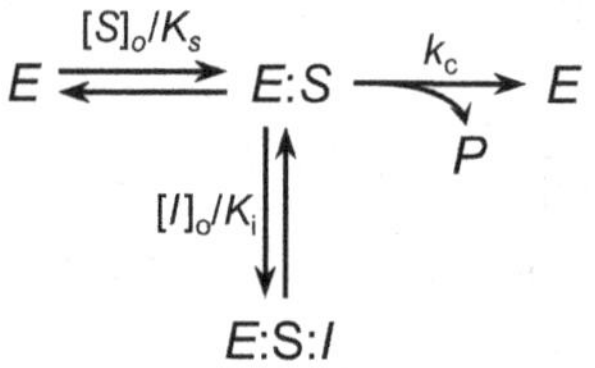

$$v_o = \frac{\dfrac{V_{max}}{1+\dfrac{[I]_o}{K_i}}[S]_o}{\dfrac{K_s}{1+\dfrac{[I]_o}{K_i}}+[S]_o} \quad (5.13)$$

$$\left(\frac{V_{max}}{K_s}\right)_{obs} = \frac{V_{max}}{K_s} \quad (5.14)$$

$$(V_{max})_{obs} = \frac{V_{max}}{1+\dfrac{[I]_o}{K_i}} \quad (5.15)$$

Noncompetitive

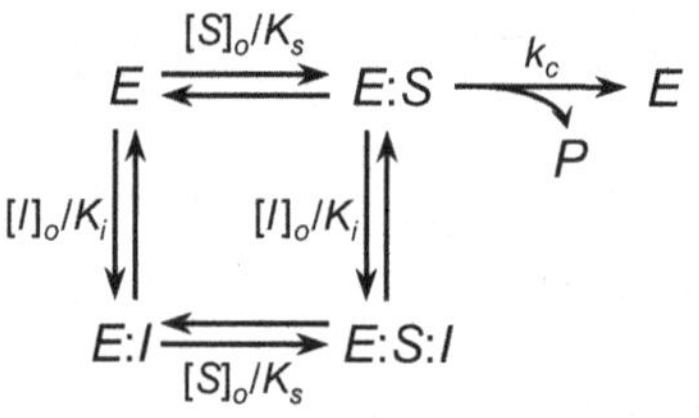

$$v_o = \frac{\dfrac{V_{max}}{1+\dfrac{[I]_o}{K_i}}[S]_o}{K_s+[S]_o} \quad (5.16)$$

$$\left(\frac{V_{max}}{K_s}\right)_{obs} = \frac{V_{max}/K_s}{1+\dfrac{[I]_o}{K_i}} \quad (5.17)$$

$$(V_{max})_{obs} = \frac{V_{max}}{1+\dfrac{[I]_o}{K_i}} \quad (5.18)$$

Figure 5.3. The three classical mechanisms of enzyme inhibition, and their rate laws as derived using the rapid equilibrium assumption.

From first principles of inhibition that we learned in Chapter 4, we know that inhibitors exert their effect by reversibly binding to a particular form of enzyme, to produce a catalytically inactive complex, thereby reducing the overall flux through the reaction cycle. We also know, from Chapter 3, that each enzyme form is the reactant state for a particular steady-state kinetic parameter; free enzyme E is reactant state for

V_{max}/K_s, and E:S is reactant state for V_{max}. Putting these two concepts together we see that if an inhibitor binds to E, the magnitude of V_{max}/K_s will decrease, and if an inhibitor binds to E:S, the magnitude of V_{max} will decrease. This is reflected in the three classical mechanisms of inhibition. Competitive inhibitors bind only to E and effect only V_{max}/K_s. Likewise, uncompetitive inhibitors bind only to E:S and effect only V_{max}. Finally, noncompetitive and mixed inhibitors bind to both E and E:S, effecting V_{max}/K_s and V_{max}.

5.2.2 Kinetic Mechanism of an Inhibitor Does Not Necessarily Indicate Where the Inhibitor Binds

We must be cautious in extending our interpretation of modes of inhibition to structure. That is, a given mode of inhibition does not necessarily tell us where on the enzyme surface an inhibitor binds. For example, it is commonly assumed that a competitive inhibitor binds to the active site of the enzyme. However, all we know from kinetic data is that it binds to free enzyme. Kinetics cannot give us any information about *where* the inhibitor binds on free enzyme. Inhibitor could very well be binding to some site on the backside of the enzyme and inducing a conformational change that closes down the active site. Likewise, an uncompetitive inhibitor need not be binding to the Michaelis complex. If the enzyme reaction proceeds through several intermediates, as we discussed in Chapter 3, the inhibitor could be binding to one of these rather than to the Michaelis complex. There is an important lesson here—*kinetics can never reveal structure*.

5.2.3 The Meaning of K_i for Simple, One-Substrate Reactions

In Chapter 4, the two measures of inhibitor potency we considered were IC_{50} and $K_{i,app}$. We saw that both of these constants were dependent on the mechanism of inhibition, the mechanism of substrate turnover, and the substrate concentration. In contrast to these parameters, the K_i values that emerge for kinetic mechanistic studies of one-substrate reactions with a single intermediate species (i.e., E:S) are true equilibrium constants for dissociation of inhibitor from the enzyme form to which it is bound. In this subsection, we consider two potentially complicating factors in the interpretation of K_i values for such systems: multiple, sequentially formed E:I species and nonequilibrium substrate turnover. At the end of this chapter, we consider one-substrate reactions that have more than a single, intermediate species (see Section 5.4) and see the difficulty in actually getting at the true K_i value in systems that are more representative of "real-world" enzymology.

5.2.3.1 Multiple, Sequentially Formed E:I Species. It is common to interpret K_i values as if they reflect the $[I]_o$-dependent equilibrium between two species. For a competitive inhibitor, this equilibrium is assumed to be between E and E:I, as illustrated in the mechanism of Figure 5.4A. This need not be the case, however; there can be any number of species that form subsequent to the initial encounter complex between enzyme and inhibitor.

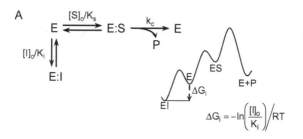

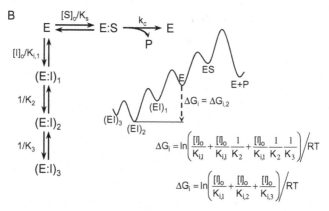

Figure 5.4. Free energy diagram inhibition. *Panel A:* Standard mechanism of competitive inhibition involving a single E:I complex. *Panel B:* A mechanism of competitive inhibition involving three consequtively formed E:I complexes.

Consider the competitive inhibition mechanism of Figure 5.4B. The interaction between enzyme and inhibitor generates three sequentially formed E:I species. It is easy to imagine plausible chemical and/or structural explanations for multiple intermediates. Perhaps the unstable encounter complex, $(E:I)_1$, undergoes a conformational change to form the more stable $(E:I)_2$, which can then conformationally isomerize into $(E:I)_3$. Or perhaps the encounter complex undergoes some sort of reversible chemistry to form $(E:I)_2$, which then isomerizes to $(E:I)_3$.

The rate expression for the mechanism depicted in Figure 5.4 can be derived using the rapid equilibrium assumption and is expressed in Equations 5.19 and 5.21:

$$v_o = \frac{V_{\max}[S]_o}{K_s\left\{1+[I]_o\left(\dfrac{1}{K_{i,1}}+\dfrac{1}{K_{i,1}K_2}+\dfrac{1}{K_{i,1}K_2K_3}\right)\right\}+[S]_o} \tag{5.19}$$

or, more simply,

$$v_o = \frac{V_{\max}[S]_o}{K_s\left(1+\dfrac{[I]_o}{K_i}\right)+[S]_o} \tag{5.20}$$

where

$$K_i = \left(\frac{1}{K_{i,1}} + \frac{1}{K_{i,1}K_2} + \frac{1}{K_{i,1}K_2K_3} \right)^{-1}. \qquad (5.21)$$

We see that for an inhibition mechanism involving sequentially formed E:I complexes, the K_i value is equal to the reciprocal of the sum of reciprocals of the net dissociation constant for each of the intermediates:

$$K_i = \left(\frac{1}{K_{i,1}} + \frac{1}{K_{i,2}} + \frac{1}{K_{i,3}} \right)^{-1}. \qquad (5.22)$$

As we saw in Chapter 3, it is often instructive to translate concepts such as these into free energy diagrams. In doing so, the first thing to note is that free energy values can only be calculated from unitless equilibrium constants, according to $\Delta G = -lnK/RT$. In the case of enzyme inhibition constants, the term $[I]_o/K_i$ is used as the unitless constant, where the inhibitor concentration is either $[I]_o$ of an experiment, or a standard state concentration that one chooses for consistency among several experiments. Note that by using $[I]_o/K_i$ one has effectively converted a dissociation constant, K_i, into an *association* constant.

This is illustrated in the free energy diagram of Figure 5.4A. In this diagram, $[I]_o > K_i$ so the calculated free energy change is negative, meaning that formation of E:I from E is an exothermic process, thus rendering E:I more stable than E. If $[I]_o$ would have been less than K_i, then a positive free energy would have been calculated, meaning E:I is less stable than E.

To illustrate these concepts for the more complex mechanism of Figure 5.4B, let us consider a specific case for the mechanism of Figure 5.4 in which formation of the encounter complex is governed by $K_{i,1}$ with value 100 µM. (E:I)$_1$ then undergoes a conformational isomerization to (E:I)$_2$ with equilibrium constant $K_2 = $ (E:I)$_1$/ (E:I)$_2 = 10^{-6}$. Finally, (E:I)$_2$ can form the final complex (E:I)$_3$ with $K_2 = $ (E:I)$_2$/ (E:I)$_3 = 10^2$. The net dissociation constants (see Eqs. 5.21 and 5.22) for (E:I)$_1$, (E:I)$_2$, and (E:I)$_1$ are 100 µM, 0.01 µM, and 1 µM, respectively. The reciprocal of the sum of reciprocals equals (101.01)$^{-1}$ or 0.0099 µM. This value is essentially identical to $K_{i,2}$, indicating that K_i reflects the accumulation of (E:I)$_2$, as suggested by inspection of the free energy diagram for this mechanism, where $[I]_o > K_{i,1}$. (E:I)$_1$ and (E:I)$_3$ can be said to be "kinetically invisible," and not accessible by steady-state kinetic techniques.

If the three net dissociation constants would have been closer in magnitude, K_i would have reflected accumulation of all three species. In such cases, a "virtual" E:I complex could be said to exist, analogous to the virtual transition state that exists in cases where the rate is partially rate limited by a number of steps (Schowen 1978; Stein 1985).

5.2.3.2 K_i Values and Nonequilibrium Mechanisms of Substrate Turnover. The topics we have thus far covered in this chapter, theory of enzyme

inhibition and meaning of K_i, have been developed in the context of rapid equilibrium kinetics. A natural question to ask is if these concepts still hold true for situations in which the steady state for substrate turnover does not attain equilibrium (the case of nonequilibrium kinetics for inhibitor binding) is covered in Chapter 6).

Such a situation is illustrated in the mechanism of Figure 5.5 for competitive inhibition of a one-substrate, one-intermediate reaction. The rate law of Equation 5.28 is precisely the same as Equation 5.10, except for the substitution of K_m for K_s, where the former equals the kinetic expression $(k_{-1} + k_2)/k_1$.

For this simple reaction, the meaning of K_i is independent of whether or not substrate turnover is an equilibrium process. For more complex cases, this need not be the case (see Section 5.4).

5.3   ANALYSIS OF INITIAL VELOCITY DATA FOR ENZYME INHIBITION

We see from the rate laws for the various mechanisms of inhibition (see Figures 5.2 and 5.3) that initial velocities for inhibition of an enzymatic reaction are dependent on $[S]_o$ and $[I]_o$. Therefore, to determine the kinetic mechanism of inhibition, v_o must be determined as a function of both these variables. In practice, an investigator will measure initial velocities for reactions in which $[S]_o$ and $[I]_o$ have been systematically varied. For example, a kinetic experiment might involve initial velocity determinations for the reaction of six concentrations of substrate and inhibitor, for a total of 36 ($[S]_o$, $[I]_o$) pairs. Once this data have been generated, the problem shifts to analysis.

5.3.1   Analysis of Data with Two Independent Variables

The solution to many important mechanistic problems in enzymology are found in the outcome of kinetic experiments that depend on more than a single independent variable. You will recall from the beginning of this chapter, to assess how changes in inhibitor structure effect the binding affinity to each of the various enzyme forms, the kinetic mechanism of inhibition must be determined for the structurally modified inhibitors. For one-substrate reactions, this requires an experiment having two independent variables (i.e., $[S]_o$ and $[I]_o$), while for two-substrate reactions, experiments having three independent variables must be conducted (Chapter 8). The data generated from these experiments can be analyzed by one of two general methods: the global fit or the method of replots. These methods are discussed below in the context of the specific problem at hand—the inhibition of a single-substrate reaction.

5.3.1.1   Method of Replots.   This is a two stage process—the construction of *primary plots* and, from these, *secondary plots*. For inhibition of a single-substrate reaction, there are two possible primary plots: plot I_I^o, the dependence of v_o on $[I]_o$ at several fixed $[S]_o$, and plot I_S^o, the dependence of v_o on $[S]_o$ at several fixed $[I]_o$. For I_I^o, each dependence of v_o on $[I]_o$ is fit to the four-parameter inhibition function of Equation 4.1, while for I_S^o, each dependence of v_o on $[S]_o$ is individually fit to the Michaelis–

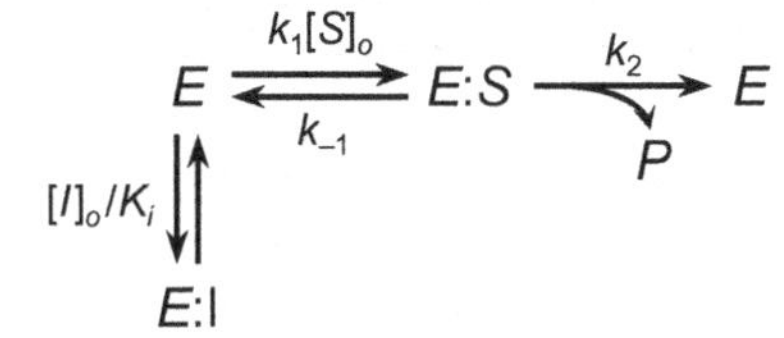

$$E \underset{k_{-1}}{\overset{k_1[S]_o}{\rightleftharpoons}} E{:}S \overset{k_2}{\longrightarrow} \underset{P}{\overset{E}{\searrow}}$$

$$[I]_o/K_i \updownarrow$$

$$E{:}I$$

Experimental condition:	$[S]_o \gg [E]_o$
Conservation equations:	$[E]_o = [E] + [ES] + [EI]$
	$[S]_o = [S] + [ES] = [S]$
	$[I]_o = [I] + [EI]\}$
Statement of rate dependence:	$v_o = k_2[ES]$

Steady-state condition:

$$\frac{d[ES]}{dt} = 0$$

$$k_1[E][S] = (k_{-1} + k_2)[ES]$$

$$[ES] = \frac{k_1}{k_{-1}+k_2}[E][S] \; ; \quad \frac{[E]}{[ES]} = \frac{\dfrac{k_{-1}+k_2}{k_1}}{[S]}$$

Derivation of rate equation:

$$\frac{v_o}{[E]_o} = \frac{k_2[ES]}{[E]+[ES]+[EI]} \tag{5.23}$$

$$v_o = \frac{V_{max}}{\dfrac{[E]}{[ES]}+1+\dfrac{[EI]}{[ES]}} \tag{5.24}$$

$$v_o = \frac{V_{max}}{\dfrac{[E]}{[ES]}+1+\dfrac{[EI]}{[E]}\dfrac{[E]}{[ES]}} \tag{5.25}$$

$$v_o = \frac{V_{max}}{\dfrac{\dfrac{k_{-1}+k_2}{k_1}}{[S]_o}+1+\dfrac{[I]_o}{K_i}\dfrac{\dfrac{k_{-1}+k_2}{k_1}}{[S]_o}} \tag{5.26}$$

$$v_o = \frac{V_{max}[S]_o}{\dfrac{k_{-1}+k_2}{k_1}+[S]_o+\dfrac{k_{-1}+k_2}{k_1}\dfrac{[I]_o}{K_i}} \tag{5.27}$$

$$v_o = \frac{V_{max}[S]_o}{K_m\left(1+\dfrac{[I]_o}{K_i}\right)+[S]_o} \tag{5.28}$$

Figure 5.5. Steady-state derivation of the rate law for competitive inhibition.

Menten equation. Each curve from the two primary plots is then carefully inspected to judge whether it adequately fits the data and thus supports the use of the simple rate laws that were used to fit them. If the curves can be described simply, then secondary plots are constructed. An example is given below illustrating what is done if primary plots are not simple.

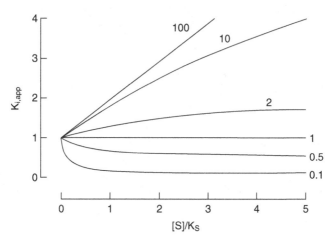

Figure 5.6. Dependence of $K_{i,app}$ on $[S]_o/K_s$ according to Equation 5.28 at various values of β.

Secondary plots are constructed as the dependence of each parameter from the primary plot on the fixed independent variable for that plot. It is from the analysis of the secondary plots that we discover the mechanism of inhibition and estimate K_i values.

For I_I^o, there is a single secondary plot: $II_{Ki,obs}^o$, $K_{i,obs}$ versus $[S]_o$, while for I_S^o, there are two: $II_{(V)obs}^o$, $(V_{max})_{obs}$ versus $[I]_o$ and $II_{(V/K)obs}^o$, $(V_{max}/K_s)_{obs}$ versus $[I]_o$. In principle, analysis of $II_{Ki,obs}^o$ or the combination of $II_{(V)obs}^o$ and $II_{(V/K)obs}^o$ can lead to the kinetic mechanism and estimates of K_i values. However, I advise proceeding with the analysis using $II_{(V)obs}^o$ and $II_{(V/K)obs}^o$, due to the ambiguity that can accompany mechanism assignment using $II_{K,iobs}^o$.

This is illustrated in the plot of Figure 5.6, where I used Equation 5.29 to simulate the

$$K_{i,app} = K_i \left(\frac{1 + \dfrac{[S]_o}{K_s}}{1 + \dfrac{[S]_o}{\beta K_s}} \right) \tag{5.29}$$

dependence of $K_{i,app}$ on $[S]_o$ for mixed inhibition at the indicated values of β. If a sufficiently wide range of $[S]_o$ is not used, it may be impossible to distinguish a competitive mechanism of inhibition ($\beta \gg 1$) from mixed inhibition $\beta \sim 10$. Likewise, uncompetitive inhibition ($\beta \ll 1$) can easily be mistaken for mixed inhibition unless β is greater than about 0.1.

On the other hand $II_{(V)obs}^o$ and $II_{(V/K)obs}^o$ can usually be interpreted with far less ambiguity. As shown in Figure 5.7, the classical mechanisms of inhibition produce distinct pairs of $II_{(V)obs}^o$ and $II_{(V/K)obs}^o$ plots. So, from inspection of $II_{(V)obs}^o$ and $II_{(V/K)obs}^o$, one can assign the mechanism.

Once the kinetic mechanism is established, K_i values are estimated by fitting the data for the relevant secondary plot to equations of Figure 5.3.

- **Competitive Inhibition.** Data of $II^o_{(V/K)obs}$ is fit to Equation 5.11.
- **Uncompetitive Inhibition.** Data of $II^o_{(V)obs}$ is fit to Equation 5.15.
- **Noncompetitive/Mixed Inhibition.** Data of $II^o_{(V/K)obs}$ is fit to Equation 5.17 and data of $II^o_{(V)obs}$ to Equation 5.18.

To illustrate these concepts, consider the plots of Figure 5.8, which contain the analysis of the inhibition of the ATPase activity of cdk5/p25 kinase by N4-(6-aminopyrimidin-4-yl)-sulfanilamide (APS) (Liu et al. 2008). Primary plot I^o_I was constructed (not shown here) and demonstrates that APS inhibition curves, at each substrate concentration, could be fit to the four parameter function with $n = 1$ and $v_{bkg} = 0$. Primary plot I^o_S was also constructed (Figure 5.8A) and shows that the dependence of v_o on $[ATP]_o$ at each inhibitor concentration obeys Michaelis–Menten kinetics and thus allowed secondary plots $II^o_{(V)obs}$ and $II^o_{(V/K)obs}$ to be constructed (Figure 5.8B). It is evident from inspection of the replots that both $(V_{max})_{obs}$ and $(V_{max}/K_m)_{obs}$ titrate with APS, indicating a mixed mechanism of inhibition. Nonlinear least squares analysis of these two plots to the equations for mixed inhibition of Figure 5.7 yield best-fit parameters, $K_i = 3.6\,\mu M$ and $\beta K_i = 94\,\mu M$.

It should be noted that the inhibition curve for the secondary plot $(V_{max})_{obs}$ does not go to zero at infinite $[APS]_o$, but rather goes to a value of about $0.1V_{max}$. This suggests that the ternary complex of cdk5/p25, ATP, and APS still retains catalytic activity. The ability to pick up this subtle feature of mechanism is one of the advantages of the method of replots.

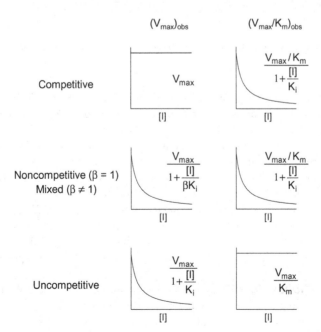

Figure 5.7. $[I]_o$ dependence of $(V_{max})_{app}$ and $(V_{max}/K_m)_{app}$ for the three classical mechanisms of inhibition.

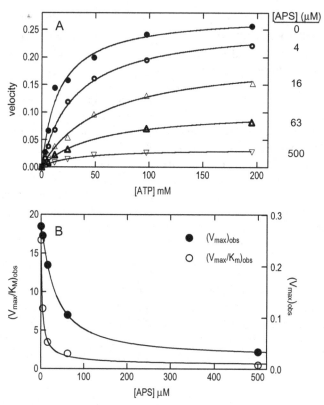

Figure 5.8. Method of replots—inhibition by APS of the ATPase activity of cdk5/p25 kinase.

As another example of these concepts, consider Figure 5.9, which shows secondary plots for the inhibition of ubiquitin C-terminal hydrolase-L1 (UCH-L1) by LDN-091946. Recall that in Chapter 4, we saw that the plot of v_o versus [LDN-091946]$_o$ at a single concentration of substrate obeyed a simple binding law with $n = 1$ and $v_{bkg} = 0$. Additional experiments confirmed that this simple behavior was observed over a range of substrate concentrations. Also, primary plot I_S^o was constructed and demonstrates that the dependence of v_o on $[S]_o$ at each inhibitor concentration obeys Michaelis–Menten kinetics and thus allowed secondary plots $II_{(V)obs}^o$ and $II_{(V/K)obs}^o$ to be constructed (Figure 5.9). Here we see that while $(V_{max})_{obs}$ titrates with inhibitor, $(V_{max}/K_m)_{obs}$ does not, and is a clear case of uncompetitive inhibition (see Figure 5.7). From the data for $(V_{max})_{obs}$ versus $[I]_o$, a K_i value of 2.8 µM by best-fit analysis to Equation 5.15.

5.3.1.2   Global Fit.

This method is much simpler to execute than the method of replots. According to this method, the entire data set representing the dependence of v_o on independent variables $[S]_o$ and $[I]_o$ is fit to a rate law for a candidate inhibition mechanism. The results from such an exercise are usually presented as a plot of the $[S]_o$ dependence of v_o, at the various fixed values of $[I]_o$, with the best-fit parameters

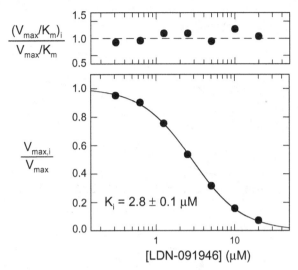

Figure 5.9. Method of replots—inhibition by LDN-091946 of the hydrolase activity of UCH-L1.

used to draw lines for the dependencies of v_o on $[S]_o$ at each value of $[I]_o$. Typically, rate laws for several mechanisms are used to fit the data. The mechanism that best fits the data is judged by evaluating the fitting statistics for each mechanism, as well as by visual inspection.

To illustrate this, I will once again consider the inhibition by APS of cdk5's ATPase activity. The data of Figure 5.8A were fit to the three classical mechanisms of inhibition: competitive, mixed, and uncompetitive. Graphs of these fits are shown in Figure 5.10 and the best-fit parameters and values of root mean square[1] are summarized in Table 5.1.

Statistically and by inspection, the fit of the data to uncompetitive inhibition is the poorest of the three fits. But while the statistics for the fit to mixed inhibition is better than that for competitive inhibition, visual inspection leaves one wondering which is the actual mechanism of inhibition. And, of course, this is a crucial mechanistic feature. Uncompetitive inhibition tells us that APS can bind to the complex of ATP and enzyme, while competitive inhibition tells us that no such binding mode exists.

We see from this example that while the global fit provides overall statistics for the fit of a data set to a particular model, it does have some problems associated with it.

- **Visual judgment of goodness-of-fit.** It is often difficult to visually judge the goodness of fit at low substrate concentration, where data points can be "bunched

---

[1] The root mean square, rms, equals $\sqrt{SS/(N-P)}$, where SS is the sum-of-squares of residuals, N is number of data points, and P is the number of parameters (N-P is the degree of freedom).

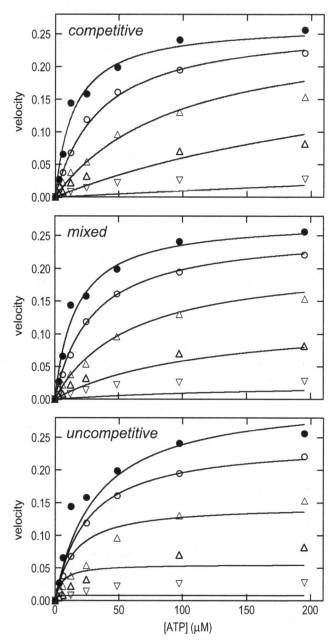

Figure 5.10. Global fit—inhibition by APS of the ATPase activity of cdk5/p25 kinase. ●, 0; ○, 4; ▲, 16; △, 63; and ▼, 500 μM.

TABLE 5.1. Parameter Estimates Using the Rate Laws for the Three Standard Inhibition Mechanisms for the Data of Figure 5.8A

Mechanism	V_{max}	K_m (μM)	K_i (μM)	βK_i (μM)	rms
Competitive	0.27 ± 0.01	15 ± 2	2.8 ± 0.4	—	0.0108
Mixed	0.28 ± 0.01	17 ± 2	4.4 ± 0.7	58 ± 21	0.0029
Uncompetitive	0.31 ± 0.02	31 ± 5	14 ± 2	—	0.0182

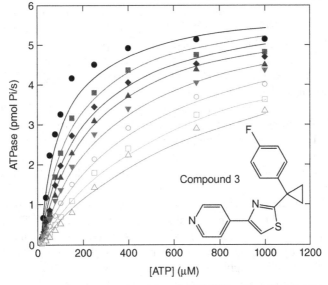

Figure 5.11. Inhibition of the ATPase activity of Eg5[1-367] by thiazole **3**. Inhibitor concentrations from low to high are: 0, 50, 100, 150, 250, 400, 600, and 800 nM.

up." This is clearly seen in Figure 5.11, which is the next example we will consider, where it is impossible to visually assess the fit of data at the low range of substrate concentration.

- **Ambiguity in the assignment of mechanism.** We saw in the above example that fit of the same data set to competitive and mixed inhibition give very nearly the same fit. We are in the uncomfortable position of being left only with statistics to assign mechanism.

- **Departures from standard mechanisms are hidden.** An investigator will often fit a data set to the rate laws for the three classical inhibition mechanisms: competitive, noncompetitive, and uncompetitive. One of these will fit the data better than the other two, and it will be this mechanism that will be said to be the mechanism of inhibition. Related to both this practice and the difficulty to visually assess the data and fits is the likelihood that an investigator will miss mechanistically significant departures from the standard mechanism. For example, we saw that the analysis of the inhibition of cdk5/p25 by APS suggests that classical mixed inhibition (see Figure 5.1) may not adequately describe this case. Rather it is likely that the complex cdk5/p25:APS:ATP retains some catalytic activity.

This next example illustrates many of the points I have been making about the relative merits of the global fit and the method of replots.

5.3.1.3 Thiazole Inhibitors of Monomeric Eg51-367. Eg5, or kinesin spindle protein, is a motor protein that uses the energy available from its hydrolysis of ATP to drive the separation of spindle poles during mitosis. This protein is essential for cell mitosis, thus making its ATPase activity a prime target for oncology drug discovery. Eg5[1-367] is a truncated form of Eg5 that is monomeric in solution and apparently more "user-friendly" for the development of assays that might be used in high-throughput screens (HTS) to identify inhibitors.

In the course of such a screening campaign, investigators at Merck identified a new class of thiazole inhibitors of Eg5[1-367] (Rickert et al. 2008). To determine the kinetic mechanism of inhibition of these compounds, studies of the sort we have been discussing were conducted. Figure 5.11 is taken from a paper by the Merck group and is a plot of the steady-state inhibition data for their Compound **3**. The authors state that they initially fit the data for inhibition by Compound **3** to a model for mixed inhibition but found that β was poorly defined. They were, however, able to fit the data to the rate law for competitive inhibition and calculated a K_i value of 125 nM. The authors conclude that "the thiazoles described here show ATP competitive kinetic behavior" (Rickert et al. 2008; abstract). However, inspection of the data and the best-fit lines suggests that the situation might not be so straightforward.

The first feature to note in Figure 5.11 is that the best-fit lines for competitive inhibition really do not fit the data very well at all. This calls into question the assignment of a competitive mode of inhibition. Close inspection of Figure 5.11 reveals a systematic departure of the best-fit lines from the data points that is suggestive of some form of cooperativity (see Chapter 3). Rather than a hyperbolic dependence of v_o on $[ATP]_o$, the data are better described by a sigmoidal shape that is characteristic of positive cooperativity.

The model I will use to discuss Eg5's cooperativity is shown in Figure 5.12, a simplified version of the mechanism of Figure 3.12. The rate law for this mechanism is given in Equation 5.30. As we learned in Chapter 3, if $K_{s,1}/K_{s,2} > 10$, $K_{s,1}$ and $K_{s,2}$ cannot be individually

$$v_o = \frac{V_{\max}[S]_o}{K_{s,2}\left(1+\dfrac{K_{s,1}}{[S]_o}\right)+[S]_o} \tag{5.30}$$

determined, an expression of the form shown in Equation 5.31 must be used to fit the data:

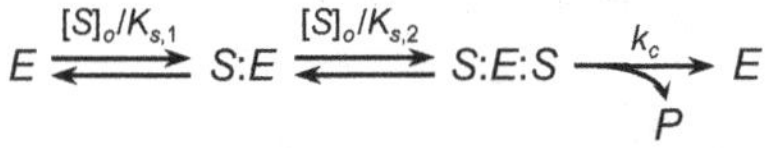

$$E \overset{[S]_o/K_{s,1}}{\rightleftharpoons} S{:}E \overset{[S]_o/K_{s,2}}{\rightleftharpoons} S{:}E{:}S \overset{k_c}{\longrightarrow} E \\ \searrow \\ P$$

Figure 5.12. Model for cooperative binding of substrate to enzyme. Positive cooperativity is observed when $K_{s,1} > K_{s,2}$.

$$v_o = \frac{V_{max}}{\dfrac{K_{s,1}K_{s,2}}{[S]_o^2} + 1}. \tag{5.31}$$

According to this minimal mechanism for cooperativity, free enzyme binds ATP at a non-catalytic, allosteric site, perhaps the same site to which the inhibitor monastrol binds, to form S:E. Binding of this first molecule of ATP allows binding of the second at the active site to form catalytically competent S:E:S. In mechanisms such as this, positive cooperativity will be observed if $K_{s,1} > K_{s,2}$. It should be noted that for enzymes in which the catalytic step is not rate limiting, such as Eg5 (Cochran and Gilbert 2005; Krysiak and Gilbert 2006), sigmoidal kinetics can have purely kinetic origins (see Chapter 3) Nonetheless, for the sake of simplicity and illustrative purposes, I will use the rapid equilibrium mechanism of Figure 5.12.

We will proceed in our analysis using the method of replots.[2] Attempts at fitting the eight v_o versus $[ATP]_o$ dependencies to Equation 5.30 resulted in values of $K_{s,1}$ and $K_{s,2}$ that had error limits as large as 200%. Thus, the data sets were fit to Equation 5.31. In all cases, the fits were excellent. The data and best-fit lines are shown in Figure 5.13A.

Next, replots were constructed of the $[I]_o$ dependencies of $(V_{max})_{app}$ and $(V_{max}/K_{s,1}K_{s,2})_{app}$ (see Fig. 5.13B). We see from the replots that Compound 3 titrates both $(V_{max})_{app}$ and $(V_{max}/K_{s,1}K_{s,2})_{app}$, with $K_{i,app}$ values of 1–2 μM and 38 ± 3 nM, respectively, indicating that Compound 3 is a mixed inhibitor of Eg5, binding to free enzyme and the S:E:S complex. What this data do not tell us is whether the inhibitor can bind to S:E.

This example illustrates how global fitting to standard models can result in the assignment of incorrect mechanisms. In the present case, several important mechanistic features were missed due to global fitting of the data to standard inhibition models.

5.3.1.4 *Emergence of Mechanism from Data.* The previous discussion of global fits and the method of replots illustrates an important feature of my philosophy of data analysis, a feature that runs throughout this book. When possible, we should avoid imposing preconceived theoretical constructs or mechanisms on data. Rather we should employ methods of data reduction that all allow the *mechanism to emerge from the data.*

---

[2] I would like to thank Dr. Keith Rickert for supplying this data for analysis.

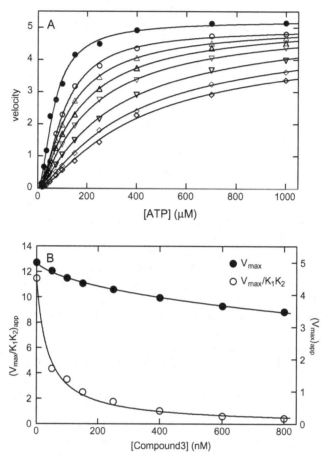

Figure 5.13. Methods of replots—inhibition of the ATPase activity of Eg5[1-367] by Compound 3.

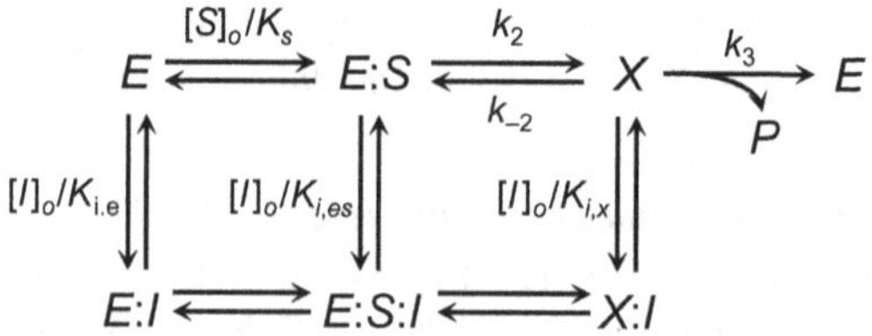

Figure 5.14. Mechanism of inhibition of a single substrate reaction that proceeds through intermediate species X.

Experimental condition: $[S]_0, [I]_0 \gg [E]_0$

Conservation equations: $[E]_0 = [E] + [ES] + [X] + [EI] + [ESI] + [XI]$

$[S]_0 = [S] + [ES] + [ESI]$

$[I]_0 = [I] + [EI] + [ESI] + [XI]$

Statement of rate dependence: $v_0 = k_3[X]$

Steady-state condition: $\dfrac{d[X]}{dt} = 0 \Rightarrow [X] = \dfrac{k_2}{k_{-2} + k_3}[ES]$

Derivation of rate equation:

$$\frac{v_0}{[E]_0} = \frac{k_3[X]}{[E] + [ES] + [X] + [EI] + [ESI] + [XI]}$$

$$\frac{v_0}{[E]_0} = \frac{k_3}{\dfrac{[E]}{[X]} + \dfrac{[ES]}{[X]} + 1 + \dfrac{[EI]}{[X]} + \dfrac{[ESI]}{[X]} + \dfrac{[XI]}{[X]}}$$

$$\frac{v_0}{[E]_0} = \frac{k_3}{\dfrac{K_s}{[S]_0}\dfrac{k_{-2}+k_3}{k_2} + \dfrac{k_{-2}+k_3}{k_2} + 1 + \dfrac{[I]_0}{K_{i,e}}\dfrac{K_s}{[S]_0}\dfrac{k_{-2}+k_3}{k_2} + \dfrac{[I]_0}{K_{i,es}}\dfrac{k_{-2}+k_3}{k_2} + \dfrac{[I]_0}{K_{i,x}}}$$

$$\frac{v_0}{[E]_0} = \frac{k_3}{\dfrac{K_s}{[S]_0}\dfrac{k_{-2}+k_3}{k_2} + \dfrac{k_{-2}+k_3+k_2}{k_2} + \dfrac{[I]_0}{K_{i,e}}\dfrac{K_s}{[S]_0}\dfrac{k_{-2}+k_3}{k_2} + \dfrac{[I]_0}{K_{i,es}}\dfrac{k_{-2}+k_3}{k_2} + \dfrac{[I]_0}{K_{i,x}}}$$

$$\frac{v_0}{[E]_0} = \frac{\dfrac{k_2 k_3}{k_2 + k_{-2} + k_3}}{\dfrac{K_s}{[S]_0}\dfrac{k_{-2}+k_3}{k_2+k_{-2}+k_3} + 1 + \dfrac{[I]_0}{K_{i,e}}\dfrac{K_s}{[S]_0}\dfrac{k_{-2}+k_3}{k_2+k_{-2}+k_3} + \dfrac{[I]_0}{K_{i,es}}\dfrac{k_{-2}+k_3}{k_2+k_{-2}+k_3} + \dfrac{[I]_0}{K_{i,x}}\dfrac{k_2}{k_2+k_{-2}+k_3}}$$

$$\frac{v_0}{[E]_0} = \frac{\dfrac{k_2 k_3}{k_2 + k_{-2} + k_3}[S]_0}{K_s\dfrac{k_{-2}+k_3}{k_2+k_{-2}+k_3}\left(1+\dfrac{[I]_0}{K_{i,e}}\right) + [S]_0\left(1 + \dfrac{[I]_0}{K_{i,es}}\dfrac{k_{-2}+k_3}{k_2+k_{-2}+k_3} + \dfrac{[I]_0}{K_{i,x}}\dfrac{k_2}{k_2+k_{-2}+k_3}\right)}$$

$$\frac{v_0}{[E]_0} = \frac{\dfrac{k_2 k_3}{k_2 + k_{-2} + k_3}[S]_0}{K_s\dfrac{k_{-2}+k_3}{k_2+k_{-2}+k_3}\dfrac{1+\dfrac{[I]_0}{K_{i,e}}}{1+\dfrac{[I]_0}{K_{i,es}\left(1+\dfrac{k_2}{k_{-2}+k_3}\right)} + \dfrac{[I]_0}{K_{i,x}\left(+\dfrac{k_{-2}+k_3}{k_2}\right)}} + [S]_0} \cdot \left(1+\dfrac{\dfrac{[I]_0}{K_{i,es}\left(1+\dfrac{k_2}{k_{-2}+k_3}\right)} + \dfrac{[I]_0}{K_{i,x}\left(1+\dfrac{k_{-2}+k_3}{k_2}\right)}}{}\right)$$

(5.32)

$$\frac{v_0}{[E]_0} = \frac{\dfrac{k_c}{1+\dfrac{[I]_0}{\beta K_i}}[S]_0}{K_m\dfrac{1+\dfrac{[I]_0}{K_i}}{1+\dfrac{[I]_0}{\beta K_i}} + [S]_0}$$

(5.33)

Figure 5.15. Derivation of the rate law for the mechanism of Figure 5.14.

The method of replots, with its stepwise analytical procedure, is such a method. At each stage of the analysis, there is the opportunity for appraisal and judgment, and reflection on what this stage has just revealed. At each stage, we can look for anomalies that may alert us to important mechanistic features.

5.4   INHIBITION OF ONE-SUBSTRATE, TWO-INTERMEDIATE REACTIONS

In Chapter 3, we examined the kinetic behavior of the one-substrate, two-intermediate reaction shown in Figure 3.4. We learned that the meaning of k_c and K_m is dependent on the relative values of k_2, k_{-2}, and k_3 as expressed in Equations 3.23–3.26. Here we extend our examination to the inhibition of this reaction.

The inhibition reaction we will consider is shown in Figure 5.14. In it, inhibitor binds to three enzyme forms: E, ES, or X, where X can be one of a great many types of intermediary species including a conformational isomer of ES, a product of active-site chemistry (e.g., acyl- or phospho-enzyme), or the complex of enzyme and product. The question that we hope to answer in this section is how do we then interpret $[S]_o$-independent inhibition constants that emerge from mechanism of inhibition studies?

We start by deriving the rate expression for the mechanism of Figure 5.14. Using the steady-state assumption for intermediate X, the derivation proceeds in the usual way as shown in Figure 5.15, resulting finally in the rate law of Equation 5.32, which can be recast as Equation 5.33 using the definitions of Equations 5.34–5.37:

$$k_c = \frac{k_2 k_3}{k_2 + k_{-2} + k_3} \tag{5.34}$$

$$K_m = K_s \left(\frac{k_{-2} + k_3}{k_2 + k_{-2} + k_3} \right) \tag{5.35}$$

$$K_i = K_{i,e} \tag{5.36}$$

$$\beta K_i = \frac{\left\{ K_{i,es} \left(1 + \dfrac{k_2}{k_{-2} + k_3} \right) \right\} \left\{ K_{i,x} \left(1 + \dfrac{k_{-2} + k_3}{k_2} \right) \right\}}{K_{i,es} \left(1 + \dfrac{k_2}{k_{-2} + k_3} \right) + K_{i,x} \left(1 + \dfrac{k_{-2} + k_3}{k_2} \right)}. \tag{5.37}$$

Equations 5.36 and 5.37 are based on the kinetic equivalence of the mechanisms of Figures 5.1 and 5.14. These two equations tell us several things about enzymes that follow the mechanism of Figure 5.14:

- **Competitive inhibitors**, as determined by the standard methods described earlier in this chapter, bind to E and have a $[S]_o$-independent inhibition constant K_i that is equal to $K_{i,e}$.

- **Uncompetitive inhibitors** can bind either to $E:S$ or to X, and βK_i will equal $K_{i,es}\left(1 + \dfrac{k_2}{k_{-2} + k_3}\right)$ or $K_{i,x}\left(1 + \dfrac{k_{-2} + k_3}{k_2}\right)$, respectively. If inhibitor binds to both $E:S$ and X, βK_i will equal the weighted average of Equation 5.37. Whether these constants actually equal the true dissociation constants $K_{i,es}$ and $K_{i,x}$, depends on the relative magnitudes of k_2, k_{-2}, and k_3.
- **Noncompetitive inhibitors** bind to E and $E:S$ or X, or to all three species.

To summarize the above, we see that for inhibitors that are determined to be competitive, there is no ambiguity in assignment of either the species to which inhibitor binds, which is free enzyme, or the meaning of K_i, the inhibition constant determined from initial velocity studies, which is $K_{i,e}$. Things are not so simple with uncompetitive inhibitors. Here, standard mechanism of inhibition studies reveal neither the identity of the species to which inhibitor binds, which could be $E:S$, X, or both, nor the meaning of βK_i. For noncompetitive inhibition, the value of $K_{i,e}$ can still be determined unambiguously. Binding to complexed forms of enzyme is analyzed as shown below for uncompetitive inhibitors.

The meaning of βK_i can be better understood if Equation 5.37 is expressed in reciprocal form, in which case βK_i becomes an association equilibrium constant $\beta K_{i,assoc}$:

$$\beta K_{i,assoc} = \frac{\left(K_{i,es}\right)_{assoc}}{1 + \dfrac{k_2}{k_{-2} + k_3}} + \frac{\left(K_{i,x}\right)_{assoc}}{1 + \dfrac{k_{-2} + k_3}{k_2}}. \tag{5.38}$$

Figure 5.16A is a redrawing of Figure 5.15 emphasizing that $\beta K_{i,assoc}$ reflects the two parallel reaction paths that $E:S$ can take. This concept is shown in simplified form in Figure 5.16B, where it is seen that the parallel reactions of A to produce B and C is governed by rate constant k_{obs}, which is equal to the sum of the rate constants for the two parallel paths.

The reaction path that $E:S$ takes, resulting in the steady-state accumulation of $E:S:I$ or $X:I$, is governed by the magnitudes of the two association constants $(K_{i,es})_{assoc}$ and $(K_{i,x})_{assoc}$, and the relative magnitudes of k_2, k_{-2}, and k_3.

Equation 5.38 can be simplified to Equation 5.39 for enzymes in which k_{-2} is zero:

$$\beta K_{i,assoc} = \frac{\left(K_{i,es}\right)_{assoc}}{1 + \dfrac{k_2}{k_3}} + \frac{\left(K_{i,x}\right)_{assoc}}{1 + \dfrac{k_3}{k_2}}. \tag{5.39}$$

Such enzymes are hydrolases that proceed through a covalently bound intermediate, and include phosphatases, serine, and cysteine proteases, esterases, amidases, acyl-transferases operating in the absence of acyl-acceptor, and certain glycohydrolases. Uncompetitive inhibition of these enzymes can only be interpreted if independent kinetic studies have revealed the relative magnitude of k_2 and k_3.

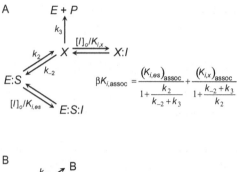

Figure 5.16. Reactions involving parallel pathways. *Panel A:* Redrawn mechanism of Figure 5.14 emphasizing how E:S can potentially associate with inhibitor via two different pathways. *Panel B:* Reaction of compound A by parallel paths to form B and C.

A case that is relevant to drug discovery is the inhibition of acetylcholinesterase (AchE). Inhibitors of this enzyme, which act to extend the lifetime of acetylcholine in neuronal synapses, are used widely in clinical practice to treat the dementia of Alzheimer's disease. One such inhibitor is tetrahydroaminoacridine, or tacrine which is marketed as Cognex™. Berman and Leonard reported that the mechanism of inhibition of AchE depends on substrate identity, being noncompetitive for acetylthiocholine, and competitive for 7-acetoxy-4-aminocoumarin (7AMC) (Berman and Leonard 1991).

Now, it was known that for acetylthiocholine, $k_2/k_3 > 10$, while for 7AMC, $k_2/k_3 < 0.1$. Thus, according to Equation 5.39, for acetylthiocholine, $\beta K_{i,assoc} = (K_{i,x})_{assoc}$, and for 7AMC, $\beta K_{i,assoc} = (K_{i,es})_{assoc}$. Given these kinetics, Berman and Leonared explained the substrate-dependent inhibition mechanisms by proposing a mechanism in which tacrine is able to bind to free enzyme and the acyl-enzyme, but not the Michaelis complex. We see then that for both substrates, binding to free enzyme is evidenced by the inhibition of V_{max}/K_m, but only for hydrolysis of acetylthiocholine, where $k_2/k_3 \gg 1$ and the reactant state for V_{max} is the acyl-enzyme, will inhibition of V_{max} be observed. For hydrolysis of 7AMC, where $k_2/k_3 \ll 1$ and the reactant state for V_{max} is the Michaelis complex, V_{max} will be unaffected by inhibitor.

The upshot of this section is that for inhibition of an enzyme whose reaction proceeds through more than a single intermediate, the meaning of βK_i for an uncompetitive or noncompetitive inhibitor will be dependent on the internal kinetics of the mechanism and thus on the identity of the substrate. Thus, we should not be surprised to see different values of βK_i when different substrates are used. Finally, we see that the only way to fully understand βK_i is through a thorough understanding of the kinetic, and perhaps chemical, mechanism of the enzyme.

5.5 INHIBITION BY DEPLETION OF SUBSTRATE

Many enzymes of interest to drug discovery researchers catalyze reactions of macro-molecular substrates. Examples of these enzymes, and their substrates and reactions, are summarized in Table 5.2.

When designing screening assays for these enzymes, one of the first decisions the investigator must make is whether to use a synthetic, low-molecular-weight substrate or a natural, macromolecular substrate. As we discussed in Section 2.5.1.1, this decision will be based on practical considerations (e.g., which substrates afford the most precise, reliable, and cost-effective assay) and whether the macromolecule makes contacts with enzyme exosites that are mechanistically important. If a macromolecular substrate is chosen, one must be aware of a factor that can confound interpretation of screening results—the possibility that the observed inhibition is due to the inhibitor binding to the substrate rather than the enzyme.

For example, in a screen for inhibitors for COT kinase, only such compounds were discovered (Jia et al. 2006). COT kinase is a member of the mitogen-activated protein kinase family of enzymes and plays a role in TNF-α production in macrophages, and is thus a target for anti-inflammatory drug discovery. In a screen for COT inhibitors using an assay based on the COT-catalyzed phosphorylation of the kinase MEK1, compounds were identified whose inhibitory activity was not due to interaction with COT kinases, but rather was due to the ability of the compounds to bind to MEK1, thereby preventing its interaction with COT.

Perhaps the best documented case of enzyme inhibition being caused by the binding of inhibitor to substrate is the inhibition of cell wall biosynthesis in gram$^+$ bacteria by the glycopeptide antibiotic vancomycin (see structure in Fig. 5.17). Vancomycin is used to treat infections caused by gram$^+$ bacteria, including methicillin-resistant *Staphylococcus aureus* (MRSA) and strains of penicillin-resistant *Streptococcus pneumonia*. As just mentioned, the mode of action of this antibiotic is through the inhibition of cell wall biosynthesis. But instead of exerting its inhibitory effect by binding to one of the biosynthetic enzymes, vancomycin binds to the D-alanyl-D-alanine

TABLE 5.2. Enzymes that Catalyze Reactions of Macromolecular Substrates

Enzyme Class	Substrate	Reaction
Polymerases	RNA & DNA	Nucleic acid formation
Nucleases	RNA & DNA	Hydrolysis
Glycohydrolases	Polysaccharides	Hydrolysis of glycosidic linkages
Proteases	Proteins	Hydrolysis of peptide bonds
Kinases	Proteins	ATP-dependent phosphorylation of Ser, Thr, Tyr
Phosphatases	Proteins	Hydrolysis of phosphorylated residues
Acetyltransferases	Proteins	AcCoA-dependent acetylation of Lys
Deacetylases	Proteins	Hydrolysis of Lys(Ac)
Methyltransferases	Proteins	SAM-dependent methylation of Lys

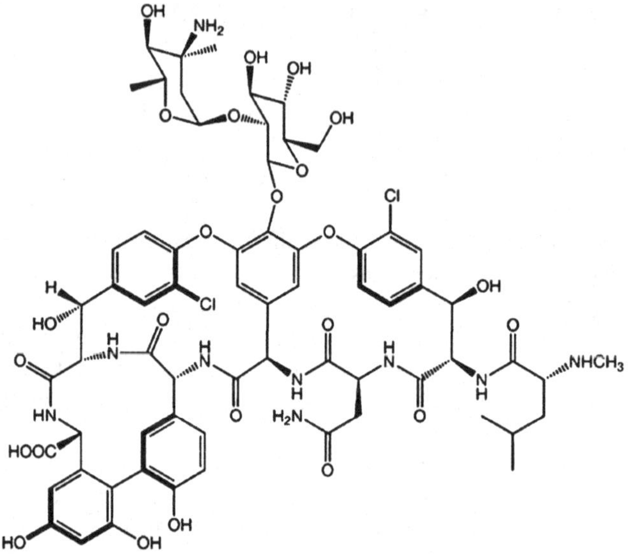

Figure 5.17. Structure of vancomycin.

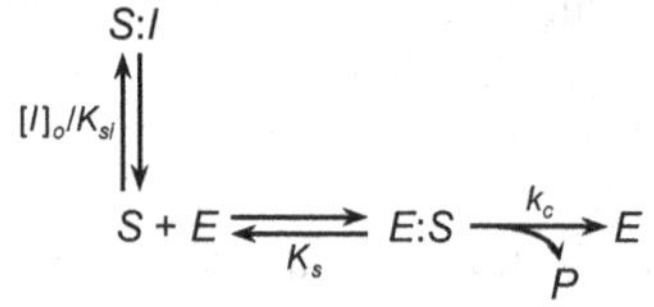

Figure 5.18. Mechanism for inhibition by substrate depletion.

dipeptide on the peptide side chain of newly synthesized peptidoglycan subunits, preventing them from being incorporated into the cell wall.

The minimal mechanism for such inhibition is shown in Figure 5.18. According to this mechanism, inhibitor is seen to reversibly bind to substrate, forming a complex that cannot bind to enzyme. Simple inspection of this mechanism provides immediate insights into this mechanism of inhibition. We see that the three limiting cases for the relative magnitudes of K_{si}, $[S]_o$, and $[I]_o$ are accompanied by limiting extents of inhibition:

- $K_{si} \gg [S]_o$, $[I]_o \Rightarrow [S:I] \cong 0$, no inhibition is observed.
- $[S]_o \gg K_{si}$, $[I]_o \Rightarrow [S:I] \cong 0$, no inhibition is observed.
- $[I]_o \gg K_{si}$, $[S]_o \Rightarrow [S:I] \cong [S]_o$, 100% inhibition is observed.

However, in non-limiting cases, where K_{si}, $[S]_o$, and $[I]_o$ are within an order of magnitude of one another, a rate law will be required to predict the extent of inhibition. This rate law must take into account that $[S:I]$ is a significant fraction of $[S]_o$. Thus, the assumption we have made throughout this chapter that $[S] = [S]_o$ no longer holds. In this case, the rate equation for the mechanism of Figure 5.18 must be written as

$$v_o = \frac{k_c[E]_o[S]}{K_s + [S]} \qquad (5.40)$$

where initial velocity depends on $[S]$, and not on $[S]_o$.

Given this, we now must derive an expression for the concentration of free substrate. This derivation starts with the definition of K_{si}:

$$K_{si} = \frac{[S][I]}{[SI]}. \qquad (5.41)$$

From the conservation of matter relationship $[I]_o = [I] + [SI]$, the numerator of Equation 5.41 becomes

$$K_{si} = \frac{[S]\{[I]_o - [SI]\}}{[S]_o - [S]}. \qquad (5.42)$$

Replacing $[SI]$ in the numerator of Equation 5.42 with $[S]_o - [S]$ yields

$$K_{si} = \frac{[S]\{[I]_o - [S]_o + [S]\}}{[S]_o - [S]}. \qquad (5.43)$$

Multiplying through by $[S]_o - [S]$ and rearranging yields the quadratic equation

$$[S]^2 + (K_{si} + [I]_o - [S]_o)[S] - K_{si}[S]_o = 0, \qquad (5.44)$$

the solution for which is:

$$[S] = \frac{-(K_{si} + [I]_o - [S]_o) + \sqrt{(K_{si} + [I]_o - [S]_o)^2 + 4K_{si}[S]_o}}{2} \qquad (5.45)$$

Due to the form of the rate law for the mechanism of Figure 5.18, the shapes of inhibitor titration curves will vary with substrate concentration. In Figure 5.19 are simulated titration curves drawn using Equations 5.40 and 5.45. We see that IC_{50} values increase with increased $[S]_o$, which at first blush suggests competitive inhibition. However, closer inspection of these curves reveals a situation that is not so simple. As $[S]_o$ increases the curves become sigmoidal. Due this sigmoidal behavior, IC_{50} values will not depend on $[S]_o$ in the simple way that IC_{50} depends on $[S]_o$ for competitive inhibition. Rather, IC_{50} values for inhibition by substrate depletion will have a much steeper dependence on $[S]_o$.

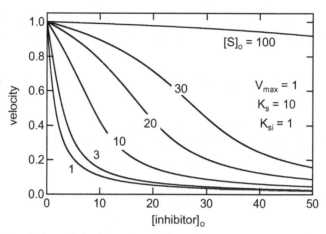

Figure 5.19. Simulations of the dependence of initial velocity on inhibitor concentration for the substrate depletion mechanism of Figure 5.18.

REFERENCES

Berman, H. A. and K. Leonard (1991). "Interaction of tetrahydroaminoacridine with acetylcholinesterase and butyrylcholinesterase." *Mol. Pharmacol.* **41**: 412–418.

Cochran, J. C. and S. P. Gilbert (2005). "ATPase mechanism of Eg5 in the absence of microtubules." *Biochemistry* **44**: 16633–16648.

Jia, Y., et al. (2006). "Comparative analysis of various in vitro COT kinase assay formats and their applications in inhibitor identification and characterization." *Anal. Biochem.* **350**: 268–276.

Krysiak, T. C. and S. P. Gilbert (2006). "Dimeric Eg5 maintains processivity through alternating-site catalysis with rate-limiting ATP hydrolysis." *J. Biol. Chem.* **281**: 39444–39454.

Liu, M., et al. (2008). "Kinetic studies of Cdk5/p25 kinase: Phosphorylation of tau and complex inhibition by two prototype inhibitors." *Biochemistry* **47**: 8367–8377.

Rickert, K. W., et al. (2008). "Discovery and biochemical characterization of selective ATP competitive inhibitors of the human mitotic kinesin KSP." *Arch. Biochem. Biophys.* **469**: 220–231.

Schowen, R. L. (1978). Catalytic power and transition state stabilization. In *Transition States of Biochemical Processes*, R. D. Gandour and R. L. Schowen, eds. New York, Plenum Press: 77–144.

Stein, R. L. (1985). "Catalysis by human leukocyte elastase. 5. Structural features of the virtual transition state for acylation." *J. Am. Chem. Soc.* **107**(25): 7768–7769.

6

TIGHT-BINDING, SLOW-BINDING, AND IRREVERSIBLE INHIBITION

In Chapters 4 and 5, the kinetic expressions we derived for analyzing enzyme inhibition data apply only in those cases were the following three conditions are met:

- Total enzyme concentration is much less than the experimental range of inhibitor concentration.
- The interaction of inhibitor with enzyme is reversible.
- The attainment of equilibrium between enzyme and the enzyme:inhibitor (E:I) complex is rapid relative to the timescale of the kinetic experiment.

In practice, however, these conditions are sometimes not met, leading to *tight-binding inhibition*, when $[E]_o < [I]_{expt}$, *irreversible inhibition*, when $k_{dissociation} = 0$ for E:I, or *slow-binding inhibition*, when $t_{equilibrium} > 5\,min$. The goal of this chapter is to develop methods to accurately and precisely estimate inhibitor potency for these cases.

6.1 IMPORTANCE IN DRUG DISCOVERY

Reliable estimates of inhibitor potency are the foundation of every drug discovery program that has an enzyme as its therapeutic target. These estimates not only guide

Kinetics of Enzyme Action: Essential Principles for Drug Hunters, First Edition. Ross L. Stein.
© 2011 John Wiley & Sons, Inc. Published 2011 by John Wiley & Sons, Inc.

the medicinal chemist toward greater inhibitor potency, but also the biologist in the choice of inhibitor dose for cellular and animal models of disease, and even the clinical pharmacologist in explaining various aspects of pharmacodynamics. These diverse activities all suffer when inhibitor potency is not determined accurately.

Given this, it is essential that the team's enzymologist know if the above conditions are met for the inhibitor he is characterizing, and, if not, that he use the appropriate methods to estimate the inhibitor's potency. For treating a tight-binding or slow-binding inhibitor as if it were classical can lead to IC_{50} or $K_{i,app}$ estimates that are orders of magnitude higher than the actual K_i value, and treating an irreversible inhibitor as if it were classical leads to IC_{50} or $K_{i,app}$ values that are meaningless, since enzyme and the complex formed between enzyme and irreversible inhibitor can never attain equilibrium.

6.2 TIGHT-BINDING INHIBITION

6.2.1 Estimation of K_i Values for Tight-Binding Inhibitors

As we will see in this and the next section, the analysis of potent enzyme inhibitors presents special challenges that, if not adequately dealt with, can lead to large underestimates of potency. In this section, we will consider the situation in which the inhibitor under analysis is of such potency that $[I]_{expt}$, the experimental concentration range needed to estimate K_i, encompasses $[E]_o$.

Recall that for classical inhibitors, where $[I]_{expt} \sim K_i \gg [E]_o$, the expression for conservation of inhibitor,

$$[I] = [I]_o - [EI] \tag{6.1}$$

simplifies to $[I] = [I]_o$. If, however, K_i is the same order of magnitude as $[E]_o$, then $[I]_{expt}$ must also be similar in magnitude to $[E]_o$ if one is to accurately estimate inhibitor potency. In such cases, no simplification of Equation 6.1 is possible. To estimate inhibitor potency for such cases, we must use a rate expression that takes into account the fact that the concentration of the EI complex is a significant fraction of the total inhibitor concentration.

Using the model of competitive inhibition shown in Figure 5.3, we begin this derivation in the usual way, by dividing the statement of rate dependence by the expression for conservation of enzyme:

$$\frac{v_o}{[E]_o} \frac{k_c[ES]}{[E]+[ES]+[EI]} \tag{6.2}$$

Since $[EI]$ is a significant fraction of $[I]_o$, we must use an expression for $[EI]$ that takes this into account. To derive such an expression, we start with Equation 6.3,

$$K_i = \frac{[E][I]}{[EI]} \tag{6.3}$$

and substitute $[I]$ with the term, $[I]_o - [EI]$:

$$K_i = \frac{[E]\{[I]_o - [EI]\}}{[EI]}. \tag{6.4}$$

Rearranging, we obtain the expression we need for $[EI]$:

$$[EI] = \frac{[I]_o[E]}{K_i + [E]}. \tag{6.5}$$

But this is still not quite what we need. For this expression to be useful for substitution into Equation 6.2, we must express $[E]$ in terms of $[ES]$. We accomplish this by rearranging the expression for K_s to obtain $[E] = (K_s[ES])/[S]_o$. When we substitute this into Equation 6.5 and rearrange we obtain,

$$[EI] = \frac{[I]_o[ES]}{\dfrac{K_i[S]_o}{K_s} + [ES]}. \tag{6.6}$$

Substituting this expression for $[EI]$ into Equation 6.2 and dividing through by $[ES]$, yields

$$v_o = \frac{k_c[E]_o}{\dfrac{K_s}{[S]_o} + 1 + \dfrac{[I]_o}{\dfrac{K_i[S]_o}{K_s} + \dfrac{v_o}{k_c}}} \tag{6.7}$$

where v_o/k_c has been substituted for $[ES]$.

Multiplying the numerator and denominator by $[S]_o/(K_s + [S]_o)$, we obtain

$$v_o = \frac{v_c}{1 + \dfrac{[I]_o}{\dfrac{K_i[S]_o}{K_s} + \dfrac{v_o}{k_c}} \dfrac{[S]_o}{K_s + [S]_o}}, \tag{6.8}$$

where v_c is the control velocity, which equals $k_c[E][S]_o/(K_s + [S]_o)$. Carefully rearranging Equation 6.8, yields

$$v_o = \frac{v_c}{1 + \dfrac{[I]_o}{K_{i,app} + \dfrac{v_o}{k_c}\left(1 + \dfrac{K_s}{[S]_o}\right)}} \tag{6.9}$$

where $K_{i,app} = K_i(1 + [S]_o/K_s)$. Finally, multiplying through by the denominator of Equation 6.9 and collecting terms yields the quadratic equation,

$$\left(\frac{[E]_o}{v_c}\right)v_o^2 + v_o\left(K_{i,app} + [I]_o - [E]_o\right) - v_c K_{i,app} = 0. \tag{6.10}$$

Solving this equation for v_o yields

$$v_o = \frac{-\left(K_{i,app} + [I]_o - [E]_o\right) + \sqrt{\left(K_{i,app} + [I]_o - [E]_o\right)^2 + 4K_{i,app}[E]_o}}{2\left(\dfrac{[E]_o}{v_c}\right)} \tag{6.11}$$

or, in its more familiar form,

$$\frac{v_o}{v_c} = \frac{1}{2[E]_o}\left\{-\left(K_{i,app} + [I]_o - [E]_o\right) + \sqrt{\left(K_{i,app} + [I]_o - [E]_o\right)^2 + 4K_{i,app}[E]_o}\right\}. \tag{6.12}$$

The dependence of v_o/v_c on $[I]_o$ for a tight-binding inhibitor can be fit to the expression of Equation 6.12 to arrive at best-fit values of $K_{i,app}$ and $[E]_o$. This is illustrated in the example below.

6.2.1.1 Case Study: Inhibition of the Prolyl cis-trans Isomerase Activity of FKBP by FK-506.
To illustrate how treating an inhibitor as classical, when in fact it is tight-binding, can lead to underestimates of potency, consider Figure 6.1 which shows two analyses of data for the inhibition of FKBP's prolyl *cis-trans* isomerase activity by FK-506, a natural product immunosuppressive agent. In the late 1980s, it was discovered that the immunosuppressive activity of FK-506 is mediated through the complex it forms with FKBP, or FK-506 binding protein (Stein 1991; Stein 1993). Around the same time, it was found that FKBP has enzymatic activity, catalyzing the *cis-trans* isomerization of Pro-Xaa bonds in proteins and peptides (Stein 1991; Stein 1993).

Figure 6.1A contains data for a titration of FKBP with FK-506 (Harrison and Stein 1992). These data were fit to a two parameter function, in which $v_{control}$ and v_{bkg} of Equation 4.1 are constrained to one and zero, respectively. The best fit K_i value is 23 nM and roughly agrees with a literature IC_{50} estimate of 50 nM (Harding et al. 1989). However, the slope factor of this fit is 2. Recall from Chapter 4 (Section 4.3.2) that one reason for slope factors being larger than one is K_i being on the same order of magnitude as $[E]_o$. And this is precisely what is occurring here. In this experiment, the nominal enzyme concentration was 50 nM and $[I]_{expt}$ ranged from 10 to 100 nM. Fitting this data to Equation 6.12 yielded $K_i = 1.9 \pm 0.1$ nM and $[E]_o = 45 \pm 3$ nM.

The actual K_i value is an order of magnitude lower than the value estimated by treating FK-506 as a classical inhibitor. The lesson here is that the enzymologist must always be alert to $[E]_o$ relative to $[I]_{expt}$.

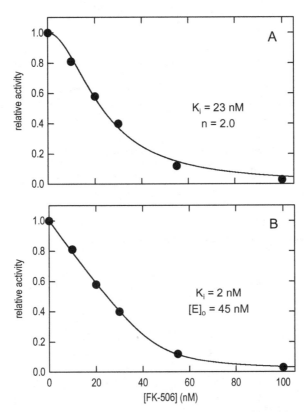

Figure 6.1. Inhibition of the prolyl *cis-trans* isomerase activity of FKBP by KF-506. *Panel A:* Solid line through the data points was drawn using the best-fit paramaters from nonlinear

least-squares fit to: $\dfrac{v_o}{v_c} = \dfrac{1}{1 + \left(\dfrac{[I]_o}{K_{i,app}}\right)^n}$; $K_{i,app} = 23 \pm 1\,\text{nM}$, $n = 2.0 \pm 0.2$. *Panel B:* Solid line through

the data was drawn using the best-fit paramaters from nonlinear least-squares fit to Equation 6.12; $K_{i,app} = 1.7 \pm 0.2\,\text{nM}$, $[E]_o = 45 \pm 5\,\text{nM}$.

6.2.2 Use of a Tight-Binding Inhibitor as Active-Site Titrant

Figure 6.2 contains simulations of inhibitor titrations using Equation 6.12, an enzyme concentration of 1 nM, and the indicated values of $K_{i,app}$. What is clear from this figure is that when $K_{i,app} < [E]_o$, titration curves are characterized by a distinctive linear region at low inhibitor concentrations. The form of this linear dependence of v_o/v_c on $[I]_o$ can be derived from Equation 6.12 under the condition $K_i \ll [I]_o \sim [E]_o$:

$$\frac{v_o}{v_c} = -\frac{[I]_o}{[E]_o} + 1. \tag{6.13}$$

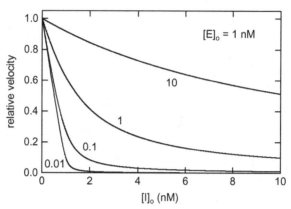

Figure 6.2. Use of tight-binding inhibitors as active-site titrants. Simulations of binding isotherms according to Equation 6.12 for $[E]_o = 1$ nM and the indicated $K_{i,app}$ values (nM units).

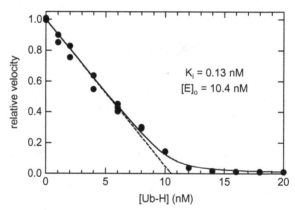

Figure 6.3. Titration of *UCH-L1* by ubiquitin C-terminal aldehyde (*Ub-H*). In this experiment the nominal enzyme concentration was 10 nM. Solid line through the data was drawn using the best-fit parameters from nonlinear least-squares fit to Equation 6.12; $K_{i,app} = 0.13 \pm 0.04$ nM, $[E]_o = 10.4 \pm 0.3$ nM.

Significantly, the x-axis intercept is $[E]_o$.

From the above analysis, one can see that a potent inhibitor can be used to determine the concentration of active enzyme. Knowing the concentration of active enzyme is of course essential for converting values of V_{max} to k_c.

6.2.2.1 Case Study: Inhibition of UCH-L1 by Ubiquitin C-Terminal Aldehyde (Ub-H).

Figure 6.3 contains a plot of the dependence of v_o/v_c on $[Ub-H]_o$ for the inhibition of *UCH-L1* (Case and Stein 2006). Fitting this data to Equation 6.12 yielded $K_{i,app} = 0.13 \pm 0.04$ nM and $[E]_o = 10.4 \pm 3$ nM. This enzyme concentration agrees with the nominal concentration of 10 nM. While the $K_{i,app}$ value has a relative

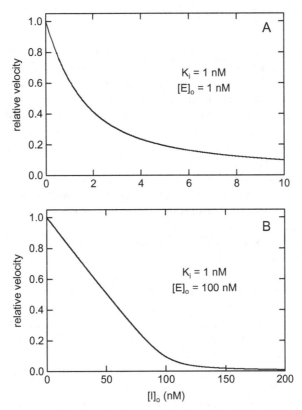

Figure 6.4. Simulations of binding isotherms using Equation 6.12 and the indicated values of $K_{i,app}$ and $[E]_o$.

large error associated with it, the enzyme concentration is determined well enough that it can serve as an accurate estimate of the concentration of active *UCH-L1*.

6.2.3   Different $[E]_o$ for Different Purposes

Once preliminary studies of a tight-binding inhibitor provide a tentative $K_{i,app}$ value, an accurate estimate of $K_{i,app}$ can be obtained by determining the $[I]_o$ dependence of v_o/v_c at an enzyme concentration that has been set to a value similar to the tentative $K_{i,app}$ value. As shown in the simulation of Figure 6.4A, an inhibitor concentration range running from around 0.5 to 10 μM would be appropriate for an inhibitor with a $K_{i,app}$ of 1 nM.

With knowledge of the tight-binding inhibitor's $K_{i,app}$ value, an experiment can be designed to use the inhibitor as an active-site titrant. The key factor to bear in mind here is that $[E]_o$ must be much larger than $K_{i,app}$ if an accurate estimate of $[E]_o$ is to be obtained. This can be seen in Figures 6.3 and 6.4B where a significant region of linearity in the v_o/v_c versus $[I]_o$ plot is observed under the condition $[E]_o = 100K_{i,app}$.

6.3 SLOW-BINDING INHIBITION

A feature of enzyme inhibition that we have not yet discussed is its time dependence. In typical kinetic experiments of the sort simulated in Figure 6.5A, the full effect of an inhibitor is expressed immediately after addition of the inhibitor to a reacting solution of enzyme and substrate. However, there is a class of inhibitor, so-called "slow-binding inhibitors," in which full inhibitory potency is not immediate but rather develops slowly (see Fig. 6.5B). The interaction of a slow-binding inhibitor with its target enzyme is characterized by a pre-steady-state phase with a duration that is on the timescale of conventional kinetic experiments, that is, minutes to hours.

Recognizing that an inhibitor is a slow-binding inhibitor and then correctly determining its interaction kinetics and K_i value for the target enzyme is critical for establishing structure–activity relationships that accurately reflect the changes in inhibitory potency that attend systematic variation of inhibitor structure. The consequences of failing to notice that an inhibitor is slow-binding can be appreciated upon inspection of the inhibition progress curve of Figure 6.5B. It is clear that estimates of potency based on initial velocity measurements will be grossly underestimated. The biochemical literature has many examples of this. For example, IC_{50} values based on initial velocity measurements for the inhibition of chymotrypsin by the natural product chymostatin range from 10 nM to 100 nM (Umezawa et al. 1970; Feinstein et al. 1976), while the analysis of this system taking into account that chymostatin is a slow-binding inhibitor of chymotrypsin yields a K_i value of 0.4 nM (Stein and Strimpler 1987). Perhaps the most dramatic example of the underestimate of the potency of a slow-binding inhibitor is the inhibition of 5α-reductase by finasteride.

Case Study: Inhibition of 5α-Reductase by Finasteride. 5α-reductase is an NADPH-dependent enzyme that catalyzes the reduction of testosterone to dihydrotestosterone (Bull et al. 1996) and has been implicated in diseases in which the latter more potent androgen hormone is thought to play a role. Finasteride is one of the 5α-reductase inhibitors developed by Merck in the early 1990s to treat such diseases and is marketed

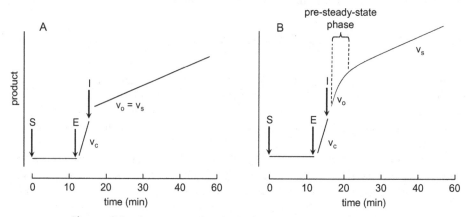

Figure 6.5. Time courses for classical and slow-binding inhibition.

today as Procepia™ and Proscar™ for the treatment of male pattern baldness and benign prostatic hyperplasia, respectively.

Initial estimates of it as a 26 nM inhibitor were at odds with the extremely small doses that were required in initial clinical trials of the drug. This was finally reconciled when it was realized that finasteride is a slow-binding inhibitor of 5α-reductase and that the initial estimates of potency, based on a single time point assay involving no preincubation of enzyme and inhibitor (Rasmusson et al. 1986), underestimated the true potency of the inhibitor. Careful kinetic studies (Bull et al. 1996) revealed that the rate constant for the association of 5α-reductase and finasteride is an unremarkable $10^6 M^{-1} s^{-1}$, but the rate constant for dissociation of the E:I complex is about $10^{-7} s^{-1}$, corresponding to a dissociation half-time of four weeks! The ratio of these rate constants is the K_i value and is less than 0.1 pM. Thus, the 26 nM IC_{50} value initially determined underestimates the potency of finasteride by five orders of magnitude.

In what follows, we will learn how to determine the kinetic constants and K_i values for slow-binding inhibitors. We start with a trivial, yet often forgotten truth in enzymology.

6.3.1 All Inhibition Is Time-Dependent Inhibition

Equation 6.14 expresses the dissociation constant for the interaction of enzyme and inhibitor in the fashion we have seen in previous chapters:

$$K_i = \frac{[E][I]}{[EI]} = \frac{[E][I]_o}{[EI]}. \tag{6.14}$$

This expression for K_i can be recast as its kinetic equivalent,

$$K_i = \frac{k_{off}}{k_{on}}, \tag{6.15}$$

where, k_{on} is the second-order rate constant for the combination of E and I to form E:I, and k_{off} is the first-order rate constant for dissociation of E:I, for the simple mechanism.

$$E + I \underset{k_{off}}{\overset{k_{on}}{\rightleftharpoons}} E{:}I$$

It is often convenient to express Equation 6.15 as

$$K_i = \frac{k_{off}}{k'_{on}} \tag{6.16}$$

where k'_{on} is the pseudo-first-order rate constant,

$$k'_{on} = k_{on}[I]_o. \tag{6.17}$$

The mechanism can now be written as the reversible interconversion of two species:

$$E \underset{k_{off}}{\overset{k_{on}[I]_o}{\rightleftharpoons}} E{:}I$$

For this mechanism, the differential equation of Equation 6.18 expresses the time-dependent

$$\frac{\partial [E]}{\partial t} = -k'_{on}[E] + k_{off}[EI] \tag{6.18}$$

change in the concentration of free enzyme. Integration of this expression yields

$$[E] = [E]_o e^{-(k'_{on}+k_{off})t+} + \frac{k_{off}}{k'_{on}+k_{off}}[E]_o e^{-(k'_{on}+k_{off})t+} + \frac{k_{off}}{k'_{on}+k_{off}}[E]_o, \tag{6.19}$$

which can be rearranged to

$$[E] = ([E]_o - [E]_{eq})e^{-k_{obs}t} + [E]_{eq}, \tag{6.20}$$

where k_{obs} is the first-order rate constant for the approach of the system to equilibrium and equals

$$k_{obs} = k'_{on} + k_{off}, \tag{6.21}$$

and $[E]_{eq}$ is the concentration of enzyme when the system has reached equilibrium, and equals

$$[E]_{eq} = \frac{[E]_o}{1 + \dfrac{[I]_o}{K_i}}. \tag{6.22}$$

Equation 6.20 predicts that in an experiment in which an enzyme is introduced into a solution of inhibitor, $[E]_o$ will exponentially decrease with time to its equilibrium concentration, $[E]_{eq}$, as it complexes with the inhibitor.

A useful metric is the time to reach equilibrium (or steady state, as discussed below), which can be estimated as 5τ, where

$$\tau = (k'_{on} + k_{off})^{-1}. \tag{6.23}$$

When 5τ has elapsed, the system's attainment of equilibrium is about 98% complete.

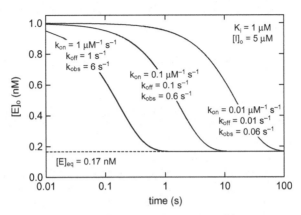

Figure 6.6. Simulations of progress for the disappearance of free enzyme as it attains equilibrium with E:I. Solid lines were drawn using Equation 6.20.

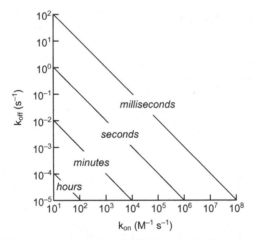

Figure 6.7. All inhibition is time-dependent inhibition. k_{off} is plotted as a function of k_{on} for an inhibitor of $K_i = k_{off}/k_{on} = 1\,\mu M$. Each line has a slope of $-1\,\mu M$ and demarcates time zones for the attainment of inhibition, with $[I]_o \sim 1\,\mu M$.

Examples of progress curves for the disappearance of enzyme with time are shown in Figure 6.6. In these simulations, while K_i and $[I]_o$ are held constant at 1 and $5\,\mu M$, respectively, k'_{on}, k_{off}, and of course k_{obs}, vary over three orders of magnitude. For the k_{obs} values 6, 0.6, and $0.06\,s^{-1}$, approximately 1, 10, and $100\,s$ are required for the system to attain equilibrium. A more global picture is shown in Figure 6.7, which is a plot of k_{off} versus k_{on} for an inhibitor of K_i equal to $1\,\mu M$. Each line has a slope of $-1\,\mu M$ and demarcates time zones for the attainment of inhibition, with $[I]_o \sim 1\,\mu M$. We see that as the individual terms of k_{off}/k_{on} both increase from $(10^{-4}\,s^{-1})/(10^2\,M^{-1}\,s^{-1})$ to $(10^2\,s^{-1})/(10^8\,M^{-1}\,s^{-1})$, the time to reach equilibrium changes from hours to milliseconds.

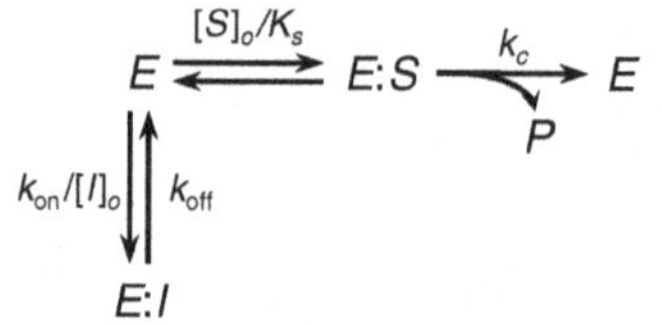

$$E \underset{}{\overset{[S]_o/K_s}{\rightleftharpoons}} E{:}S \xrightarrow{k_c} \begin{array}{c} E \\ P \end{array}$$

$$k_{on}/[I]_o \Big\updownarrow k_{off}$$

$$E{:}I$$

Figure 6.8. Mechanism for competitive inhibition.

Classical inhibitors are those for which the equilibrium between E and E:I is said to be reached "instantaneously," with no time dependence. As just illustrated, all inhibition is time-dependent inhibition. What differs among inhibitors is the timescale for the observation of the time dependence. Using conventional sorts of kinetics instrumentation, inhibitors with $k_{obs} > 10^{-2}\,s^{-1}$ ($t_{1/2} < 100\,s$) will promote what appears to be an "instantaneous" attainment of equilibrium between E and E:I. For $k_{obs} > 10^{-2}\,s^{-1}$ ($t_{1/2} > 100\,s$), attainment of equilibrium can be observed my monitoring E or E:I as they approach their equilibrium concentrations.

6.3.2   Kinetics of Inhibition

While in certain favorable cases one can determine the kinetics for the approach to equilibrium by monitoring some signal associated with either E or E:I, it is of course more common to use substrate turnover by enzyme as a reflection of free enzyme concentration. Thus, for the mechanism of Figure 6.8, the reaction velocity for the enzyme-catalyzed reaction decreases according to the following equation:

$$v_t = (v_o - v_s)e^{-k_{obs}t} + v_s \tag{6.24}$$

where v_t is the velocity at time t, v_o is the velocity at time zero, which for this simple mechanism is equal to v_c, the control velocity in the absence of inhibitor,

$$v_o = v_c = \frac{k_c[E]_o[S]_o}{K_s + [S]_o} \tag{6.25}$$

v_s is the final, steady-state velocity in the presence of inhibitor,

$$v_s = \frac{v_o}{1 + \dfrac{[I]_o}{K_{i,app}}} \tag{6.26}$$

and

$$k_{obs} = k'_{on}[I]_o + k_{off}. \tag{6.27}$$

In Equations 6.26 and 6.27, $K_{i,app}$ and k'_{on} are dependent on the mechanism of inhibition. In Figure 6.8, the mechanism is competitive, so

$$K_{i,app} = K_i \left(1 + \frac{[S]}{K_s} \right) \tag{6.28}$$

and

$$k'_{on} = \frac{k_{on}}{1 + \frac{[S]}{K_s}}. \tag{6.29}$$

This theory can be productively reduced to practice for inhibitor/enzyme systems for which 5τ is an hour or longer. In these cases, an experimental protocol, illustrated in Figure 6.9, is available to determine the kinetics of free enzyme disappearance as reflected in velocity measurements. In this illustration, a spectrophotometric assay is used to measure initial velocities of reactions solutions constructed by the addition of a small aliquot of concentrated substrate solution to a much larger volume of a solution of enzyme and inhibitor that had been allowed to incubate for a predetermined time. The dependence of velocity on incubation time can be fit to Equation 6.24 to arrive at best-fit values of k_{obs}, v_o, and v_s. As we will discuss below, if this experiment is done at several inhibitor concentrations, then the $[I]_o$ dependence of k_{obs} and v_{ss} will allow the calculation of k_{on} and k_{off} and $K_{i,app}$, respectively.

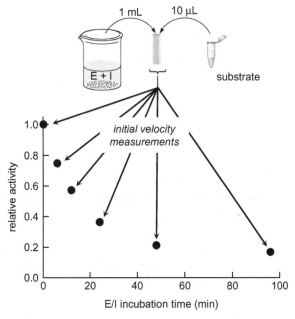

Figure 6.9. Design of an experiment to measure the kinetics of interaction of enzyme and inhibitor.

As suggested above, this method works well for slow-binding inhibitors for which $5\tau > 1\,h$, but for E/I systems that attain equilibrium more rapidly, another method is more convenient. For these cases, we will want to measure product formation, or substrate depletion, as a function of time. Such progress curves are described by Equation 6.30 which is

$$[P] = \left(\frac{v_o - v_s}{k_{obs}}\right)\left(1 - e^{-k_{obs}t}\right) + v_s t \tag{6.30}$$

the integrated form of Equation 6.24.

A simulated progress curve, drawn with Equation 6.30, is shown in Figure 6.10. In this simulated experiment, enzyme was added to a solution of inhibitor and substrate, and product was then monitored as a function of time. This curve is characterized by three phases: an initial phase that extrapolates to time = 0 with slope v_o,

$$[P] = v_o t \tag{6.31}$$

a final phase with a slope equal to v_s (see Eq. 6.26), and an exponential phase that connects the two linear phases with a pseudo-first rate constant of k_{obs}.

Note that at infinite time, Equation 6.30 simplifies to:

$$[P] = \left(\frac{v_o - v_s}{k_{obs}}\right) + v_s t. \tag{6.32}$$

This equation describes a linear dependence of product concentration on time with slope equal to v_s and an intersect of $(v_o - v_s)/k_{obs}$ (see inset of Fig. 6.10). Equations 6.31 and 6.32 intersect at a point with a time coordinate equal to $1/k_{obs} = \tau$.

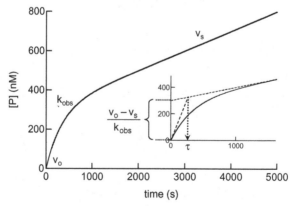

Figure 6.10. Progress curve for slow-binding inhibition. The curve was drawn using Equation 6.30 and the paramater values: $v_o = 1.0\,nM\,s^{-1}$, $v_s = 0.1\,nM\,s^{-1}$, and $k_{obs} = 0.003\,s^{-1}$.

While the above analysis provides a graphical means to determine values of v_o, v_s, and k_{obs} from reaction progress curve for a slow-binding inhibitor, it is more common to fit the data points of the curve to Equation 6.30 by nonlinear least squares analysis. In subsequent sections, it will be assumed that parameter estimates are obtained by the latter means.

6.3.3 Mechanisms of Slow-Binding Inhibition

To this point, the mechanism of inhibition we have been considering involves formation of the E:I complex without formation of intermediate species. This is the mechanism of Figure 6.8, which is shown again in Figure 6.11 as Mechanism A. Another mechanism that is commonly encountered involves formation of an intermediate on the pathway to the final E:I complex. This is depicted as Mechanism B of Figure 6.11.

According to Mechanism B, $(E:I)_1$ rapidly forms, with dissociation constant $K_{i,1}$, and then slowly undergoes transformation into $(E:I)_1$. The overall dissociation constant is expressed as

$$K_i = K_{i,1} \frac{k_{-i}}{k_i + k_{-1}}. \tag{6.33}$$

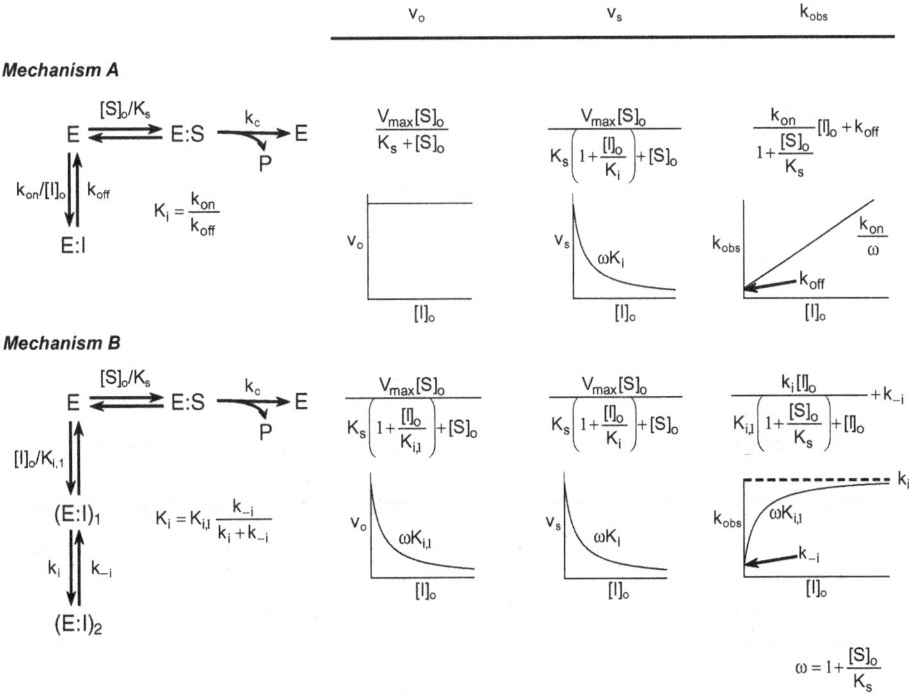

Figure 6.11. Mechanisms of slow-binding inhibition and corresponding expressions defining v_o, v_s, and k_{obs}. For the plots of v_o and v_s vs. $[I]_o$, the parameter in the plot (i.e., $\omega K_{i,1}$ or ωK_i) is the apparent inhibitor dissociation constant.

If the dissociation of $(E:I)_2$ is slow relative to its formation from $(E:I)_1$ (i.e., $k_{-i} \ll k_i$), then

$$K_i = K_{i,1} \frac{k_{-i}}{k_i} = K_{i,1} K_{i,2} \qquad (6.34)$$

and $(E:I)_2$ is the species that accumulates in the steady state. If $k_{-i} \gg k_i$, then $K_i = K_{i,1}$, and while $(E:I)_2$ may form, it is kinetically invisible (see Section 5.2.2).

While both mechanisms manifest progress curves of the sort shown in Figure 6.10, they can be distinguished by how the three progress curve parameters, v_o, v_s, and k_{obs}, depend on $[I]_o$ (see Fig. 6.11).

- **Dependence of v_o on $[I]_o$.** For Mechanism A, v_o is independent of $[I]_o$ since at very early times $E:I$ will not have formed. In contrast, for Mechanism B, since $(E:I)_1$ forms rapidly (i.e., within the dead time of the experiment), the dependence of v_o on $[I]_o$ follows the simple binding isotherm we first learned about in Chapter 4. Thus,

$$v_o = \frac{v_c}{1 + \dfrac{[I]_o}{K_{i1,app}}} \qquad (6.35)$$

where

$$v_c = \frac{V_{max}[S]_o}{K_m + [S]_o} \qquad (6.36)$$

and

$$K_{i1,app} = K_{i,1}\left(1 + \frac{[S]_o}{K_s}\right). \qquad (6.37)$$

- **Dependence of v_s on $[I]_o$.** For both mechanisms, the dependence of v_s on $[I]_o$ follows the form of a simple binding curve defined by ωK_i, where K_i is defined by Equation 6.33 and reflects the accumulation of the most stable $E:I$ complex.
- **Dependence of k_{obs} on $[I]_o$.** For Mechanism A, k_{obs} is linearly dependent on $[I]_o$, with slope and y-intercept equal to $k_{on}/(1 + [S]_o/K_s)$ and k_{off}, respectively. In contrast, the dependence of k_{obs} on $[I]_o$ for Mechanism B is hyperbolic, obeying the expression

$$k_{obs} = \frac{k_i[I]_0}{K_{i,1}\left(1 + \dfrac{[S]_o}{K_s}\right) + [I]_o} + k_{-i}. \qquad (6.38)$$

This function is similar to the Michaelis–Menten equation, with its familiar saturation kinetics. In this case, as $[I]_o$ goes to infinity, k_{obs} asymptotically approaches k_i, doing so with an apparent dissociation constant equal to $K_{i,1}(1 + [S]_o/K_s)$. However, unlike the Michaelis–Menten equation, which has an intercept of zero, the hyperbolic dependence of k_{obs} on $[I]_o$ intersects the y-axis at k_{-i}.

Note that for a system in which $K_{i,1} \gg [I]_o$, Equation 6.38 simplifies to

$$k_{obs} = \frac{(k_i / K_{i,1})[I]_0}{1 + \dfrac{[S]_o}{K_s}} + k_{-i} \qquad (6.39)$$

which is of course identical to Equation 6.27, where $k_{on} = k_i/K_{i,1}$. This is analogous to kinetics of substrate turnover for cases in which substrate concentration is much less than K_s. For such cases, the Michaelis–Menten equation simplifies to $v_o = (k_c/K_s)[S]_o$, and v_o will be linearly dependent of $[S]_o$ with a slope equal to k_c/K_s. We see that for inhibition and substrate turnover, a linear dependence of v_o on ligand concentration does not necessarily rule out the existence of an intermediary species, it just means that the ligand concentration range was not sufficiently broad to reveal the species.

6.3.4 Experimental Approaches

6.3.4.1 Initial Experiments: Titration Curves and Time-Dependent $K_{i,app}$ Values.
The initial hint that an inhibitor might be slow-binding will usually be the observation of steep titration curves with slope factors greater than 1. This occurs because the slow kinetics of inhibition do not allow establishment of equilibrium at low concentrations of inhibitor. The suspicion of slow-binding inhibition can be confirmed by determining titration curves as a function of preincubation time between enzyme and inhibitor. For a slow-binding inhibitor, longer preincubation times will be accompanied by the observation of curves with smaller values of $K_{i,app}$ and slope factors that approach 1. This is illustrated in Figure 6.12 for an inhibitor with $K_{i,app}$, k_{on}, and k_{off} values of 0.1 μM, $10^4 M^{-1} s^{-1}$, and $10^{-3} s^{-1}$, respectively. We see that as the preincubation increases from 10 s to 1000 s, the slope factor decreases from 1.5 to 1.1, and the $K_{i,app}$ value decreases 50-fold from 6.4 μM to 0.12 μM.

6.3.4.2 Progress Curves.
As we saw above, the evaluation of the kinetics of slow-binding inhibition is based on the analysis of inhibition progress curves. Several points must be kept in mind if the investigator is going acquire reliable titration curves:

- **Start data collection immediately after reaction initiation.** Unlike kinetic analysis of simple first-order reactions in which sampling data at early times is not critical for the calculation of accurate estimates of first-order rate constants, these early data point are critical for accurate estimates of inhibition progress curve parameters, especially v_o and k_{obs}.

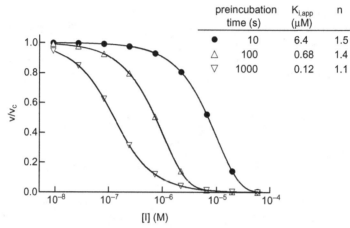

Figure 6.12. The dependence of titration curve shape preincubation time for a slow-binding inhibitor.

Consider the progress curve of Figure 6.13A for slow-binding inhibition, by LDN-27219, of the transglutaminase-catalyzed transamidation of Gln residues N, N-dimethylated casein (NMC) by the fluorogenic Lys-containing peptide αN-Boc-Lys-NH-CH$_2$-CH$_2$-NH-dansyl (KXD) (Case and Stein 2005). (See Chapter 4 for introductory comments about this enzyme.) Formation of isopeptide bonds between KXD and the Gln residues of NMC is followed by the change in fluorescence intensity that occurs when the dansyl moiety of KXD is incorporated into the protein substrate (Case and Stein 2005; Case et al. 2005).

When the full data set (i.e., data points from time 0s to 1000s) is fit to Equation 6.30, the following best-fit parameters are calculated: $v_o = 4.5 \, FI \, s^{-1}$, $v_s = 0.38 \, FI \, s^{-1}$, and $k_{obs} = 0.011 \, s^{-1}$. The inset summarizes the kinetic consequences of fitting truncated data sets in which initial data points are deleted. We can see that as more and more of the initial data points are deleted, simulating the initiation of data collection at times farther and farther away from the actual t_o of the reaction, v_o and k_{obs} are estimated with decreasing accuracy. As might be expected, estimates of v_s values are little effected by early data point truncation.

Capturing early data points can be difficult for inhibition systems with k_{obs} values greater than about $0.03 \, s^{-1}$, or $t_{1/2} \leq 20 \, s$. While rapid kinetic instrumentation (e.g., stopped flow apparatus) will give the best results for such systems, reliable progress curves can be acquired with conventional equipment if care is taken. What will not work in these situations are microplate readers, which should only be used for inhibition progress curves with k_{obs} values less than $0.003 \, s^{-1}$, or $t_{1/2} > 200 \, s$.

- **Continue data collection sufficiently long to ensure accurate parameter estimates.** For inhibition systems with $t_{1/2}$ values much greater than about 5 min, there is a natural tendency to terminate the reaction before sufficient data have

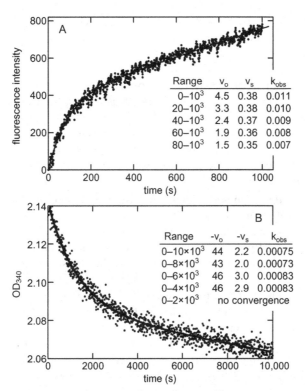

Figure 6.13. Progress curve analysis for the inhibition of tissue transglutaminase (TGase) by LDN-27219. *Panel A:* Inhibition of the TGase-catalyzed transamidation of NMC by KXD. The line through the points is the best fit line to all the data (i.e., t = 0–1000 s). The inset summarizes best-fit parameters to Equation 6.30 for full and truncated data sets. Velocities have units of FI s^{-1}. [NMC] = 100 μM, [KXD] = 20 μM, [LDN-21219] = 4 μM, and [$TGase$] = 0.15 μM. *Panel B:* Inhibition of the TGase-catalyzed hydrolysis of the Gln residue of Z-Pro-Gln-Gly-Trp. The line through the points is the best fit line to all the data (i.e., t = 0–10,000 s). The inset summarizes best-fit parameters to Equation 6.30 for full and truncated data sets. Velocities have units of μOD s^{-1}. [Z-Pro-Gln-Gly-Trp] = 630 μM, [LDN-21219] = 20 μM, and [$TGase$] = 0.15 μM.

been collected to accurately define v_s. The consequences of this can be seen in Figure 6.13B, which contains a progress curve for inhibition of the TGase-catalyzed hydrolysis of the Gln residue of Z-Pro-Gln-Gly-Trp (Case and Stein 2005). The hydrolysis of the Gln residue was monitored using a glutamate dehydrogenase-coupled assay that allows continuous measurement of product ammonia. In this assay, ammonia serves as substrate for glutamate dehydrogenase to promote the α-ketoglutarate-dependent oxidation of NADH to NAD$^+$ ($\Delta\varepsilon_{340}$ = −6220) (Case and Stein 2003).

When the full data set (i.e., data points from time 0 s to 10,000 s) is fit to Equation 6.30, the following best-fit parameters are calculated: v_o = 44 μOD s^{-1},

$v_s = 2.2 \, \mu\text{OD s}^{-1}$, and $k_{obs} = 7.5 \times 10^{-4} \text{s}^{-1}$. The inset summarizes the kinetic consequences of fitting truncated data sets in which initial data points are deleted. Nonlinear least-square analysis of the first four data sets yielded parameter estimates in which v_o is seen to be the least sensitive parameter to truncation and v_s the most sensitive. However, for the most truncated data set, the analysis did not converge, indicating that this was too extensive of a truncation to be approximated by Equation 6.30.

For accurate parameter estimates, progress curve data should be collected for at least 10 half-times. At this point, the progress curve will have a very clear and well-defined steady-state region, which will ensure accurate estimates of all three inhibition progress curve parameters.

- **Terminate data collection before substrate depletion.** The derivation of Equation 6.30 assumes that in the course of collecting the data of an inhibition progress curve significant substrate depletion has not occurred, so that $[S] \sim [S]_o$. From a practical point of view, this means that in the course of collecting the progress curve, no more than 10% of the substrate should have been consumed.

This is illustrated in Figure 6.14A which contains progress curves for the inhibition of human leukocyte elastase by the trifluoromethyl ketone MeOSuc-Val-Pro-Val-CF$_3$ (Stein et al. 1987; $K_i = 13 \, \text{nM}$). In the control run, $10 \, \mu\text{M}$ product has been produced, which corresponds to a consumption of only 6% of the substrate, so that $[S] \sim [S]_o$. In this experiment, the control curve is linear, as expected from the substrate consumption, and care was taken to ensure that the amount of product formed during inhibition reactions never exceeded this limiting value of $10 \, \mu\text{M}$.

The consequences of allowing inhibition progress curves to exceed low levels of substrate consumption are illustrated in the simulations of Figure 6.14B, which model a system in which 10% of substrate consumption corresponds to a product concentration of $6 \, \mu\text{M}$. All three inhibition progress curves exceed this value, by amounts that increase with decreasing inhibitor concentration. Each of these curves was fit to Equation 6.30, with best-fit values summarized in Table 6.1. The numbers in parentheses are the "actual" parameter values calculated using the various constants for this system (see legend of Fig. 6.14b).

Regardless of the inhibitor concentration and the extent to which the data of the progress curve exceeds 10% substrate consumption, the estimate of v_o is essentially identical to the nominal value of $23 \, \text{nM s}^{-1}$. This is not the case for v_s and k_o. While best-fit values for these parameters vary with $[I]_o$ in the anticipated direction, they differ from the nominal values, and this difference increases with decreasing inhibitor concentration.

6.3.4.3 *Independent Determination of* k_{off}. As outlined above, plots of k_{obs} versus $[I]_o$ allow the determination of both k_{on} and k_{off}. However, if $k_{off} \ll k_{on}[I]_{expt}$, accurate estimates of k_{off} from a plot of k_{obs} versus $[I]_o$ will be impossible. This can be seen in Figure 6.15A, which shows the $[I]_o$ dependence of k_{obs} for inhibition by γboroGlu

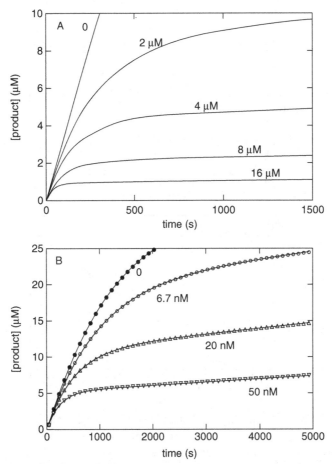

Figure 6.14. Inhibition progress curve analysis and substrate depletion. *Panel A:* Progress curves for the inhibition human leuckocyate elastase by MeOSuc-Val-Pro-Val-CF$_3$ (MeOSuc, methoxysuccinyl; pNA, *p*-nitroaniline). Substrate was MeOSuc-Ala-Ala-Pro-Val-pNA and was present at a concentration of 160 μM, which is equal to about 3K_m. Inhibitor was present at the indicated concentration. *Panel B:* Simulations of slow-binding inhibition progress under the following conditions: $K_s = 100\,\mu M$, $k_c = 1000\,s^{-1}$, $k_{on} = 1 \times 10^5\,M^{-1}s^{-1}$, $k_{off} = 1 \times 10^{-4}\,s^{-1}$, $[S]_o = 30\,\mu M$, $[E]_o = 0.1\,nM$, and the indicated $[I]_o$ in μM units.

TABLE 6.1. Parameter Estimates Using the Rate Las for Slow-Binding Inhibition for the Simulated Data of Figure 6.14B

$[I]$ (nM)		v_s	k_{obs}
6.7	22 (23)	0.96 (4.5)	1.1 (0.6)
20	22 (23)	0.80 (1.4)	2.0 (1.6)
50	23 (23)	0.46 (0.6)	4.2 (4.0)

Units: v_o and v_s, $nM\,s^{-1}$; k_{obs}, $10^{-3}\,s^{-1}$.

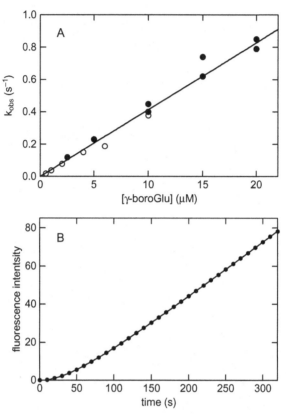

Figure 6.15. Inhibition of the γGTase catalyzed hydrolysis of γGlu-AMC by γboroGlu. *Panel A:* Dependence of k_{obs} on [γboroGlu]. Data represented by open circles was collected on a standard fluorometer, while data reperesented by closed circles were collected on a stopped-flow fluorometer. Solid line drawn with parameters: *slope* = $k_{on.obs}$ = 41 ± 1mM^{-1}s^{-1} and *y-intercept* = k_{off} = –0.0015 ± 0.028s^{-1}. *Panel B:* Determination of k_{off}. 15 μM enzyme and 20 μM inhibitor were incubated for 10 min and then diluted 1,000-fold into assay buffer containing 20 μM (10K_m) γGlu-AMC. Solid line was drawn using Equation 6.30 and best fit parameters: v_o = 0.003 ± 0.007 Fl s^{-1}, v_s = 0.285 ± 0.008 Fl s^{-1}, and k_{obs} = k_{off} = 0.0219 ± 0.0007 s^{-1}.

of the γ-glutamyl transpeptidase (γGTase)-catalyzed hydrolysis of γGlu-AMC (Stein et al. 2001). These plots allow a $k_{on,app}$ value of 41 mM^{-1} s^{-1} to be determined, or a k_{on} of 400 mM^{-1} s^{-1}. In the inhibitor concentration range shown in this figure, k_{off} cannot be distinguished from zero. To determine k_{off} from the dependence of k_{obs} versus [γboroGlu]$_o$, lower concentrations of inhibitor would have to be used. However, at these low values of [I]$_o$, steady-state velocities cannot be reached before substrate depletion exceeds 10%, thus rendering these progress unusable for k_{obs} estimation.

To determine k_{off} for this system, the experiment of Figure 6.15B was conducted, involving the dilution of a solution of γGTase and γboroGlu into a solution of substrate γGlu-AMC. Key features of this experiment, that must be attended to in all determinations of k_{off} using this dilution method, include the following:

- In the preincubation solution, enzyme concentration must be high enough to ensure that when diluted into the substrate solution, $[E]_{final}$ is still high enough to generate reliable reaction rates.
- In the preincubation solution, inhibitor concentration must be greater than the enzyme concentration to ensure that all the enzyme exists as E : I.
- Dilution of the preincubation solution into the substrate solution must be of such a magnitude that $[I]_{final} < K_i$, so that inhibitor will not rebind to enzyme.
- To further ensure that diluted inhibitor cannot rebind to enzyme, the substrate concentration should be much greater than K_m.

If these conditions are met, the progress curve will be characterized by an initial velocity of zero, a final velocity equal to $k_c[E]_{final} = V_{max}$, and a k_{obs} value equal to k_{off}, thus adhering to a simplified form of Equation 6.30:

$$[P] = -\frac{V_{max}}{k_{off}}\left(1 - e^{-k_{off}t}\right) + V_{max}t. \tag{6.40}$$

When the progress curve of Figure 6.15B was fit to Equation 6.40, a k_{obs} value of $0.022\,s^{-1}$ was determined. If this value approximates k_{off}, the ratio k_{obs}/k_{on} should closely approximate the K_i value of 35 nM that was determined from analysis of v_s on $[I]_o$. This ratio equals $(0.022\,s^{-1})/(4 \times 10^5\,M^{-1}\,s^{-1})$ or 55 nM, which is in good agreement with K_i.

6.4 IRREVERSIBLE INHIBITION

Irreversible inhibitors are molecules that bind to the active site of their target enzyme and chemically modify it in such a way as to render the enzyme catalytically inactive. Methods for kinetic analysis of irreversible inhibition are similar to those for analysis of slow-binding inhibition, with two important differences: (i) inhibition progress curves have final velocities equal to zero, and thus are described by Equation 6.41 or its integrated form, Equation 6.42, and (ii) plots of k_{obs} versus $[I]_o$ have an intercept of zero, and can be described by

$$v_t = v_o e^{-k_{obs}t} \tag{6.41}$$

$$[P] = \frac{v_o}{k_{obs}}\left(1 - e^{-k_{obs}t}\right) \tag{6.42}$$

either Equation 6.43,

$$k_{obs} = \frac{k_{inact}}{1 + \dfrac{[S]_o}{K_m}}[I]_o \tag{6.43}$$

where k_{inact} is the second-order rate constant for irreversible interaction of inhibitor with enzyme, or Equation 6.44

$$k_{obs} = \frac{k_i [I]_o}{K_{i,1}\left(1 + \dfrac{[S]_o}{K_m}\right) + [I]_o} \qquad (6.44)$$

where $K_{i,1}$ is the dissociation complex of the initial E:I complex and k_i is the first-order rate constant for irreversible interaction of inhibitor with enzyme. Equation 6.44 simplifies to Equation 6.43 when $K_{i,1} > [I]_{expt}$.

Since enzyme and the final complex formed between enzyme and irreversible inhibitor can never attain equilibrium, the only methods that are appropriate to estimate the potency of an irreversible inhibitor are kinetic methods. IC_{50} values determined for irreversible inhibitors have no meaning and will decrease in magnitude as preincubation time between enzyme and inhibitor is lengthened.

Kinetic parameters for enzyme inactivation by an irreversible inhibitor are determined by one of two methods:

- A discontinuous method can be used in which residual enzyme activity is measured after enzyme and inhibitor have incubated for a preset time. These velocities can then be plotted versus time and fit to Equation 6.41 to arrive at a value of k_{obs} for that particular inhibitor concentration. Or,
- Progress curves can be generated by monitoring product formation after addition of enzyme to a solution of substrate and inhibitor. The progress curve can then be fit to Equation 6.42 to estimate k_{obs}.

In either of the above two cases, k_{obs} is determined at various concentrations of inhibitor to determine the mechanism of inactivation and then the kinetic parameters.

Case Study: Inactivation of Human Leukocyte Elastase (HLE) by the Peptide-Based Chloromethyl Ketone MeoSuc-Ala-Ala-Pro-Val-CH₂Cl. Peptide-based chloromethyl ketones are a class of serine protease irreversible inhibitors that inactivate their target enzyme by alkylating the imidazole moiety of the active-site histidine. A detailed kinetic study of the inactivation of HLE by MeoSuc-Ala-Ala-Pro-Val-CH₂Cl was performed to understand mechanistic details of how this compound inactivates HLE (Stein and Trainor 1986).

The initial experiment reported in this study was the determination of progress curves for the inactivation of HLE by MeoSuc-Ala-Ala-Pro-Val-CH₂Cl. In Figure 6.16A, I have plotted this data and fit each progress curve to Equation 6.42 to arrive at best-fit values for k_{obs}, which are plotted versus $[I]_o$ in Figure 6.16B. The dependence of k_{obs} on $[I]_o$ is clearly not linear, but rather is hyperbolic, suggesting a mechanism involving rapid formation of a reversible E:I complex, with a $K_{i,1}$ value 11 μM, prior to the inactivation step, which occurs by a process with rate constant $0.04\,s^{-1}$.

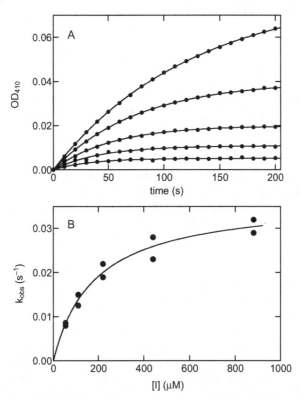

Figure 6.16. Irreversible inhibition of human leuckocyte elastase (HLE) by MeoSuc-Ala-Ala-Pro-Val-CH$_2$Cl. *Panel A:* Inactivation progress curves were generated by adding enzyme to a solution of substrate MeOSuc-Ala-Ala-Pro-Val-pNA and inhibitor, and monitoring the release of product *p*-nitroraniline at 440 nm. $[I]_o$ = 55, 110, 220, 440, 880 µM. *Panel B:* Hyperbolic dependence of k_{obs}, calculated from the curves of Panel A, on $[I]_o$. The best-fit parameters according to Equation 6.44, with the term $(1 + [S]/K_m)$ constrained to 16, are k_i = 0.037 ± 0.002 s^{-1} and $K_{i,1}$ = 11 ± 2 µM.

REFERENCES

Bull, H. G., M. Garcia-Calvo, S. Andersson, W. Baginsky, H. K. Chan, D. E. Ellsworth, R. R. Miller, R. A. Sterns, R. K. Bakshi, G. R. Rasmussen, R. L. Tolman, R. W. Myers, J. W. Kozarich, and G. S. Harris (1996). "Mechanism-based inhibition of human 5α-reductase by finasteide." *J. Am. Chem. Soc.* **118**: 2359–2365.

Case, A. and R. L. Stein (2003). "Kinetic analysis of the action of tissue transglutaminase on peptide and protein substrates." *Biochemistry* **42**: 9466–9481.

Case, A. and R. L. Stein (2005). "Mechanistic studies of the interaction of tissue transglutaminase with a novel, non-peptidic slow-binding inhibitor." *Biochemistry* **46**: 1106–1115.

Case, A. and R. L. Stein (2006). "Mechanistic studies of ubiquitin C-terminal hydrolase L1." *Biochemistry* **45**: 2443–2452.

Case, A., J. Ni, L.-A. Yeh, and R. L. Stein (2005). "Development of a mechanism-based assay for tissue transglutaminase—Results of a high-throughput screen and discovery of inhibitors." *Anal. Biochem.* **338**: 237–244.

Feinstein, G., C. J. Malemud, and A. Janoff (1976). "The inhibition of human leukocyte elastase and chymotrypsin-like protease by elastatinal and chymostatin." *Biochim. Biophys. Acta* **429**: 925–932.

Harding, M. W., A. Galat, D. E. Uehling, and S. L. Schreiber (1989). "A receptor for the immunosuppressant FK506 is a cis-trans peptidyl-prolyl isomerase." *Nature* **341**: 758–760.

Harrison, R. K. and R. L. Stein (1992). "Mechanistic studies of enzymatic and nonenzymatic prolyl cis-trans isomerization." *J. Am. Chem. Soc.* **114**: 3464–3471.

Rasmusson, G., G. F. Reynolds, N. G. Steinberb, E. Walton, G. Patel, T. Liang, M. A. Cascieri, A. H. Cheung, J. R. Brooks, and C. Berman (1986). "Azasteroids: Structure-activity relationships for inhibition of 5α-reductase and of androgen receptor binding." *J. Med. Chem.* **29**: 2298–2315.

Stein, R. L. (1991). "Exploring the catalytic activity of immunophilins." *Curr. Biol.* **1**: 234–236.

Stein, R. L. (1993). "Mechanisms of enzymic and non-enzymic prolyl isomerization." *Adv. Protein Chem.* **44**: 1–24.

Stein, R. L. and A. M. Strimpler (1987). "Slow-binding inhibition of chymotrypsin and cathepsin G by the peptide aldehyde chymostatin." *Biochemistry* **26**(9): 2611–2615.

Stein, R. L. and D. A. Trainor (1986). "Mechanism of inactivation of human leukocyte elastase by a chloromethyl ketone: Kinetic and solvent isotope effect studies." *Biochemistry* **25**(19): 5414–5419.

Stein, R. L., A. M. Strimpler, P. D. Edwards, J. J. Lewis, R. C. Mauger, J. A. Schwartz, M. A. Stein, and D. A. Trainor (1987). "Mechanism of slow-binding inhibition of human leukocyte elastase by trifluoromethyl ketones." *Biochemistry* **26**: 2682–2689.

Stein, R. L., C. DeCicco, and B. Thomas (2001). "Slow-binding inhibition of γ-glutamyl transpeptidase by γ-boroGlu." *Biochemistry* **40**: 5804–5811.

Umezawa, H., Y. Aoyagi, H. Morishma, S. Kunimoto, and M. Matsuzari (1970). "Chymostatin, a new chymotrypsin inhibitor produced by acinomycetes." *J. Antibiot.* **23**: 425–427.

7

KINETICS OF TWO-SUBSTRATE ENZYMATIC REACTIONS

In 2005, enzymologist James Robertson published a review article containing an analysis of marketed drugs (Robertson 2005). It seems that of the 1278 marketed drugs, 317 are targeted at 71 different enzymes and, of these, 37 react with two or more substrates. This piece of pharmaceutical history, coupled with the fact that the reactions catalyzed by most enzymes involve two or more substrates, makes clear the importance of understanding the kinetics of multisubstrate enzymes.

In this chapter, I will focus exclusively on two-substrate reactions. Kinetic treatment of enzymes catalyzing reactions of three or even four substrate can be found in the excellent book of Irwin Segel (Segel 1975). The principal goals of this chapter are to outline methods for determining an enzyme's *kinetic mechanism; that is, the order in which the substrates bind to the enzyme and the associated rate and equilibrium constants.*

7.1 IMPORTANCE IN DRUG DISCOVERY

Knowledge of an enzyme's kinetic mechanism is a critical prerequisite for two activities that are central to drug discovery programs focused on enzymes as therapeutic targets.

Kinetics of Enzyme Action: Essential Principles for Drug Hunters, First Edition. Ross L. Stein.
© 2011 John Wiley & Sons, Inc. Published 2011 by John Wiley & Sons, Inc.

- **High-Throughput Screening (HTS).** The goal of a high-throughput screen is to identify enzyme inhibitors of certain mechanistic classes. For example, for two-substrate reactions, the investigator may want to identify inhibitors that are competitive with one substrate, but noncompetitive with the other. If such inhibitors are to be identified, the HTS assay must be conducted with substrate concentrations set at very specific values. To calculate these values, the investigator must know the enzyme's kinetic mechanism.[1]

- **Determining the Kinetic Mechanism of Inhibition.** The kinetic mechanism of inhibition for an enzyme inhibitor comprises knowledge of the enzyme forms to which the inhibitor binds and the dissociation constants for enzyme:inhibitor complexes that form. In Chapter 8, we will see that this is impossible to determine without prior knowledge of the enzyme kinetic mechanism for substrate turnover.

7.2 BASIC MECHANISMS

A moment's reflection reveals that there are only a limited number of ways for two substrates to productively bind to enzyme. These are shown in Figure 7.1 and comprise the three basic mechanisms for two-substrate enzymes.

- **Random Mechanism.** As the name implies, in a random mechanism, either substrate can bind first to enzyme. The random mechanism of Figure 7.1 depicts a rapid equilibrium mechanism, meaning that the chemical step for transforma-

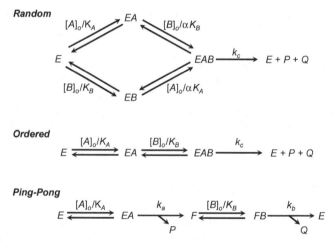

Figure 7.1. The three standard kinetic mechanisms for two-substrate reactions.

---

[1] To see how these substrate concentrations are calculated, the reader is referred to Appendix C.

tion of the $E:A:B$ complex into products is rate-limiting. For rapid equilibrium random mechanisms, K_A and K_B are true dissociation constants for the $E:A$ and $E:B$ complexes, respectively. α is the cooperativity factor and expresses how binding of the first substrates affects the affinity of enzyme for the second substrate. For example, for random mechanisms in which α is less than one, substrate X has higher affinity for $E:Y$ than it has for E.

- **Ordered Mechanism.** Here, the enzyme imposes a compulsory order for the addition of the two substrates. The ordered mechanism of Figure 7.1 is shown as a rapid equilibrium mechanism, in which turnover of $E:A:B$ is rate-limiting, and K_A and K_B are dissociation constants.
- **Ping-Pong Mechanism.** In ping-pong mechanisms, reaction occurs in two stages. In the first, enzyme binds substrate A to form a complex $E:A$, which then reacts to form product P and enzyme form F. In the second stage of a ping-pong mechanism, F binds substrate B to form $F:B$, which then turns over to yield product Q and regenerate enzyme in its original form E.

For random and ordered mechanisms, which are called sequential mechanisms, there is a single chemical step, which follows formation of the ternary complex $E:A:B$. If this chemical step is slower than substrate binding processes, all enzyme species of the system will exist in a state of equilibrium. As we will see in Sections 7.3 and 7.4, the possibility of equilibrium for sequential mechanisms allows for ready analysis using free energy diagrams, and rate equation derivation under the rapid equilibrium assumption. In contrast, for ping-pong mechanism, equilibrium cannot be established among all the enzyme species of the system. This is because the two equilibria of the system (i.e., $E + A \leftrightarrows EA$ and $F + B \leftrightarrows FB$) are separated by irreversible steps (i.e., turnover of EA and turnover of FB). This precludes the use of the rapid equilibrium assumption for the derivation of the rate equation and renders construction and interpretation of free energy diagrams a bit more tricky. For these reasons, we will consider ping-pong mechanisms apart from sequential mechanisms, in Section 7.5.

7.3 CONCEPTUAL UNDERSTANDING OF SEQUENTIAL MECHANISMS

Before deriving rate equations for sequential mechanisms, it will be useful to see if we can develop a conceptual, nonmathematical understanding of these mechanisms. Specifically, we will want to see how changes in the concentrations of the two substrates can lead to changes in the identity of the reactant state, and thus to changes in the rate-limiting step. This analysis will be made with the help of free energy diagrams.

Several preliminary points about free energy diagrams for sequential mechanisms are noteworthy:

- The rapid equilibrium assumption is operative, meaning that the highest energy transition state for any steady-state rate parameter is the transition state for the chemical step, $\neq_c$.

- Changing the concentration of a substrate will only affect the relative stabilities of the enzyme form to which it binds and the enzyme form from which it dissociates. For example, for an ordered mechanism, changes in $[A]_o$ will affect the stability of $E{:}A$ relative to E, but will not affect the relative stabilities of $E{:}A$ and $E{:}A{:}B$.
- The free energy difference between a specific enzyme form and $\neq_c$ is reflected in a specific steady-state rate parameter. While the free energy difference between $E{:}A{:}B$ and $\neq_c$ is reflected in k_c, free energy differences between other forms of the enzyme and $\neq_c$ are equal to the product of k_c and a unitless association constant for formation of $E{:}A{:}B$ from that enzyme form. For example, in an ordered mechanism, the free energy difference between $E{:}A$ and $\neq_c$ is reflected in the rate parameter $k_c([B]/K_B)$.
- Each of these rate parameters is a pseudo-first order rate constant that equals $v_o/[E]_o$ for the initial velocity measured at certain limiting concentrations of the two substrates. For example, for all sequential mechanisms, k_c is equal to $v_o/[E]_o$ for the initial velocity measured at $[A]_o > K_A$ and $[B]_o > K_B$.

With these general points in mind, we will construct and analyze free energy diagrams for each mechanism under the four possible limiting substrate concentrations:

- $[A]_o < K_A$, $[B]_o < K_B$
- $[A]_o < K_A$, $[B]_o > K_B$
- $[A]_o > K_A$, $[B]_o < K_B$
- $[A]_o > K_A$, $[B]_o > K_B$

7.3.1 Ordered Mechanism

7.3.1.1 $[A]_o < K_A$, $[B]_o < K_B$. Our starting point is the situation in which the concentrations of both A and B are less than their respective dissociation constants. Under these conditions, EA will be more stable than EAB, and E will be more stable than EA. The free energy diagram for this situation is in Figure 7.2 and tells us that the most stable species is free enzyme, and thus $v_o/[E]_o$ for this process reflects the energy difference between E and $\neq_c$. Since the reactant state is E, reaching the transition state requires equilibrium formation of EA and then EAB. Thus, $v_o/[E]_o$ equals the product of $[A]_o/K_A$, $[B]_o/K_B$, and k_c.

7.3.1.2 $[A]_o < K_A$, $[B]_o > K_B$. When A remains fixed at a concentration less than K_A, as above, and B is increased to a concentration greater than K_B, EAB becomes stabilized relative to EA. The stability of EA relative to E is insensitive to this increase in $[B]_o$. Thus, the reactant state is still free enzyme, and $v_o/[E]_o$ still equals the product of $[A]_o/K_A$, $[B]_o/K_B$, and k_c.

7.3.1.3 $[A]_o > K_A$, $[B]_o < K_B$. Now, again taking the situation $[A]_o < K_A$ and $[B]_o < K_B$ as our starting reference point, if $[B]_o$ remains less than K_B, but A is increased

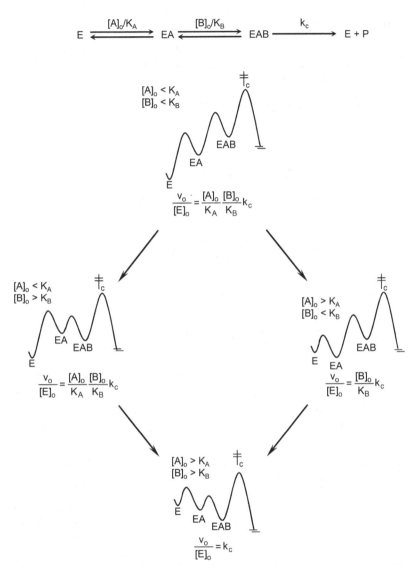

Figure 7.2. Free energy diagrams for ordered mechanisms.

to a concentration greater than K_A, the reactant state has shifted from E to EA. $v_o/[E]_o$ now equals the product of $[B]_o/K_B$, and k_c.

7.3.1.4 $[A]_o > K_A$, $[B]_o > K_B$. When the concentrations of both substrates are increased to concentrations above their respective dissociation constants, EAB is stabilized relative to EA, which is stabilized relative to E. The reactant state is EAB and $v_o/[E]_o$ now equals k_c.

7.3.2   Random Mechanism

Compared to free energy diagrams for ordered mechanisms, free energy diagrams for random mechanisms present more of a challenge to construct since there are two parallel paths to the ternary complex *EAB*. In the free energy diagrams of Figure 7.3, the

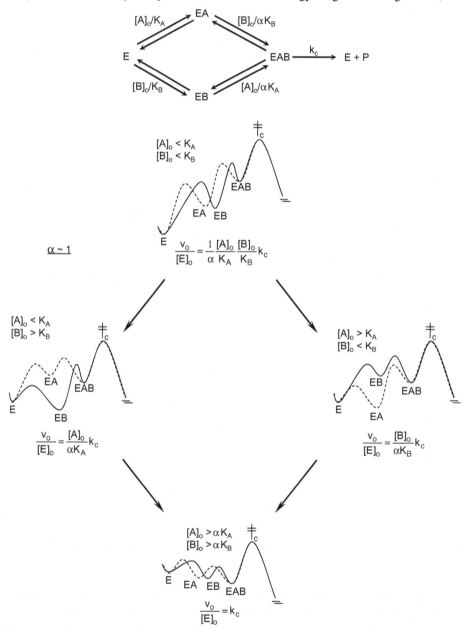

Figure 7.3. Free energy diagrams for random mechanisms.

dashed line represents the path in which A binds first to free enzyme, while the solid line reflects the path in which B binds first. Also in these diagrams, the assumption is made that α is close to unity.

7.3.2.1 ***$[A]_o < K_A$, $[B]_o < K_B$.*** Under these conditions, the most stable enzyme species is free enzyme. Thus, $v_o/[E]_o$ reflects the energy difference between E and $\neq_c$, and equals the product of $1/\alpha$, $[A]_o/K_A$, $[B]_o/K_B$, and k_c. Note that the denominator of the expression for $v_o/[E]_o$ is $\alpha K_A K_B$, which means that the two paths from E to EAB are equivalent, as chemical thermodynamics requires.

7.3.2.2 ***$[A]_o < K_A$, $[B]_o > K_B$.*** If A is held at a concentration less than K_A and B is increased to a concentration greater than K_B, EB and EAB become populated and stabilized relative to E and EA, respectively. The net result is that the reactant state has shifted from E to EB, with $v_o/[E]_o$ equaling the product of $[A]_o/\alpha K_A$ and k_c.

7.3.2.3 ***$[A]_o > K_A$, $[B]_o < K_B$.*** Given the symmetry of the random mechanism, the situation is the same as above, but with the reactant state changing from E to EA and $v_o/[E]_o$ now equal to the product of $[B]_o/\alpha K_B$ and k_c.

7.3.2.4 ***$[A]_o > K_A$, $[B]_o > K_B$.*** Under this condition, EAB is stable relative to both EA and EB, both of which are more stable than free enzyme. The reactant state is EAB and $v_o/[E]_o$ equals k_c.

7.4 DERIVATION OF RATE EQUATIONS FOR SEQUENTIAL MECHANISMS

In this section, we will consider sequential mechanisms, first deriving an overall equation in which v_o is expressed as a function of both $[A]_o$ and $[B]_o$, and then expressions of general form

$$v_o = \frac{V_{X,obs}[X]_o}{K_{X,obs} + [X]_o} \tag{7.1}$$

where

$$V_{X,obs} = f([Y]_o) \tag{7.2}$$

and

$$\left(\frac{V}{K}\right)_{X,obs} = f([Y]_o). \tag{7.3}$$

The dependence of $V_{x,obs}$ and $(V/K)_{X,obs}$ on $[Y]_o$ will be seen to be critical in establishing the kinetic mechanism of an enzyme.

7.4.1 Random Mechanism: Rapid Equilibrium Assumption

The rate equation for the random mechanism, Equation 7.4, is derived in Figure 7.4 using the

$$v_o = \frac{V_{max}}{\dfrac{\alpha K_A K_B}{[A]_o [B]_o} + \dfrac{\alpha K_B}{[B]_o} + \dfrac{\alpha K_A}{[A]_o} + 1} \qquad (7.4)$$

Experimental: $[E]_o \ll [A]_o, [B]_o$

Conservation: $[E]_o = [E]+[EA]+[EB]+[EAB]$

$[A]_o = [A]+[EA]+[EAB] = [A]$

$[B]_o = [B]+[EB]+[EAB] = [B]$

Binding constants: $K_A = \dfrac{[E][A]}{[EA]}; \quad \dfrac{[E]}{[EA]} = \dfrac{K_A}{[A]} = \dfrac{K_A}{[A]_o}$

$K_B = \dfrac{[E][B]}{[EB]}; \quad \dfrac{[E]}{[EB]} = \dfrac{K_B}{[B]} = \dfrac{K_B}{[B]_o}$

$\alpha K_A = \dfrac{[EB][A]}{[EAB]}; \quad \dfrac{[EB]}{[EAB]} = \dfrac{\alpha K_A}{[A]} = \dfrac{K_A}{[A]_o}$

$\alpha K_B = \dfrac{[EA][B]}{[EAB]}; \quad \dfrac{[EA]}{[EAB]} = \dfrac{\alpha K_B}{[B]} = \dfrac{K_B}{[B]_o}$

Rate dependence: $v_o = k_c [EAB]$

Derivation of rate equation:

$$\dfrac{v_0}{[E]_o} = \dfrac{k_c [EAB]}{[E]+[EA]+[EB]+[EAB]}$$

$$\dfrac{v_0}{[E]_o} = \dfrac{k_c}{\dfrac{[E]}{[EAB]} + \dfrac{[EA]}{[EAB]} + \dfrac{[EB]}{[EAB]} + 1}$$

$$\dfrac{v_0}{[E]_o} = \dfrac{k_c}{\dfrac{[E]}{[EA]} \dfrac{[EA]}{[EAB]} + \dfrac{\alpha K_B}{[B]_o} + \dfrac{\alpha K_A}{[A]_o} + 1}$$

$$\dfrac{v_0}{[E]_o} = \dfrac{k_c}{\dfrac{\alpha K_A K_B}{[A]_o [B]_o} + \dfrac{\alpha K_B}{[B]_o} + \dfrac{\alpha K_A}{[A]_o} + 1}$$

Figure 7.4. Derivation of rate equation of the random mechanism using the rapid equilibrium assumption.

rapid equilibrium assumption. The only difference between this derivation and others in previous chapters of this book is that total enzyme distributes among four forms, thus requiring us to track four dissociation constants. There is no conceptual difference, only more bookkeeping.

To rearrange Equation 7.4 into the form of Equation 7.1, we have only to isolate $[A]_o$ or $[B]_o$ in the denominator and collect terms. I illustrate this for substrate A.

We start by multiplying the numerator and denominator of Equation 7.4 by $[A]_o$:

$$v_o = \frac{V_{\max}[A]_o}{\alpha K_A \dfrac{K_B}{[B]_o} + [A]_o \dfrac{\alpha K_B}{[B]_o} + \alpha K_A + [A]_o}. \qquad (7.5)$$

Collecting terms produces Equation 7.6:

$$v_o = \frac{V_{\max}[A]_o}{\alpha K_A\left(1 + \dfrac{K_B}{[B]_o}\right) + [A]_o\left(1 + \dfrac{\alpha K_B}{[B]_o}\right)}. \qquad (7.6)$$

And, dividing the numerator and denominator of Equation 7.6 by the term $(1 + \alpha K_B/[B]_o)$ yields

$$v_o = \frac{\dfrac{V_{\max}}{1 + \dfrac{\alpha K_B}{[B]_o}}[A]_o}{\alpha K_A\left(\dfrac{1 + \dfrac{K_B}{[B]_o}}{1 + \dfrac{\alpha K_B}{[B]_o}}\right) + [A]_o}. \qquad (7.7)$$

By inspection of Equation 7.7 we see that

$$V_{A,obs} = \frac{V_{\max}}{1 + \dfrac{\alpha K_B}{[B]_o}} = \frac{V_{\max}[B]_o}{\alpha K_B + [B]_o} \qquad (7.8)$$

and

$$\left(\frac{V}{K}\right)_{A,obs} = \frac{V_{\max}/\alpha K_A}{1 + \dfrac{K_B}{[B]_o}} = \frac{(V_{\max}/\alpha K_A)[B]_o}{K_B + [B]_o}. \qquad (7.9)$$

We see that both $V_{A,obs}$ and $(V/K)_{A,obs}$ have hyperbolic dependencies on $[B]_o$. Due to the symmetry of the random mechanism, $V_{B,obs}$ and $(V/K)_{B,obs}$ will have hyperbolic dependencies on $[A]_o$, as shown in Equations 7.10 and 7.11, respectively:

$$V_{B,obs} = \frac{V_{max}[A]_o}{\alpha K_A + [A]_o} \tag{7.10}$$

and

$$\left(\frac{V}{K}\right)_{B,obs} = \frac{(V_{max}/\alpha K_B)[A]_o}{K_A + [A]_o}. \tag{7.11}$$

7.4.2 Ordered Mechanism: Rapid Equilibrium Assumption

The derivation of the rate equation for the ordered mechanism proceeds in a manner that is analogous to that for the random mechanism and results in Equation 7.12:

$$v_o = \frac{V_{max}}{\dfrac{K_A K_B}{[A]_o[B]_o} + \dfrac{K_B}{[B]_o} + 1}. \tag{7.12}$$

To produce an equation that expresses v_o as a function of $[A]_o$, we start by multiplying the numerator and denominator of Equation 7.12 by $[A]_o$ to produce

$$v_o = \frac{V_{max}[A]_o}{K_A \dfrac{K_B}{[B]_o} + [A]_o\left(1 + \dfrac{K_B}{[B]_o}\right)}. \tag{7.13}$$

Dividing the numerator and denominator by the term $(1 + K_B/[B]_o)$ yields

$$v_o = \frac{\dfrac{V_{max}}{1 + \dfrac{K_B}{[B]_o}}[A]_o}{K_A \dfrac{[B]_o}{1 + \dfrac{K_B}{[B]_o}} + [A]_o}. \tag{7.14}$$

Again, by inspection, we can equate the following:

$$V_{A,obs} = \frac{V_{max}}{1 + \dfrac{K_B}{[B]_o}} = \frac{V_{max}[B]_o}{K_B + [B]_o} \tag{7.15}$$

$$\left(\frac{V}{K}\right)_{A,obs} = \frac{V_{max}/K_A}{\dfrac{K_B}{[B]_o}} = \frac{V_{max}}{K_A K_B}[B]_o. \tag{7.16}$$

We see that while $V_{A,obs}$ has a hyperbolic dependence on $[B]_o$, $(V/K)_{A,obs}$ is linearly dependent on $[B]_o$.

Our next step in this analysis of the ordered mechanism is to derive an equation that expresses v_o as a function of $[B]_o$. We proceed by multiplying the numerator and denominator of Equation 7.12 through by $[B]_o$:

$$v_o = \frac{V_{max}[B]_o}{K_B\left(1 + \dfrac{K_A}{[A]_o}\right) + [B]_o}. \tag{7.17}$$

We see from this equation,

$$V_{B,obs} = V_{max} \tag{7.18}$$

$$\left(\frac{V}{K}\right)_{B,obs} = \frac{(V_{max}/K_B)[A]_o}{K_A + [A]_o}. \tag{7.19}$$

Thus, V_B is independent of $[A]_o$, and $(V/K)_{B,obs}$ has a hyperbolic dependence on $[A]_o$.

7.4.3 Ordered Mechanism: Steady-State Assumption

If the turnover of EAB to products and free enzyme is not rate-limiting, then we must express ordered mechanisms as shown in Figure 7.5. To derive the rate equation for this mechanism, which assumes rapid product release, we will use the steady-state approximation for the two intermediate species, EA and EAB.

This derivation starts in the usual way, with the quotient of the rate expression and the conservation of enzyme:

$$\frac{v_o}{[E]_o} = \frac{k_3[EAB]}{[E] + [EA] + [EAB]} \tag{7.20}$$

which upon rearrangement yields

$$\frac{v_o}{[E]_o} = \frac{k_3}{\dfrac{[E]}{[EAB]} + \dfrac{[EA]}{[EAB]} + 1}. \tag{7.21}$$

Our goal now is to express $[E]/[EAB]$ and $[EA]/[EAB]$ in terms of the various rate constants and initial substrate concentrations, $[A]_o$ and $[B]_o$.

$$E \underset{k_{-1}}{\overset{k_1[A]_o}{\rightleftarrows}} EA \underset{k_{-2}}{\overset{k_2[B]_o}{\rightleftarrows}} EAB \overset{k_3}{\longrightarrow} E + P$$

Figure 7.5. Ordered mechanism of with microscopic rate constants.

Starting with $[EA]/[EAB]$, the steady-state approximation states that

$$(k_{-2} + k_3)[EAB] = k_2[B]_o[EA] \tag{7.22}$$

or

$$\frac{[EA]}{[EAB]} = \frac{k_{-2} + k_3}{k_2[B]_o}. \tag{7.23}$$

For $[E]/[EAB]$, the steady-state approximation tells us that

$$(k_{-1} + k_2[B]_o)[EA] = k_1[A]_o[E] + k_{-2}[EAB] \tag{7.24}$$

which can be rearranged to

$$(k_{-1} + k_2[B]_o)\frac{[EA]}{[EAB]} = k_1[A]_o\frac{[E]}{[EAB]} + k_{-2}. \tag{7.25}$$

Substituting the expression for $[EA]/[EAB]$ of Equation 7.23 yields

$$(k_{-1} + k_2[B]_o)\frac{k_{-2} + k_3}{k_2[B]_o} = k_1[A]_o\frac{[E]}{[EAB]} + k_{-2}, \tag{7.26}$$

which, upon rearrangement gives the expression for $[E]/[EAB]$:

$$\frac{[E]}{[EAB]} = \frac{(k_{-1} + k_2[B]_o)\dfrac{k_{-2} + k_3}{k_2[B]_o} - k_{-2}}{k_1[A]_o} \tag{7.27}$$

or

$$\frac{[E]}{[EAB]} = \frac{k_{-1}k_2}{k_1[A]_o k_2[B]_o} + \frac{k_{-1}k_3}{k_1[A]_o k_2[B]_o} + \frac{k_3}{k_1[A]_o}. \tag{7.28}$$

Substituting these expressions for $[E]/[EAB]$ and $[EA]/[EAB]$ into Equation 7.21 yields

$$\frac{v_o}{[E]_o} = \frac{k_3}{\left(\dfrac{k_{-1}k_2}{k_1[A]_o k_2[B]_o} + \dfrac{k_{-1}k_3}{k_1[A]_o k_2[B]_o} + \dfrac{k_3}{k_1[A]_o}\right) + \left(\dfrac{k_{-2} + k_3}{k_2[B]_o}\right) + 1}. \tag{7.29}$$

If we collect terms, we arrive at

$$\frac{v_o}{[E]_o} = \frac{k_3}{\dfrac{K_A}{[A]_o}\left(\dfrac{k_{-2}+k_3}{k_2[B]_o}\right) + \dfrac{k_{-2}+k_3}{k_2[B]_o} + \dfrac{k_3}{k_1[A]_o} + 1} \tag{7.30}$$

where K_A is k_{-1}/k_1.

The relative magnitudes of these rate constants will determine which step is rate-limiting. For example, if k_3 is very large, Equation 7.30 simplifies to

$$\frac{v_o}{[E]_o} = \frac{(k_2[B]_o)[A]_o}{\left(\dfrac{k_{-1}+k_2}{k_1}\right) + [A]_o} \tag{7.31}$$

which has the form of a Michael–Menten equation for a one-substrate reaction, where k_c equals $k_2[B]_o$ and K_m for substrate A is $(k_{-1} + k_2)/k_1$.

Significantly, when k_3 is very small and rate-limiting, Equation 7.30 simplifies to

$$\frac{v_o}{[E]_o} = \frac{k_3}{\dfrac{K_A K_B}{[A]_o[B]_o} + \dfrac{K_B}{[B]_o} + 1} \tag{7.32}$$

which is precisely the equation we derived using the rapid equilibrium assumption.

7.5 PING-PONG MECHANISMS

The ping-pong mechanism, depicted in Figure 7.1, differs from the two sequential mechanisms in that it involves two irreversible chemical steps. As a consequence, the rate equation for this mechanism cannot be derived using the rapid equilibrium assumption but rather must be derived using the steady-state assumption. Once derived, we will see that k_c is a function of both k_a and k_b, and that the two dissociation K_A and K_B do not appear in the equation as isolated parameters, but as components of $K_{m,A}$ and $K_{m,B}$, each also containing the two catalytic constants, k_a and k_b. Because of these complicating factors, a clearer account of the fundamental concepts of ping-pong kinetics can be made if it follows the derivation of the rate equation for this mechanism.

7.5.1 Derivation of the Rate Equation for Ping-Pong Mechanisms

We start this derivation by noting that in the steady state, the rate of formation of P must equal the rate of formation of Q. Thus, the following equality must hold in the steady state:

$$v_o = k_a[EA] = k_b[FB]. \tag{7.33}$$

Given this,

$$\frac{[EA]}{[FB]} = \frac{k_b}{k_a}.$$

(7.34)

We now proceed in the usual way by dividing the expression of the rate dependence, which can be either $v_o = k_a[EA]$ or $v_o = k_b[FB]$, by the conservation of enzyme equation:

$$\frac{v_o}{[E]_o} = \frac{k_b[FB]}{[E]+[EA]+[F]+[FB]}.$$

(7.35)

Equation 7.35 can be rearranged to

$$\frac{v_o}{[E]_o} = \frac{k_b}{\dfrac{k_b}{k_a}\dfrac{K_A}{[A]_o} + \dfrac{k_b}{k_a} + \dfrac{K_B}{[B]_o} + 1}$$

(7.36)

or

$$\frac{v_o}{[E]_o} = \frac{k_b}{\dfrac{k_b}{k_a}\dfrac{K_A}{[A]_o} + \dfrac{K_B}{[B]_o} + \dfrac{k_a+k_b}{k_a}}.$$

(7.37)

Multiplying through by $k_a/(k_a + k_b)$, yields

$$\frac{v_o}{[E]_o} = \frac{\dfrac{k_a k_b}{k_a+k_b}}{\dfrac{K_A}{[A]_o}\dfrac{k_b}{k_a+k_b} + \dfrac{K_B}{[B]_o}\dfrac{k_a}{k_a+k_b} + 1}$$

(7.38)

or

$$v_o = \frac{k_c[E]_o}{\dfrac{K_{m,A}}{[A]_o} + \dfrac{K_{m,B}}{[B]_o} + 1}$$

(7.39)

where

$$k_c = \frac{k_a k_b}{k_a+k_b}$$

(7.40)

$$K_{m,A} = K_A\left(\frac{k_b}{k_a+k_b}\right)$$

(7.41)

$$K_{m,B} = K_B\left(\frac{k_a}{k_a+k_b}\right).$$

(7.42)

Equation 7.39 can be rearranged to yield an expression in which v_o is a function of $[A]_o$:

$$v_o = \frac{V_{\max}[A]_o}{K_{m,A} + [A]_o\left(1 + \dfrac{K_{m,B}}{[B]_o}\right)} \qquad (7.43)$$

or

$$v_o = \frac{\dfrac{V_{\max}}{1 + \dfrac{K_{m,B}}{[B]_o}}[A]_o}{\dfrac{K_{m,A}}{1 + \dfrac{K_{m,B}}{[B]_o}} + [A]_o}. \qquad (7.44)$$

By inspection we see

$$V_{A,obs} = \frac{V_{\max}[B]_o}{K_{m,B} + [B]_o} \qquad (7.45)$$

$$\left(\frac{V}{K}\right)_{A,obs} = \frac{V_{\max}}{K_{m,A}} \qquad (7.46)$$

Due to the symmetry of the ping-pong mechanism we can immediately write

$$V_{B,obs} = \frac{V_{\max}[A]_o}{K_{m,A} + [A]_o} \qquad (7.47)$$

$$\left(\frac{V}{K}\right)_{B,obs} = \frac{V_{\max}}{K_{m,B}}. \qquad (7.48)$$

We see from these equations that while $V_{X,obs}$ is hyperbolically dependent on $[Y]_o$, $(V/K)_{X,obs}$ is independent of $[Y]_o$.

7.5.2 Conceptual Understanding of Ping-Pong Mechanisms

The distinctive feature of ping-pong mechanisms is that the two substrates combine with enzyme at points in the mechanism that are separated by an irreversible reaction step. This has the effect of dividing each catalytic cycle into two independent half reactions, where the concentration of the substrate involved in one half reaction can have no influence on the rate of the other half reaction. In terms of free energy, this means that while $[A]_o$ influences the relative stabilities of E and EA, and thus the free energy difference between E and the transition state for k_a, $[A]_o$ can have no effect on free energy differences in the half reaction involving substrate B. This contrasts with sequential mechanisms in which binding of substrates occurs with no intervening species, and binding of X to EY diminishes the free energy difference between EY and the catalytic transition state.

Viewed as comprising two half reactions, a ping-pong reaction will have an initial velocity that is equal to rate of the slower half reaction. Thus, the initial velocity reflects the half reaction with the larger overall $\Delta G^{\neq}$.

As with sequential mechanisms, developing a conceptual understanding of ping-pong mechanisms will be done with the aid of free energy diagrams. The free energy diagrams that correspond to limiting cases for $[A]_o/K_A$ and $[B]_o/K_B$ are collected in Figure 7.6. Note that in these diagrams the half reactions are not connected, as they must not be since equilibria cannot exist between EA and F, or FB and E. Finally, for ease of interpretation, these diagrams are drawn with EA and FB having the same relative free energies, and $k_a/k_b > 1$.

7.5.2.1 $[A]_o < K_A, [B]_o < K_B.$ Our starting point is to consider the consequences of both substrates having concentrations less than their respective dissociation constants. This means for each of the half reactions, the reactant states are free enzyme, E or F. Since in this diagram A and B are set at concentrations such that $[A]_o/K_A = [B]_o/K_B$, and $k_a/k_b > 1$, the half reaction with larger value of $\Delta G^{\neq}$ is the one involving enzyme form F. For this case, $v_o/[E]_o$ equals the product of $[B]_o/K_B$, and k_b.

7.5.2.2 $[A]_o > K_A, [B]_o < K_B.$ If substrate B is maintained at a concentration less than K_B, but A is increased, the half reaction with larger value of $\Delta G^{\neq}$ remains the one involving enzyme form F. $v_o/[E]_o$ remains equal to the product $[B]_o/K_B$, and k_b.

7.5.2.3 $[A]_o < K_A, [B]_o > K_B.$ In contrast to the above situation, if $[A]_o/K_A$ is less than one, and the concentration of B is increased above K_B, depending on the precise value of k_a/k_b, the free energy difference between FB and $\neq_b$ may be similar in magnitude to the free energy difference between E and $\neq_a$. In this case, $v_o/[E]_o$ is a complex function of both k_b and $k_a([A]_o/K_A)$.

7.5.2.4 $[A]_o > K_A, [B]_o > K_B.$ When both substrates are at saturating conditions, the larger value of $\Delta G^{\neq}$ corresponds to $\Delta G_b^{\neq}$, and $v_o/[E]_o$ simply equals k_b.

7.6 DETERMINING THE KINETIC MECHANISM FOR TWO-SUBSTRATE REACTIONS

The previous sections provide the theoretical backdrop for investigations of the kinetic mechanism of two-substrate reactions. These investigations proceed in a manner not unlike the approach described for the inhibition of one-substrate reaction, since both systems have two independent variables. In the present case, the primary data set will comprise initial velocities determined as a function of both $[A]_o$ and $[B]_o$.

Recall that in Chapter 5, I described the two principal analytical methods for kinetic experiments that depend on two independent variables: the methods of replots and

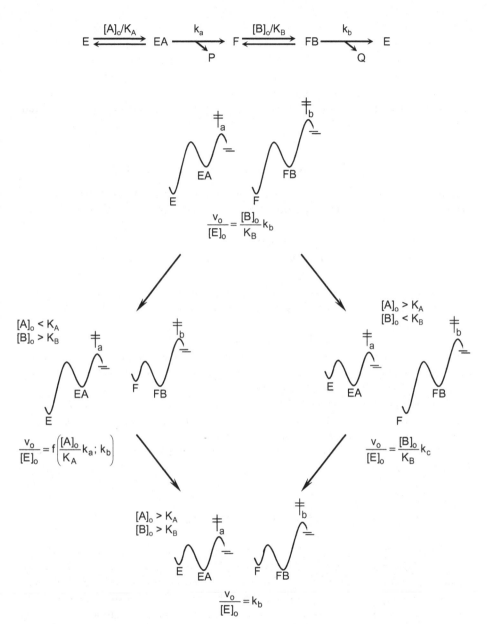

Figure 7.6. Free energy diagrams depicting ping-pong mechanisms. In all four diagrams, $\Delta G_a^{\neq} < \Delta G_{\bar{a}}^{\neq}$, thus $k_a > k_b$, and in the first diagram, $[A]_o/K_A = [B]_o/K_B < 1$.

global fitting. I made that case that the former method provides a stage-wise analysis of experimental results that allows easy visualization and multiple opportunities for appraisal. Perhaps even more importantly, the method of replots do not require us to impose preconceived mechanisms on the data, but rather provides an analytical path that allows the mechanism to emerge from the data. Because of these advantages, I will use the method of replots to illustrate how to analyze kinetic data for two-substrate reactions.

7.6.1   Method of Replots for Two-Substrate Reactions

This is a two stage process—the construction of *primary plots* and, from these, *secondary plots*. For a two-substrate reaction, there are two possible primary plots:

- Primary plot I_A^o: the dependence of v_o on $[A]_o$ at several fixed $[B]_o$
- Primary plot I_B^o: the dependence of v_o on $[B]_o$ at several fixed $[A]_o$

For both I_A^o and I_B^o, each dependence of initial velocity on substrate concentration is individually fit to the Michaelis–Menten equation, to generate value of $V_{X,obs}$ and $(V/K)_{X,obs}$ for each $[Y]_o$. Each curve is then carefully inspected to judge whether it adequately fits the data and thus supports the use of the simple Michaelis–Menten equation. If the curves can be described simply, then secondary plots are constructed. In a subsequent section, I briefly describe how to handle cases in which the primary plots do not follow Michaelis–Menten kinetics.

Secondary plots are the dependencies $V_{X,obs}$ and $(V/K)_{X,obs}$ on $[Y]_o$. It is from the analysis of the secondary plots that we discover the kinetic mechanism of the enzyme and estimate the various steady-state parameters.

There are four secondary plots for a two-substrate reaction, two from the results I_A^o, and and two from the results of I_B^o:

- $II_A^o \Rightarrow V_{A,obs}$ versus $[B]_o$ and $(V/K)_{A,obs}$ versus $[B]_o$
- $II_B^o \Rightarrow V_{B,obs}$ versus $[A]_o$ and $(V/K)_{B,obs}$ versus $[A]_o$

Figure 7.7 contains the possible secondary plots for the three standard mechanisms, as well as the rate equations that describe the various dependencies of $V_{x,obs}$ and $(V/K)_{x,obs}$ on $[Y]_o$ that were derived in the previous section. Identifying the kinetic mechanism comes from simple pattern recognition. Once the kinetic mechanism has been identified, kinetic parameters can be estimated using the appropriate equations. These concepts will become clearer when we consider the following case study.

7.6.2   cdk5/p25-Catalyzed Phosphorylation of Tau

Neurofibrillary tangles (NFTs) are one of the pathological hallmarks of Alzheimer's disease. A major component of NFTs is hyperphosphorylated forms of tau, a protein whose normal physiological role is thought to be involved in the structural integrity of

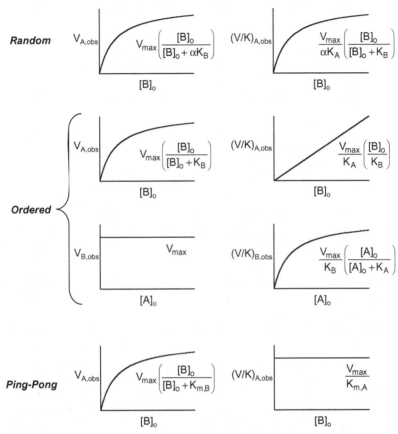

Figure 7.7. Secondary plots for the three standard mechanisms for two-substrate reactions.

microtubles. While a number of kinases are able to phosphorylate tau, cdk5/p25 has been implicated as one of the principal Alzheimer's disease (AD)-associated kinases. Cdk5/p25 is a proline-directed Ser/Thr kinase and a member of the broad family of cyclin-dependent kinases that are involved in cell cycle regulation. Unlike other family members that require association with a cyclin for activity, cdk5 associates with the activator protein p35. Neuronal-specific processing of p35 to p25 by calpain deregulates cdk5 and allows the resultant cdk5/p25 complex to hyperphosphorylate tau. Given the likely role of cdk5/p25 in the pathogenesis of AD, identifying inhibitors of this enzyme has become a priority in the search for drugs to treat AD. As a first necessary step in developing an HTS assay for this system and to allow the determination of the mechanism of inhibitors discovered during screening, our group set out to determine the kinetic mechanism of the cdk5/p25-catalyzed phosphorylation of tau.

Figure 7.8 contains the primary data of an experiment in which we determined initial velocity as a function of five concentrations each of tau and ATP (Liu et al. 2008).

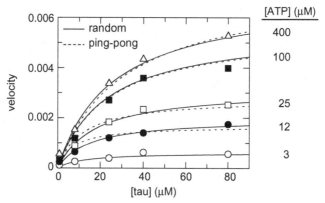

Figure 7.8. Global fits of initial velocity data for the cdk5-catalyzed phosphorylation of tau, to random and ping-pong mechanisms.

Also shown in this plot are theoretical lines drawn using the best-fit values for a global fit to the ping-pong and random mechanisms (attempts at fitting the data to the ordered mechanism failed due to nonconvergence). From a simple visual inspection of this plot, it is clear that both of these mechanisms, as different as they are from one another, fit the data equally well. However, when these data are analyzed using the method of replots, it becomes clear that the mechanism is random and not ping-pong.

Figure 7.9 contains the two primary and four secondary plots of this data. Each dependence of v_o on $[X]_o$ at a fixed concentration of $[Y]_o$ from the two primary plots obeys Michaelis–Menten kinetics. Given this simple behavior, the four secondary plots that are shown can be constructed. Since all four of these plots display hyperbolic dependences of a steady-state kinetic parameter on the alternate substrate concentration, we can conclude that the mechanism is random and not ping-pong. Analysis of the data according to Equations 7.10 and 7.11 provided the best-fit values that are summarized schematically in Figure 7.10.

7.6.3 Analysis of Two-Substrate Reactions with Non-Michaelian Kinetics

As outlined above, the kinetic analysis of a two-substrate reaction is straightforward if the primary plots follow Michaelis–Menten kinetics. Deviations from simple Michaelis–Menten kinetics do occur, and, as described in Section 3.4, will typically fall into one of three categories: substrate inhibition, positive cooperativity, and negative cooperativity.

In Section 3.4, we saw that the expression of Equation 7.49 can describe deviations for one-substrate enzymatic reactions:

$$\frac{v_o}{[E]_o} = \frac{\alpha_1[S]_o + \alpha_2[S]_o^2}{\beta_0 + \beta_1[S]_o + \beta_2[S]_o^2}. \tag{7.49}$$

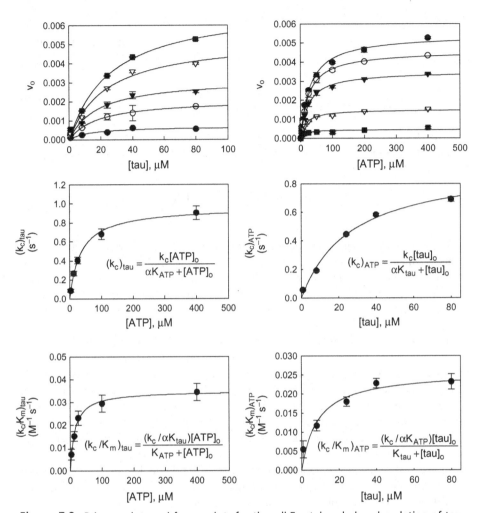

Figure 7.9. Primary plots and four replots for the cdk5-catalyzed phosphorylation of tau.

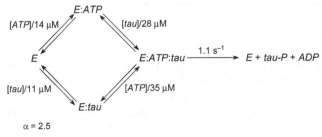

Figure 7.10. Kinetic mechanism for the cdk5-catalyzed phosphorylation of tau.

Our discussion here, for two substrate reactions, will be made somewhat simpler by assuming that for only one of the two substrates must an equation analogous to Equation 7.49 be used. That is, while Equation 7.50 must be used to

$$\frac{v_o}{[E]_o} = \frac{\alpha_{1,Y}[X]_o + \alpha_{2,Y}[X]_o^2}{\beta_{0,Y} + \beta_{1,Y}[X]_o + \beta_{2,Y}[X]_o^2} \tag{7.50}$$

describe the dependence of initial velocity on the concentration of substrate X, the simplified Equation 7.51 can be used to describe the dependence of initial velocity on the concentration of substrate Y:

$$\frac{v_o}{[E]_o} = \frac{\alpha_{1,X}[Y]_o}{\beta_{0,X} + [Y]_o} = \frac{V_{Y,obs}[Y]_o}{K_{Y,obs} + [Y]_o}. \tag{7.51}$$

All five coefficients of Equation 7.50 are functions of $[Y]_o$, while the two coefficients of Equation 7.51 are functions of $[X]_o$. Since the dependence of v_o on $[Y]_o$ follows Michaelis–Menten kinetics, the coefficients of Equation 7.50 will have simple dependencies on $[Y]_o$. On the other hand, $\alpha_{1,X}$ and $\beta_{0,X}$ will have complex dependencies on $[X]_o$, since the dependence of v_o on $[X]_o$ does not follow Michaelis–Menten kinetics. The form of all these dependencies will be dictated by the general mechanism of the enzyme, that is, sequential versus ping-pong, and steady state versus random.

The analysis of a two-substrate reaction showing this sort of kinetic complexity not only has the usual goal of ascertaining the mechanism of the enzyme, but also the goal of understanding and describing the kinetics of the non-Michaelian behavior. These two problems are best solved sequentially, perhaps first by analyzing the dependence of v_o on $[X]_o$ according to some form of Equation 7.50 that reflects the specific non-Michaelian behavior observed, and then seeing if patterns appear in secondary plots that give a clue to the general mechanism of the enzyme (i.e., sequential vs. ping-pong, and steady state vs. random). For example, if sigmoidal kinetics is observed and allows simplification of Equation 7.50 to Equation 7.52, a form analogous to Equation 3.57, then the secondary plots of $V_{X,obs}$, $(V/K_1)_{X,obs}$, and

$$v_o = \frac{V_{X,obs}\dfrac{[X]_o}{K_{1,obs}} + 2V_{X,obs}\dfrac{[X]_o^2}{K_{1,obs}K_{2,obs}}}{1 + \dfrac{[X]_o}{K_{1,obs}} + \dfrac{[X]_o^2}{K_{1,obs}K_{2,obs}}} \tag{7.52}$$

$(V/K_1K_2)_{X,obs}$ versus $[Y]_o$, and $V_{Y,obs}$ and $(V/K)_{Y,obs}$ versus $[X]_o$ should suggest the mechanism of the enzyme.

However, it is sometimes the case that previous studies or mechanistic insight allow a simultaneous solution to the problem. For example, if substrate inhibition is observed for a reaction that likely proceeds through a ping-pong mechanism, a probable mechanism involves binding of X to the enzyme to form a catalytically incompetent complex.

This is what was seen in studies of the kinetic mechanism of tissue transglutaminase (TGase) (Case and Stein 2003). Recall from previous chapters, that TGase catalyzes transamidation reactions of Gln residues of proteins or peptides with the ε-amine of Lys residues or other primary amines. In these reactions, the Gln-containing substrate binds first to TGase, with subsequent formation of an acyl–enzyme intermediate. The acyl–acceptor substrate then binds and attacks the carbonyl carbon of the acyl–enzyme to complete the transamidation reaction. In a study using the minimal substrates Z-Gln-Gly and Gly-OMe, substrate inhibition was observed when the acyl–acceptor substrate Gly-OMe was the varied substrate (Case and Stein 2003). This result was interpreted to indicate that Gly-OMe can bind to free enzyme to form a complex to which Z-Gln-Gly cannot bind. Using this general mechanistic scheme, the initial velocity pattern determined when Z-Gln-Gly and Gly-OMe were varied could be successfully analyzed. A similar situation was encountered with γ-glutamyl transpeptidase (Stein et al. 2001).

7.7 A CONCEPTUAL UNDERSTANDING OF THE SHAPES OF SECONDARY PLOTS

In the previous sections, we derived rate equations for each of the three standard mechanisms for two-substrate enzymatic reactions. We saw that these equations could be used to predict the shapes of secondary plots for the three kinetic mechanisms. Since each mechanism has a unique pattern of secondary plots, once the secondary plots are constructed for a particular enzyme system, pattern recognition reveals the mechanism. The goal of this section is develop a nonmathematical, conceptual understanding of why a specific mechanism exhibits the pattern of secondary plots that it does. We will consider each of three standard mechanisms in turn, relying on free energy diagrams to help our understanding.

7.7.1 Secondary Plots for Random Mechanisms

We start by trying to develop an understanding of why the dependence of $(V/K)_{A,obs}$ has a hyperbolic dependence on $[B]_o$. Figure 7.11 contains free energy diagrams for $(V/K)_{A,obs}$ (i.e., $[A]_o < K_A$), at $[B]_o < K_B$ and $[B]_o > K_B$. Note that in these two diagrams, as well as the diagrams for the dependence of $V_{A,obs}$ on $[B]_o$, the dashed line represents the reaction path taken when substrate A binds first, while the solid line is the path taken when B binds first. Both paths lead to the ternary complex EAB, within which rate-limiting chemisty occurs.

When $[B]_o < K_B$, the reactant state for $(V/K)_{A,obs}$ can be seen to be free enzyme, the most stable of the four enzyme species. As we consider the situation where $[B]_o > K_B$, we must first recall that the stability of only those species to which B can bind will be effected by changes in $[B]_o$. Thus, when $[B]_o > K_B$, E becomes unstable relative to EB, and EA becomes unstable relative to EAB. The overall result is that EB is now the most stable enzyme species, and thus has become the reactant state for $(V/K)_{A,obs}$.

The $[B]_o$-dependent change in reactant state manifests kinetically as a hyperbolic dependence of $(V/K)_{A,obs}$ on $[B]_o$, where at low $[B]_o$, $(V/K)_{A,obs}$ reflects the free energy difference between E and the rate-limiting transition state, while at high $[B]_o$, $(V/K)_{A,obs}$

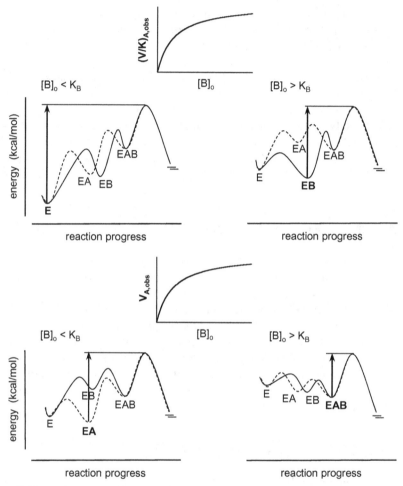

<u>Figure 7.11.</u> Explanation of replots for random mechanism by the use of free energy diagrams.

reflects the free energy difference between *EB* and the same rate-limiting transition state.

We next turn to $V_{A,obs}$ and why it is that it also has a hyperbolic dependence on $[B]_o$. At low concentrations of $[B]_o$, the most stable species is *EA*, so it is the reactant state for $V_{A,obs}$. As $[B]_o$ increases, we once more see that both *E* and *EA* are destabilized relative to *EB* and *EAB*, respectively. The result is that *EAB* becomes the most stable enzyme species at high concentrations of $[B]_o$, making *EAB* the reactant state for $V_{A,obs}$. This $[B]_o$-dependent change in reactant state is seen as a hyperbolic dependence of $V_{A,obs}$ on $[B]_o$.

Given the symmetry of the random mechanism, it is unnecessary to develop an account specifically for the hyperbolic dependencies of $(V/K)_{B,obs}$ and V_B on $[A]_o$.

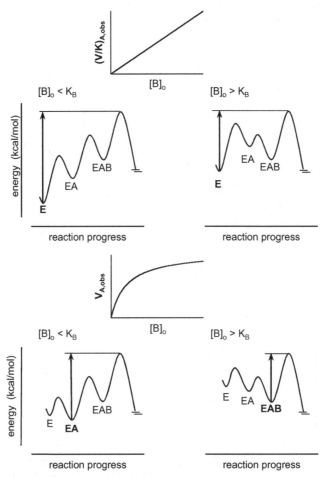

Figure 7.12. Explanation of $(V/K)_{A,obs}$ and $V_{A,obs}$ replots for the ordered mechanism by the use of free energy diagrams.

7.7.2 Secondary Plots for Ordered Mechanisms

Since ordered mechanisms lack the symmetry of random mechanisms, independent accounts will need to be developed for $(V/K)_{A,obs}$ and $V_{A,obs}$, and for $(V/K)_{B,obs}$ and $V_{B,obs}$. The relevant free energy diagrams are contained in Figures 7.12 and 7.13, respectively.

We first consider the linear dependence that $(V/K)_{A,obs}$ has on $[B]_o$. When $[B]_o < K_B$, the most stable enzyme form is free enzyme, thus making it the reactant state for $(V/K)_{A,obs}$. As B increases in concentration, the stability of EAB increases relative to EA, the only enzyme form to which B can bind. Note that increasing concentrations of B can have no influence on the stability of E relative to EA. The result is that free enzyme is the reactant state for $(V/K)_{A,obs}$ at all $[B]_o$; the effect of increasing $[B]_o$ is to decrease

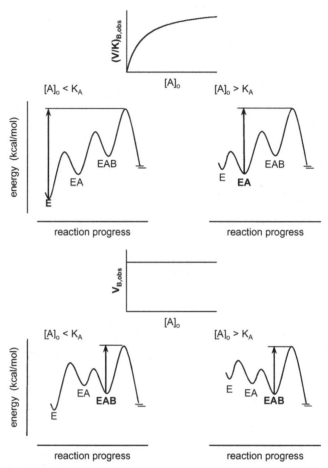

Figure 7.13. Explanation of $(V/K)_{B,obs}$ and $V_{B,obs}$ replots for the ordered mechanism by the use of free energy diagrams.

the energy barrier between free enzyme and the rate-limiting transition state. This results in a linear dependence of $(V/K)_{A,obs}$ on $[B]_o$, with no saturation kinetics.

Turning now to $V_{A,obs}$, we see that the most stable form of enzyme at low $[B]_o$ is the complex EA, which thus is the reactant state for $V_{A,obs}$ when $[B]_o < K_B$. When is increased above K_B, EA is destabilized relative to EAB, and the latter becomes the reactant for $V_{A,obs}$. Note that increasing concentrations of B have no effect on the relative stabilities E and EA. Since there is a change in reactant state with increasing concentrations of B, the dependence of $V_{A,obs}$ on $[B]_o$ is hyperbolic.

Figure 7.13 contains free energy diagrams for the dependencies of $(V/K)_{B,obs}$ and $V_{B,obs}$ on $[A]_o$. Considering first the dependence of $(V/K)_{B,obs}$ on $[A]_o$, we see that when $[A]_o < K_A$, the lowest energy species is free enzyme, and it is this species that is the reactant state for $(V/K)_{B,obs}$. When $[A]_o > K_A$, free enzyme EA is stabilized relative to free enzyme, and the former becomes the reactant state for $(V/K)_{B,obs}$. The change in reactant state with increasing concentration of A tells us to anticipate a hyperbolic dependence of $(V/K)_{B,obs}$ on $[A]_o$.

Finally, we examine the dependence of $V_{B,obs}$ on $[A]_o$. Here we see that the lowest energy species is EAB, and that the energy difference between it and the rate-limiting transition state does not change with increased concentrations of $[A]_o$. This is because while high concentrations of A stabilize EA relative to free enzyme, it can have no effect on the stability of EAB relative to EA. So, $V_{B,obs}$ is independent of $[A]_o$.

7.7.3 Ping-Pong Mechanisms

Figure 7.14 summarizes the replots and relevant free energy diagrams for ping-pong mechanisms. We start with the independence of $(V/K)_{A,obs}$ on concentration of substrate B.

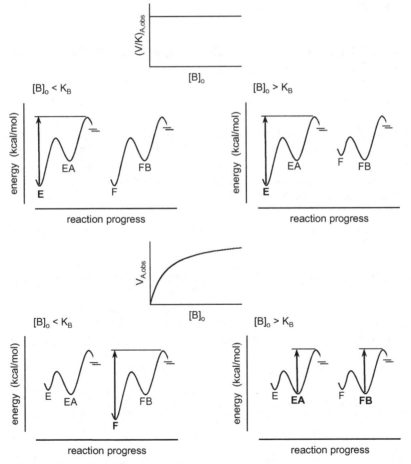

Figure 7.14. Explanation of replots for the ping-pong mechanism by the use of free energy diagrams.

When $[B]_o < K_B$, the half reaction with the greater energy barrier will be the one for turnover of A. Thus, $(V/K)_{A,obs}$ reflects the energy difference between E and the transition state for this half reaction. As $[B]_o$ is increased to values above K_B, FB will be stabilized relative to F. This will have no effect on the energetics of the half reaction for A, and $(V/K)_{A,obs}$ still reflects the energy difference between E and the transition state for this half reaction.

Turning now to $V_{A,obs}$, where E is saturated with substrate making EA stable relative to E, we see that at $[B]_o < K_B$, the larger value of $\Delta G^\neq$ corresponds to the half reaction for B. Thus, $V_{A,obs}$ reflects the energy difference between F and the transition state for this half reaction. Now, as the concentration of B is increased until F is saturated, $V_{A,obs}$ will be a function of the $\Delta G^\neq$ values for both half reactions. Since the reactant state for $V_{A,obs}$ changes with increasing concentration of B, $V_{A,obs}$ will have a hyperbolic dependences on $[B]_o$.

Due to the symmetry of ping-pong mechanisms, the same analysis holds when A is the variable substrate.

7.8 MISTAKEN IDENTITY: RAPID EQUILIBRIUM RANDOM VERSUS STEADY-STATE ORDERED

The astute reader will have noticed that Equation 7.4, for rapid equilibrium random mechanisms, has precisely the same form as Equation 7.30, for steady-state ordered mechanisms. Both mechanisms yield four hyperbolic secondary plots, indicating that initial velocity experiments of the type described in this chapter cannot distinguish between to types of sequential mechanisms. We will see at the end of the next chapter that they can be distinguished by the characterization of the mechanism of inhibition by products or dead-end substrate analogs.

REFERENCES

Case, A. and R. L. Stein (2003). "Kinetic analysis of the action of tissue transglutaminase on peptide and protein substrates." *Biochemistry* **42**: 9466–9481.

Liu, M., et al. (2008). "Kinetic studies of Cdk5/p25 kinase: Phosphorylation of tau and complex inhibition by two prototype inhibitors." *Biochemistry* **47**: 8367–8377.

Robertson, J. G. (2005). "Mechanistic basis of enzyme-targeted drugs." *Biochemistry* **44**: 5561–5571.

Segel, I. H. (1975). *Enzyme Kinetics*. New York, John Wiley & Sons.

Stein, R. L., et al. (2001). "Slow-binding inhibition of γ-glutamyl transpeptidase by γ-boroGlu." *Biochemistry* **40**: 5804–5811.

8

KINETIC MECHANISM OF INHIBITION OF TWO-SUBSTRATE ENZYMATIC REACTIONS

Analyzing the inhibition of two-substrate enzymatic reactions requires the derivation and utilization of rate laws with three independent variables: $[A]_o$, $[B]_o$, and $[I]_o$. While this level of complexity may at first seem daunting, we will see in this chapter that elucidating kinetic mechanisms for such reactions relies on the same protocols of experimental design and methods of data analysis that were developed in previous chapters for systems with two independent variables.

Given this new level of complexity, this chapter strives to supplement the mathematical understanding of mechanism that is inherent in rate laws, with a *conceptual understanding* of mechanism. Such an understanding is gained by recognizing the connection that exists between a steady-state rate parameter and a specific enzyme form, and then seeing how binding of a molecule to that enzyme form can manifest as a diminution of the associated rate parameter.

This chapter begins with a discussion on how the success of medicinal chemical optimization programs can depend on a thorough understanding of the kinetic mechanism of inhibition of two-substrate reactions. Next, general inhibition mechanisms are examined for the three standard two-substrate reactions, with a view to moving beyond rate laws to a conceptual understanding of mechanism. The focus is then narrowed to a consideration of enzyme inhibition by substrate analogs, and the mechanistic insights

Kinetics of Enzyme Action: Essential Principles for Drug Hunters, First Edition. Ross L. Stein.
© 2011 John Wiley & Sons, Inc. Published 2011 by John Wiley & Sons, Inc.

that can be gained by determining the mechanism by which these compounds inhibit their targeted enzyme. Finally, we see how the knowledge gained from these mechanistic studies can be used to drive structure–activity relationship (SAR) programs for two-substrate reactions.

8.1   IMPORTANCE IN DRUG DISCOVERY

Recall in Chapter 5, which focused on determining the inhibition of one-substrate enzymatic reactions, a principal goal of early-stage drug discovery programs that are developing enzyme inhibitors is establishing SARs that are predictive, both of enzyme inhibition, and cellular and *in vivo* activity. To maximize the chances of being able to construct such an SAR, the measure of inhibitory activity should be actual K_i values rather than IC_{50} values. This is, of course, still the case for two-substrate reactions. Without a "kinetic dissection" of $K_{i,app}$ into K_i values for the various steady-state complexes of enzyme and inhibitor, the "A" of SAR will be confusing at best and misleading at worst.

Since knowledge of kinetic mechanism provides estimates of dissociation constants for the various complexes of inhibitor and enzyme, data would be available to drive an SAR program based on optimization of one particular binding mode of the inhibitor. This strategy is discussed in more detail at the end of this chapter.

Mechanistic studies on the inhibition of two-substrate reactions involve determining the binding mode for each of the two substrates. This is important in programs in which a particular binding mode is sought. For example, for protein kinases, it might be desirable to identify inhibitors that do not bind to the ATP site. In such cases, compounds would be sought that are not competitive with ATP. However, it should be remembered that the observation of competitive inhibition does not necessarily mean active site binding (Roskoski 2003).

8.2   MECHANISMS OF INHIBITION OF TWO-SUBSTRATE REACTIONS

In this section, I start with a discussion of how we can arrive at a conceptual understanding of inhibition, using one-substrate enzymatic reactions as the model. We then move on and consider inhibition of the three standard enzymatic reactions, from both quantitative and conceptual points of view. These three sections are preceded by a brief description of the general quantitative methods that are used in determining the kinetic mechanisms of two-substrate enzymatic reactions.

8.2.1   A Conceptual Understanding of Inhibition: Connecting Steady-State Enzyme Forms to Rate Parameters

The rate laws for all enzymatic reactions reflect connections between the derived steady-state rate parameters and specific enzyme forms. Unfortunately, this connection is not readily apparent in rate laws as they are usually expressed. Consider the one-substrate reaction shown below and its rate law in Equation 8.1:

$$E + A \underset{K_A}{\rightleftharpoons} E{:}A \xrightarrow{k_c} E + P$$

$$v_o = \frac{V_{max}[A]_o}{K_A + [A]_o}$$

(8.1)

This mechanism has two enzyme forms, E and EA, that are connected to the two steady-state rate parameters that govern the reaction, V_{max}/K_A and V_{max}. While this connection is not immediately apparent in rate law of Equation 8.1, it becomes clearer when the rate law is rearranged into its reciprocal form:

$$\frac{1}{v_o} = \frac{K_A}{V_{max}} \frac{1}{[A]} + \frac{1}{V_{max}}$$

(8.2)

$$v_o = \left\{ \left(\frac{V_{max}}{K_A}[A]_o \right)^{-1} + (V_{max})^{-1} \right\}^{-1}$$

(8.3)

We see that the initial velocity can be expressed as the reciprocal of the sum of reciprocals of the two steady-state rate parameters. As we learned in Chapter 3, each of the two rate terms of Equation 8.3 reflects the free energy difference between a specific reactant state and the catalytic transition state. For V_{max}/K_A, this reactant state is E, while for V_{max}, the reactant state is EA.

A useful feature of expressing rate laws in their reciprocal form is that corresponding inhibition rate laws can be written by inspection. If an inhibitor binds to an enzyme form, the corresponding rate term will be diminished according to the general $[I]_o$-dependent expression of Equation 8.4:

$$V_{obs} = \frac{V}{1 + \dfrac{[I]_o}{K_{i,ex}}}$$

(8.4)

where V corresponds to one of the steady-state rate parameters of the mechanism, and $K_{i,ex}$ is the equilibrium constant for dissociation of I from general enzyme species EXI.

So, if an inhibitor binds to both E and EA, Equation 8.3 becomes

$$v_o = \left\{ \left(\frac{\dfrac{V_{max}}{K_A}}{1 + \dfrac{[I]_o}{K_{i,e}}}[A]_o \right)^{-1} + \left(\frac{V_{max}}{1 + \dfrac{[I]_o}{K_{i,ea}}} \right)^{-1} \right\}^{-1}$$

(8.5)

which can be rearranged into the more familiar form for mixed inhibition:

$$v_o = \frac{\dfrac{\dfrac{V_{max}}{K_A}}{1 + \dfrac{[I]_o}{K_{i,ea}}}[A]_o}{K_A\left(\dfrac{1 + \dfrac{[I]_o}{K_{i,e}}}{1 + \dfrac{[I]_o}{K_{i,ea}}}\right) + [A]_o} \tag{8.6}$$

As we have seen in previous chapters, one-substrate reactions are seldom as simple as the one we have been considering. For more complex mechanisms that proceed through more than a single steady-state intermediate, ambiguities can arise in the correlation of a rate parameter with a single enzyme form. As we saw in Chapter 3, Section 3.3, V_{max} for these reactions may reflect more than a single reactant state. Thus, non-competitive or uncompetitive inhibitors, which effect V_{max}, may be binding to more than a single enzyme species (see Section 5.4).

While such complexity may or may not arise for one-substrate reaction, this is the situation we face with *all* two-substrate reactions, since there are always two or more steady-state intermediates. For two-substrate reactions, ambiguity in the correlation of rate parameters with reactant states can occur with both V_X and $(V/K)_X$. However, there are two general rules that help us correlate rate parameters with reactant states:

- $(V/K)_X$ reflects all enzyme species to which substrate X is not bound.
- V_X reflects all enzyme species to which substrate X is bound.

These rules are illustrated in Figure 8.1 in the context of the three standard mechanisms for two substrate reactions. Ambiguity arises since the observed effect of inhibition on a measured rate parameter may reflect binding of the inhibitor to more than a single enzyme form. So, for example, in an ordered mechanism, while the observation of $[I]_o$-dependent $(V/K)_{A,obs}$ tells us, without ambiguity, that inhibitor binds to free enzyme, the observation of $[I]_o$-dependent $V_{A,obs}$ leaves us with uncertainty. Does inhibitor bind to EA or to EAB or to both?

In the sections that follow, I outline methods for elucidating the kinetic mechanism of inhibition for the three standard two-substrate mechanisms. For the sections discussing sequential mechanisms, I begin by developing a conceptual framework for connecting rate parameters to the enzyme forms to which inhibitor binds. For ping-pong mechanisms, I begin by deriving the rate law for inhibition, and then move to concepts that emerge from the rate law. But first, we discuss the general approach that will be used for the analysis of inhibition of two-substrate reactions.

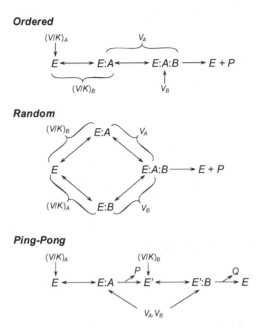

Figure 8.1. Identity of the reactant states for $(V/K)_X$ and V_X for the three standard two-substrate mechanisms. Note that there is ambiguity in the identity of $(V/K)_X$ for ordered mechanisms, and ambiguity in the identity of both $(V/K)_X$ and V_X for random mechanisms. In ping-pong mechanisms, V_A and V_B are identical and share the same two reactant states, $E{:}A$ and $E'{:}B$.

8.2.2 Elucidation of Mechanism: General Analytical Method

To elucidate the mechanism of inhibition of two-substrate reactions, we use protocols of experimental design and methods of data analysis that are an extension of what we have seen in previous chapters.

Briefly, initial velocities are first measured for two 2-dimensional matrices: v_o as a function of $[A]_o$, $[I]_o$, at fixed $[B]_o$, and v_o as a function of $[B]_o$, $[I]_o$ and at fixed $[A]_o$. The data from each of these experiments is then analyzed according to the method outlined in Chapter 5 for the inhibition of one-substrate reactions.

Each of these two experiments generates a primary plot, and from each of these, two replots:

v_o versus $[A]_o$, at various $[I]_o$ ([B]$_o$ fixed)		v_o versus $[B]_o$, at various $[I]_o$ ([A]$_o$ fixed)	
$V_{A,obs}$ versus $[I]_o$	$(V/K)_{A,obs}$ versus $[I]_o$	$V_{B,obs}$ versus $[I]_o$	$(V/K)_{B,obs}$ versus $[I]_o$

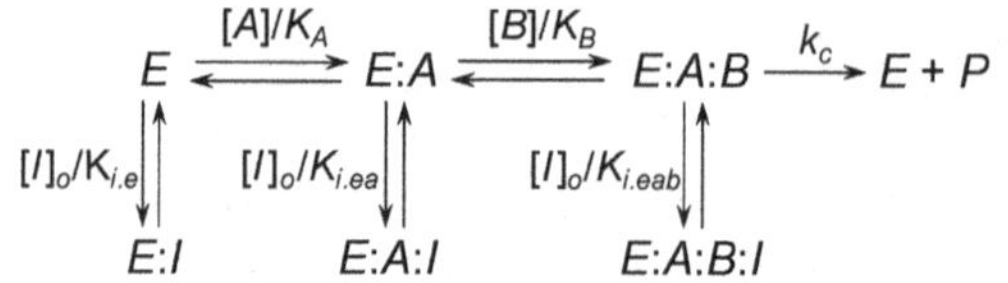

$$E \underset{}{\overset{[A]/K_A}{\rightleftharpoons}} E{:}A \underset{}{\overset{[B]/K_B}{\rightleftharpoons}} E{:}A{:}B \xrightarrow{k_c} E + P$$

$$\big\downarrow [I]_o/K_{i,e} \qquad \big\updownarrow [I]_o/K_{i,ea} \qquad \big\updownarrow [I]_o/K_{i,eab}$$

$$E{:}I \qquad E{:}A{:}I \qquad E{:}A{:}B{:}I$$

$$v_o = \frac{\dfrac{V_{max}}{1 + \dfrac{[I]}{K_{i,eab}}}}{\dfrac{K_A}{[A]_o}\dfrac{K_B}{[B]_o}\left(\dfrac{1 + \dfrac{[I]}{K_{i,e}}}{1 + \dfrac{[I]}{K_{i,eab}}}\right) + \dfrac{K_B}{[B]_o}\left(\dfrac{1 + \dfrac{[I]}{K_{i,ea}}}{1 + \dfrac{[I]}{K_{i,eab}}}\right) + 1} \tag{8.7}$$

$$v_o = \frac{\dfrac{V_{max}}{\left(1 + \dfrac{[I]}{K_{i,eab}}\right) + \dfrac{K_B}{[B]_o}\left(1 + \dfrac{[I]}{K_{i,ea}}\right)}[A]_o}{K_A\left(\dfrac{\dfrac{K_B}{[B]_o}\left(1 + \dfrac{[I]}{K_{i,e}}\right)}{\left(1 + \dfrac{[I]}{K_{i,eab}}\right) + \dfrac{K_B}{[B]_o}\left(1 + \dfrac{[I]}{K_{i,ea}}\right)}\right) + [A]_o} \tag{8.8}$$

$$V_{A,obs} + \frac{V_{max}}{\left(1 + \dfrac{[I]}{K_{i,eab}}\right) + \dfrac{K_B}{[B]_o}\left(1 + \dfrac{[I]}{K_{i,ea}}\right)} \tag{8.9}$$

$$(V/K)_{A,obs} = \frac{\dfrac{V_{max}}{K_A}}{\dfrac{K_B}{[B]_o}\left(1 + \dfrac{[I]}{K_{i,e}}\right)} \tag{8.10}$$

$$v_s = \frac{\dfrac{V_{max}[B]_o}{1 + \dfrac{[I]}{K_{i,eab}}}}{[B]_o + K_B\left(\dfrac{\dfrac{K_A}{[A]_o} + 1 + \dfrac{[I]}{K_{i,e}}\dfrac{K_A}{[A]_o} + \dfrac{[I]}{K_{i,ea}}}{1 + \dfrac{[I]}{K_{i,eab}}}\right)} \tag{8.11}$$

$$(V_{max})_B = \frac{V_{max}}{1 + \dfrac{[I]}{K_{i,eab}}} \tag{8.12}$$

$$(V_{max}/K_m)_B = \frac{V_{max}/K_B}{\dfrac{K_A}{[A]_o} + 1 + \dfrac{[I]}{K_{i,e}}\dfrac{K_A}{[A]_o} + \dfrac{[I]}{K_{i,ea}}} \tag{8.13}$$

Figure 8.2. Inhibition scheme and rate equations for an ordered mechanism, in which inhibitors bind to all forms of enzyme. Equation 8.7 is the general rate equation for this mechanism, and Equations 8.8–8.10 and Equations 8.11–8.13 are the substrate-specific rearrangements of Equation 8.7 for substrates A and B, respectively.

Just as with the inhibition of a one-substrate reaction, the shapes of these replots tells us the kinetic mechanism of inhibition:

- Inhibitors affecting only $V_{X,obs}$ are said to be uncompetitive versus X.
- Inhibitors affecting only $(V/K)_{X,obs}$ are said to be competitive versus X.
- Inhibitors affecting both $V_{X,obs}$ and $(V/K)_{X,obs}$ are said to be noncompetitive versus X.

With knowledge of the kinetic mechanism, we can move to quantitative analysis of the replots with a view to extracting the various inhibition constants. The analysis of replots is specific to each mechanism, and is discussed for each below.

8.2.3 Ordered Mechanisms

8.2.3.1 Connecting Rate Parameters to the Enzyme forms to which Inhibitor Binds.
Figure 8.2 portrays the mechanism of inhibition of an ordered mechanism in which inhibitor binds to all three steady-state enzyme forms. The rate equation for this mechanism can readily be derived assuming rapid establishment of equilibrium among all enzyme species and is given in Equation 8.7. While this equation, and the substrate specific rearrangement of Equations 8.8–8.13, are essential in the quantitative analysis of inhibition, Equation 8.7 does not make clear the connection between the enzyme form to which inhibitor binds and the affected rate parameter. As we saw for one-substrate reactions, this can be seen when the rate law of Equation 8.7 is expressed in reciprocal form.

Equation 8.14 is the reciprocal form of the rate equation for an ordered mechanism. Each of these

$$v_o = \left\{ \left(\frac{V_{max}}{K_A K_B}[A]_o[B]_o \right)^{-1} + \left(\frac{V_{max}}{K_B}[B]_o \right)^{-1} + (V_{max})^{-1} \right\}^{-1} \tag{8.14}$$

rate terms reflects the reaction of a specific steady-state form: $V_{max}/K_A K_B$ for reaction of E, V_{max}/K_B for reaction of EA, and V_{max} for reaction of EAB. From our discussion above, we know that if an inhibitor binds to one of these forms, this binding will manifest as the $[I]_o$-dependent titration of the rate parameter corresponding to that form. If inhibitor binds to all three forms, Equation 8.14 becomes

$$v_o = \left\{ \left(\frac{\frac{V_{max}}{K_A K_B}}{1 + \frac{[I]_o}{K_{I,e}}}[A]_o[B]_o \right)^{-1} + \left(\frac{\frac{V_{max}}{K_B}}{1 + \frac{[I]_o}{K_{I,ea}}}[B]_o \right)^{-1} + \left(\frac{V_{max}}{1 + \frac{[I]_o}{K_{I,eab}}} \right)^{-1} \right\}^{-1} \tag{8.15}$$

8.2.3.2 Quantitative Analysis.
Of course, steady-state inhibition studies do not provide direct access to the three rate terms and the three inhibitor dissociation

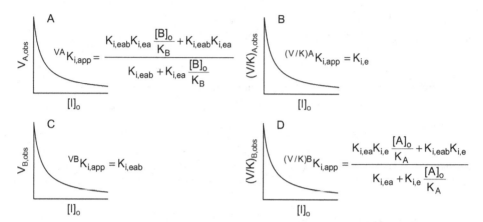

Figure 8.3. Inhibition replots for the titration of the four observed steady-state kinetic parameters for the ordered mechanism of Figure 8.2. In each plot, the expression is given for the apparent inhibition constant for that particular parameter.

constants of Equation 8.15. Rather, these studies lead us to the four replots of $(V/K)_{A,obs}$, $V_{A,obs}$, $(V/K)_{B,obs}$, and $V_{B,obs}$ versus $[I]_o$, and the four apparent inhibitor dissociation constants, $^{(V/K)A}K_{i,app}$, $^{VA}K_{i,app}$, $^{(V/K)B}K_{i,app}$, and $^{VB}K_{i,app}$, respectively, which are calculated from the replots (see Fig. 8.3).

Note in the replot of Figure 8.3B, that the dependence of $(V/K)_{A,obs}$ on $[I]_o$, which is described by Equation 8.10, provides an estimate of an apparent inhibition constant, $^{(V/K)A}K_{i,app}$, which is equal to a true dissociation constant:

$$^{(V/K)A}K_{i,app} = K_{i,e}. \tag{8.16}$$

Likewise, the dependence of $V_{B,obs}$ on $[I]_o$, (see Fig. 8.3C), which is described by Equation 8.12, provides an estimate of an apparent inhibition constant, $^{VB}K_{i,app}$, which is equal to a true dissociation constant:

$$^{VB}K_{i,app} = K_{i,eab}. \tag{8.17}$$

The situation is not so simple when we analyze the dependence of $V_{A,obs}$ on $[I]_o$ (see Fig. 8.3A) or $(V/K)_{B,obs}$ on $[I]_o$ (see Fig. 8.3D). For the former, this analysis, according to Equation 8.9, provides an apparent dissociation constant $^{VA}K_{i,app}$ that is a complex function of two dissociation constants:

$$^{VA}K_{i,app} = \frac{K_{i,eab}K_{i,ea}\dfrac{[B]_o}{K_B} + K_{i,eab}K_{i,ea}}{K_{i,eab} + K_{i,ea}\dfrac{[B]_o}{K_B}}. \tag{8.18}$$

Similarly, the dependence of $(V/K)_{B,obs}$ on $[I]_o$, according to Equation 8.13, yields an apparent dissociation constant, $^{(V/K)B}K_{i,app}$, which is also a complex function of two dissociation constants:

$$^{(V/K)B}K_{i,app} = \frac{K_{i,ea}K_{i,e}\dfrac{[A]_o}{K_A} + K_{i,ea}K_{i,e}}{K_{i,ea} + K_{i,e}\dfrac{[A]_o}{K_A}}. \tag{8.19}$$

We see then that the apparent inhibition constants calculated from the dependencies of $V_{A,obs}$ on $(V/K)_{B,obs}$ on $[I]_o$, refer to *two* steady-state enzyme species that can bind inhibitor: EA and EAB for $^{VA}K_{i,app}$, and E and EA for $^{(V/K)B}K_{i,app}$. In both of these cases, the enzyme species that prevails in the steady state, and thus the enzyme species to which inhibitor binds, is substrate-concentration dependent. For example, inspection of Equation 8.18 tells us that if $[B]_o/K_B \ll 1$, $^{VA}K_{i,app} = K_{i,ea}$. This of course results from the fact that if $[B]_o/K_B \ll 1$, then $[EA] \gg [EAB]$.

A method is now apparent for calculating the dissociation constants for an inhibitor that binds to all three steady-state enzyme forms of an ordered mechanism. First, calculate $K_{i,e}$ and $K_{i,eab}$ from the replots of $(V/K)_{A,obs}$ versus $[I]_o$ and $V_{B,obs}$ versus $[I]_o$, respectively. Then, with these results in hand, calculate $K_{i,ea}$ using Equations 8.18 or 8.19.

8.2.4 Random Mechanisms

8.2.4.1 Connecting Rate Parameters to the Enzyme Forms to Which Inhibitor Binds. Equation 8.20

$$v_o = \left\{\left(\frac{V_{max}}{\alpha K_A K_B}[A]_o[B]_o\right)^{-1} + \left(\frac{V_{max}}{\alpha K_B}[B]_o\right)^{-1} + \left(\frac{V_{max}}{\alpha K_A}[A]_o\right)^{-1} + (V_{max})^{-1}\right\}^{-1} \tag{8.20}$$

is the reciprocal form of the rate expression for random mechanisms (see Fig. 8.4 for the mechanism and rate law). As above, each rate term reflects a specific steady-state enzyme form: $V_{max}/(\alpha K_A K_B)$, $V_{max}/(\alpha K_B)$, $V_{max}/(\alpha K_A)$, V_{max}, reflecting E, EA, EB, and EAB, respectively. If inhibitor binds to all four enzyme forms, Equation 8.20 becomes

$$v_o = \left\{\left(\frac{\dfrac{V_{max}}{\alpha K_A K_B}}{1 + \dfrac{[I]_o}{K_{I,e}}}[A]_o[B]_o\right)^{-1} + \left(\frac{\dfrac{V_{max}}{\alpha K_B}}{1 + \dfrac{[I]_o}{K_{I,ea}}}[B]_o\right)^{-1} + \left(\frac{\dfrac{V_{max}}{\alpha K_A}}{1 + \dfrac{[I]_o}{K_{I,eb}}}[A]_o\right)^{-1} + \left(\frac{V_{max}}{1 + \dfrac{[I]_o}{K_{I,eab}}}\right)^{-1}\right\}^{-1}. \tag{8.21}$$

8.2.4.2 Quantitative Analysis. Our starting point for a quantitative analysis of the inhibition of random mechanisms begins with Figure 8.4, the corresponding rate

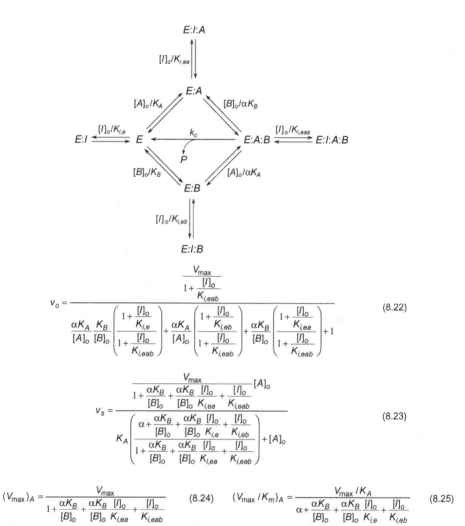

Figure 8.4. Inhibition scheme and rate equations for a random mechanism, in which inhibitors bind to all forms of enzyme. Equation 8.22 is the general rate equation for this mechanism, and Equations 8.23–8.25 are the substrate-specific rearrangements of Equation 8.22 for substrate A. Because of the symmetry of random mechanisms, the substrate-dependent rearrangements for B can be written by direct analogy to Equations 8.23–8.25.

law for this mechanism, given in Equation 8.22, and the substrate-specific rearrangements for A, in Equations 8.23–8.25. Because of the symmetry of random mechanisms, the substrate-dependent rearrangements for B can be written by direct analogy to Equations 8.23–8.25.

As indicated in Figure 8.1, both $(V/K)_{A,obs}$ and $V_{A,obs}$ have two reactant states. This means that both of these rate parameters titrate with inhibitor with apparent inhibition constants, $^{(V/K)A}K_{i,app}$ and $^{VA}K_{i,app}$, which are composites of two actual dissociation constants. The rate equations describing the dependencies of $(V/K)_{A,obs}$ and $V_{A,obs}$ on $[I]_o$ are given in Equations 8.26 and 8.27, respectively, with $^{(V/K)A}K_{i,app}$ and $^{VA}K_{i,app}$ indicated in parentheses:

$$(V/K)_{A,obs} = \cfrac{\cfrac{V_{max}}{1+\cfrac{K_B}{[B]_o}}}{1+\cfrac{[I]_o}{\left(\cfrac{K_{i,ea}K_{i,e}\cfrac{[B]_o}{K_B}+K_{i,ea}K_{i,e}}{K_{i,ea}+K_{i,e}\cfrac{[B]_o}{K_B}}\right)}} \tag{8.26}$$

$$V_{A,obs} = \cfrac{\cfrac{V_{max}}{1+\cfrac{K_B}{[B]_o}}}{1+\cfrac{[I]_o}{\left(\cfrac{K_{i,eab}K_{i,ea}\cfrac{[B]_o}{K_B}+K_{i,eab}K_{i,ea}}{K_{i,eab}+K_{i,ea}\cfrac{[B]_o}{K_B}}\right)}}. \tag{8.27}$$

Inspection of these equations reveals that without additional data, independent values for the four dissociation constants cannot be estimated. There are two sorts of additional data that would allow for a solution: (1) the $[B]_o$ dependence of $^{(V/K)A}K_{i,app}$ and $^{VA}K_{i,app}$, or (2) estimates of at least one of the four inhibitor dissociation constants by some independent means. To determine the $[B]_o$ dependence of the two apparent inhibition constants, one would have to conduct the entire experiment that we have been discussing at several concentrations of substrate B. Obtaining estimates of one or more of the four dissociation constants requires identifying an experimental approach that depends on only one of the two substrates. Such methods include biophysical techniques that lead to K_d values for an enzyme:inhibitor complex, or the analysis of the effect of the inhibitor on a partial or half reaction of the enzyme. These partial reactions include substrate hydrolysis for acyl-transferases or ATP hydrolysis for kinases. In the example below, I illustrate how we were able to use the ATPase activity of cdk5/p25 to calculate inhibition constants for the inhibition of this enzyme by N4-(6-aminopyrimidin-4-yl)-sulfanilide (APS). (Liu et al. 2008)

8.2.4.3 Analysis of the Inhibition of the Kinase Activity of cdk5/p25 Kinase.
To determine the kinetic mechanism of inhibition by APS of the cdk5/p25-catalyzed phosphorylation of tau, a microtubule-associated protein whose phosphorylation and aggregation is thought to play a role in Alzheimer's disease, we determined two initial velocity dependencies: v_o as a function of $[ATP]_o$ and $[APS]_o$, with the tau

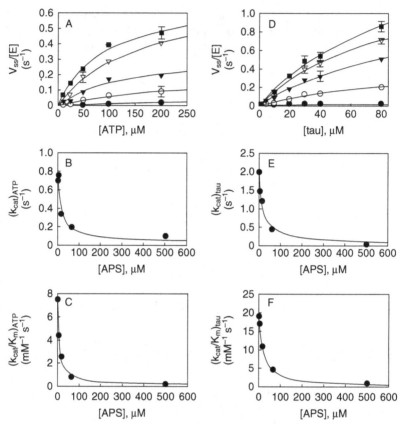

Figure 8.5. Inhibition of the cdk5/p25-catalyzed phosphorylation of tau by APS. *Panel A:* Plot of $v_{ss}/[E]$ versus $[ATP]$ at $[APS] = 0$ (●), 4 (○), 16 (▼), 63 (▽), and 500 µM (■), all at a fixed tau concentration of 80 µM. *Panels B & C:* APS concentration dependencies of $(k_{cat})_{ATP}$ and $(k_{cat}/K_m)_{ATP}$ values derived from analysis of the data of Panel A. *Panel D:* Plot of $v_{ss}/[E]$ versus $[tau]$ at $[APS] = 0$ (●), 4 (○), 16 (▼), 63 (▽), and 500 µM (■), all at a fixed *ATP* concentration of 200 µM. *Panels E & F:* APS concentration dependencies of $(k_{cat})_{tau}$ and $(k_{cat}/K_m)_{tau}$ values derived from analysis of the data of Panel D.

concentration fixed at 80 µM, and v_o as a function of $[tau]_o$ and $[APS]_o$, with the ATP concentration fixed at 200 µM. Figure 8.5A, D are the two primary plots of this data. The analysis of this data that follows is done in the context of the rapid equilibrium random mechanism for this enzyme (see Chapter 7).

As the replots of Figure 8.5 indicate, all four steady-state parameters titrate with APS. This corresponds to the "worst cases scenario" that we have been discussing, in which inhibitor can bind to all four steady-state species of enzyme. Each of the apparent inhibition constants has a form that is identical to either Equation 8.26 or 8.27, and this is a composite of two true inhibition dissociation constants.

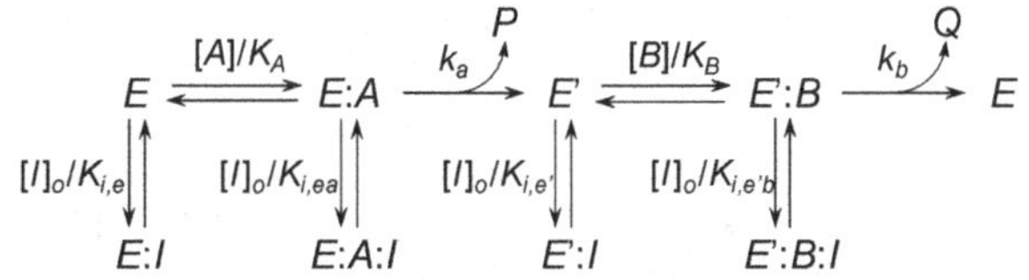

Figure 8.6. Inhibition scheme for a ping-pong mechanism, in which inhibitors bind to all forms of enzyme.

To dissect each of these apparent values into its two dissociation constants requires an independent estimate of at least one of these dissociation constants. We were able to accomplish this by taking advantage of the fact that cdk5/p25 possesses ATPase activity. This was discussed in Section 5.3, where we saw that APS is a mixed inhibitor of the ATPase activity of this enzyme, with $K_i = 4.4\,\mu M$ and $\beta K_i = 58\,\mu M$, which corresponds to $K_{i,e}$ and $K_{i,e:ATP}$, respectively. These values were then used together with the expressions for the four apparent dissociation constants to calculate $K_{i,e:tau} = 8.8\,\mu M$ and $K_{i,e:tau:ATP} = 26\,\mu M$.

8.2.5 Ping-Pong Mechanisms

Our discussion of the inhibition of enzymes that proceed through a ping-pong mechanism starts with the derivation and consideration of the rate equations for this mechanism. With this complete, we then develop a conceptual understanding of the mechanism.

8.2.5.1 Rate Equations and Quantitative Analysis. The mechanism we will be considering is that shown in Figure 8.6, in which inhibitor binds to all steady-state forms of the enzyme. The derivation of the rate law for this mechanism begins at the usual starting place, the equation in which the statement of rate dependence is divided by the conservation of enzyme:

$$\frac{v_0}{[E]_o} = \frac{k_b[FB]}{[EI]+[E]+[EAI]+[EA]+[FI]+[F]+[FBI]+[FB]}. \tag{8.28}$$

Recall from Chapter 7 that for a ping-pong mechanism, expression of the statement of rate dependence can be either $v_o = k_a[EA]$ or $v_o = k_b[FB]$.

Given that $[EA]/[FB] = k_b/k_a$ (Chapter 7), Equation 2.8 becomes

$$\frac{v_0}{[E]_o} = \frac{[FB]}{\dfrac{[I]_o}{K_{i,e}}\dfrac{K_A}{[A]_o}\dfrac{k_b}{k_a} + \dfrac{K_A}{[A]_o}\dfrac{k_b}{k_a} + \dfrac{k_b}{k_a}\dfrac{[I]_o}{K_{i,ea}} + \dfrac{[I]_o}{K_{i,f}}\dfrac{K_B}{[F]_o} + \dfrac{K_B}{[F]_o} + \dfrac{[I]_o}{K_{i,fb}} + \left(1+\dfrac{k_b}{k_a}\right)}. \tag{8.29}$$

Multiplying through by $k_b/(k_a + k_b)$ and collecting terms yields

$$\frac{v_o}{[E]_o} = \frac{\dfrac{k_a k_b}{k_a + k_b}}{\dfrac{k_b}{k_a + k_b} \dfrac{K_A}{[A]_o}\left(1+\dfrac{[I]_o}{K_{i,e}}\right)+\dfrac{k_a}{k_a + k_b} \dfrac{K_B}{[B]_o}\left(1+\dfrac{[I]_o}{K_{i,f}}\right)+1+\dfrac{k_b}{k_a + k_b}\dfrac{[I]_o}{K_{i,ea}}+\dfrac{k_a}{k_a + k_b}\dfrac{[I]_o}{K_{i,fb}}}.$$

(8.30)

Given Chapter 7's definitions for k_c, $K_{m,A}$, and $K_{m,B}$, Equation 8.3 can be expressed as

$$\frac{v_o}{[E]_o} = \frac{k_c}{\dfrac{K_{m,A}}{[A]_o}\left(1+\dfrac{[I]_o}{K_{i,e}}\right)+\dfrac{K_{m,B}}{[B]_o}\left(1+\dfrac{[I]_o}{K_{i,f}}\right)+\left(1+\dfrac{[I]_o}{\left(1+\dfrac{k_a}{k_b}\right)K_{i,ea}}+\dfrac{[I]_o}{\left(1+\dfrac{k_b}{k_a}\right)K_{i,fb}}\right)}.$$

(8.31)

Dividing through by the bracketed term of Equation 8.31 yields

$$v_o = \frac{\dfrac{V_{\max}}{1+\dfrac{[I]_o}{K_{i,ea}^*}+\dfrac{[I]_o}{K_{i,fb}^*}}}{\left(\dfrac{K_{m,A}}{[A]_o}\dfrac{1+\dfrac{[I]_o}{K_{i,e}}}{1+\dfrac{[I]_o}{K_{i,ea}^*}+\dfrac{[I]_o}{K_{i,fb}^*}}+\dfrac{K_{m,B}}{[B]_o}\dfrac{1+\dfrac{[I]_o}{K_{i,f}}}{1+\dfrac{[I]_o}{K_{i,ea}^*}+\dfrac{[I]_o}{K_{i,fb}^*}}+1\right)},$$

(8.32)

where

$$K_{i,ea}^* = \left(1+\frac{k_a}{k_b}\right)K_{i,ea}$$

(8.33)

and

$$K_{i,fb}^* = \left(1+\frac{k_b}{k_a}\right)K_{i,fb}.$$

(8.34)

Analysis now proceeds by rearranging Equation 8.32 into its substrate-specific forms. Symmetry of the ping-pong mechanism necessitates only a single rearrangement, in this case for substrate A. If we isolate substrate A, Equation 8.32 rearranges into

$$v_o = \frac{\dfrac{V_{\max}[A]_o}{1 + \dfrac{[I]_o}{K_{i,ea}^*} + \dfrac{[I]_o}{K_{i,fb}^*} + \dfrac{K_{m,B}}{[B]_o}\left(1 + \dfrac{[I]_o}{K_{i,f}}\right)}}{\dfrac{K_{m,A}}{[A]_o}\left(\dfrac{1 + \dfrac{[I]_o}{K_{i,e}}}{1 + \dfrac{[I]_o}{K_{i,ea}^*} + \dfrac{[I]_o}{K_{i,fb}^*} + \dfrac{K_{m,B}}{[B]_o}\left(1 + \dfrac{[I]_o}{K_{i,f}}\right)}\right) + [A]_o}.$$ (8.35)

Equation 8.35 can be rearranged as

$$v_o = \frac{\dfrac{V_{\max}[A]_o}{1 + \dfrac{[I]_o}{K_{i,es}} + \dfrac{K_{m,B}}{[B]_o}\left(1 + \dfrac{[I]_o}{K_{i,f}}\right)}}{\dfrac{K_{m,A}}{[A]_o}\left(\dfrac{1 + \dfrac{[I]_o}{K_{i,e}}}{1 + \dfrac{[I]_o}{K_{i,es}} + \dfrac{K_{m,B}}{[B]_o}\left(1 + \dfrac{[I]_o}{K_{i,f}}\right)}\right) + [A]_o},$$ (8.36)

where

$$K_{i,es} = \frac{K_{i,ea}^* K_{i,fb}^*}{K_{i,ea}^* + K_{i,fb}^*}.$$ (8.37)

Finally, Equation 8.36 can be expressed as

$$v_o = \frac{V_{\max}[A]_o}{\left[1 + \dfrac{[I]_o}{\left(\dfrac{K_{i,es}K_{i,f}\dfrac{K_{m,B}}{[B]_o} + K_{i,es}K_{i,f}}{K_{i,es} + K_{i,f}\dfrac{K_{m,B}}{[B]_o}}\right)}\right]}{K_{m,A}\left\{\dfrac{1 + \dfrac{[I]_o}{K_{i,e}}}{1 + \dfrac{[I]_o}{\left(\dfrac{K_{i,es}K_{i,f}\dfrac{K_{m,B}}{[B]_o} + K_{i,es}K_{i,f}}{K_{i,es} + K_{i,f}\dfrac{K_{m,B}}{[B]_o}}\right)}}\right\} + [A]_o}.$$ (8.38)

From Equation 8.38, the expressions of Equations 8.39 and 8.40 can be written for $[I]_o$-dependent $(V/K)_{A,obs}$ and $V_{A,obs}$:

$$(V/K)_{A,obs} = \frac{V_{max}/K_{m,A}}{1+\dfrac{[I]_o}{K_{i,e}}} \tag{8.39}$$

$$V_{A,obs} = \frac{V_{max}}{1+\dfrac{[I]_o}{\left(\dfrac{K_{i,es}K_{i,f}\dfrac{K_{m,B}}{[B]_o}+K_{i,es}K_{i,f}}{K_{i,es}+K_{i,f}\dfrac{K_{m,B}}{[B]_o}}\right)}}. \tag{8.40}$$

So, from replots of $(V/K)_{A,obs}$ and $(V/K)_{B,obs}$ versus $[I]_o$ we obtain

$$^{(V/K)A}K_{i,app} = K_{i,e} \tag{8.41}$$

$$^{(V/K)B}K_{i,app} = K_{i,f} \tag{8.42}$$

while in the replots of $V_{A,obs}$ and $V_{B,obs}$ versus $[I]_o$, we obtain the more complex

$$^{VA}K_{i,app} = \left(\frac{K_{i,es}K_{i,f}\dfrac{K_{m,B}}{[B]_o}+K_{i,es}K_{i,f}}{K_{i,es}+K_{i,f}\dfrac{K_{m,B}}{[B]_o}}\right) \tag{8.43}$$

$$^{VB}K_{i,app} = \left(\frac{K_{i,es}K_{i,e}\dfrac{K_{m,A}}{[A]_o}+K_{i,es}K_{i,e}}{K_{i,es}+K_{i,e}\dfrac{K_{m,B}}{[B]_o}}\right). \tag{8.44}$$

From knowledge of either $K_{i,e}$ or $K_{i,f}$ that come from Equations 8.41 and 8.42, respectively, a value of $K_{i,es}$ can be calculated. But this latter term cannot be dissected into its composite $K_{i,ea}^*$ and $K_{i,fb}^*$, and these into $K_{i,ea}$ and $K_{i,fb}$, without independent estimates of composite $K_{i,ea}^*$ or $K_{i,fb}^*$, and the ratio k_d/k_b.

8.2.5.2 Connecting Rate Parameters to the Enzyme Forms to which Inhibitor Binds.
If inhibitor binds to E, $V_{max}/K_{m,A}$ will titrate with a true dissociation constant equal to $K_{i,e}$. Likewise, if inhibitor binds to F, $V_{max}/K_{m,B}$ will titrate with a true dissociation constant equal to $K_{i,f}$. However, binding of inhibitor to EA or FB leads to apparent inhibition constants that can be analyzed only if information about the inhibition mechanism (i.e., knowledge of whether inhibitor binds to EA or FB) and enzyme mechanism (i.e., knowledge of k_d/k_b) is available from independent experiments.

8.3 INHIBITION BY SUBSTRATE ANALOGS

A substrate analog inhibitor is a compound that has a structure sufficiently similar to a substrate that it binds to the targeted enzyme in the same manner as substrate. But unlike the substrate, a substrate analog lacks a key reactive moiety so it is unable to enter into the enzyme's catalytic reaction. For example, 5′-adenylyl imidodiphosphate, a structural analog of ATP, is frequently used as a substrate analog inhibitor of kinase reactions since the nitrogen bridging the β- and γ-phosphates renders it unreactive for phosphoryl transfer. Similarly, 3-acetylpyridine adenine dinucleotide is an oxidatively unreactive analog of NADH and is used as a substrate analog inhibitor of dehydrogenases.

In this section we consider the kinetics of inhibition by this class of inhibitor and see a special use of these compounds in helping to distinguish rapid equilibrium random from steady-state ordered mechanisms (see Section 7.8 for an introduction to this problem).

8.3.1 Substrate Analog Inhibition of Ordered Mechanisms

8.3.1.1 Inhibition by an Analog of Substrate A. Suppose that compound A* is a close, but unreactive, structural analog of substrate A. One can reasonably predict that in an ordered mechanism, A* will be able to bind to E to form EA*, but will not be able to bind to either EA or to EAB. In addition, EA* may be able to bind substrate B to form EA*B. This mechanism is summarized in Figure 8.7, and its rate equation is expressed in Equation 8.45:

$$v_o = \frac{V_{max}}{\dfrac{K_A}{[A]_o}\dfrac{K_B}{[B]_o}\left(1+\dfrac{[A^*]_o}{K_{i,e}}+\dfrac{[A^*]_o}{K_{i,e}}\dfrac{[B]_o}{\phi K_B}\right)+\dfrac{K_B}{[B]_o}+1}. \tag{8.45}$$

Equation 8.45 can be rearranged into Equation 8.46, a form in which substrate A has been isolated:

$$v_o = \frac{V'_{max}[A]_o}{K'_A\left(1+\dfrac{[A^*]_o}{K'_{i,e}}\right)+[A]_o}. \tag{8.46}$$

In this equation,

$$V'_{max} = \frac{V_{max}}{1+\dfrac{[B]_o}{K'_B}}, \tag{8.47}$$

$$K'_A = K_A \frac{\dfrac{K_B}{[B]_o}}{1+\dfrac{K_B}{[B]_o}}, \tag{8.48}$$

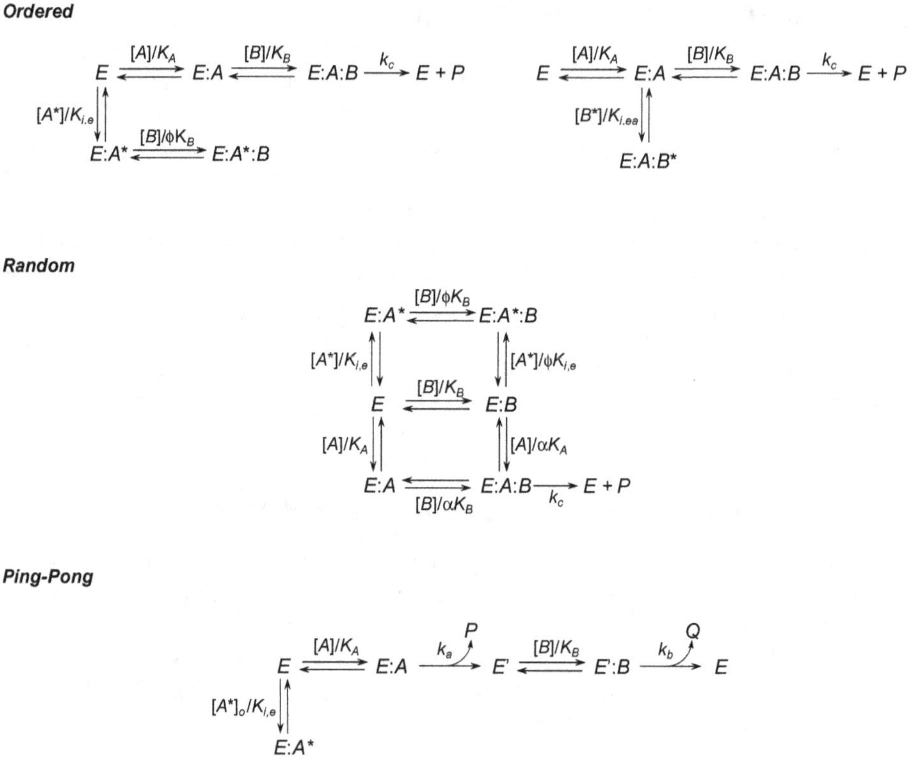

Ordered

Random

Ping-Pong

Figure 8.7. Substrate analog inhibition for the standard two-substrate mechanisms.

and

$$K'_{i,e} = \frac{K_{i,e}}{1 + \dfrac{[B]_o}{\phi K_B}}. \tag{8.49}$$

We see from Equation 8.46 that while $V_{A,obs}$ is independent of $[A*]_o$ and equal to V_{max}, $(V/K)_{A,obs}$ is dependent on $[A*]_o$ as expressed in Equation 8.50:

$$\left(\frac{V}{K}\right)_{A,obs} = \frac{\dfrac{V}{K'_B}[A]_o}{1 + \dfrac{[A*]_o}{K'_{i,e}}}. \tag{8.50}$$

Thus, $A*$ is competitive with A, with an apparent inhibition constant $K'_{i,e}$ that depends on $[B]_o/\phi K_B$.

Equation 8.51 is a rearrangement of Equation 8.45 in which substrate B has been isolated:

$$v_o = \frac{\dfrac{V_{\max}}{1 + \dfrac{K_A}{[A]_o}\dfrac{[A^*]_o}{\phi[K_{i,e}]}}[B]_o}{K_B\left(\dfrac{1 + \dfrac{K_A}{[A]_o} + \dfrac{K_A}{[A]_o}\dfrac{[A^*]_o}{\phi[K_{i,e}]}}{1 + \dfrac{K_A}{[A]_o}\dfrac{[A^*]_o}{\phi[K_{i,e}]}}\right) + [B]_o}. \tag{8.51}$$

The dependence of $V_{B,obs}$ and $(V_{\max}/K_B)_{B,obs}$ are given in Equations 8.52 and 8.53,

$$V_{B,obs} = \frac{V_{\max}}{1 + \dfrac{[A^*]}{\phi K'_{i,e}}} \tag{8.52}$$

$$\left(\frac{V}{K}\right)_{B,obs} = \frac{\left(\dfrac{V_{\max}}{K_B}[B]_o\right)[A]_o}{K_A\left(1 + \dfrac{[A^*]_o}{K_{i,e}}\right) + [A]_o} \tag{8.53}$$

where

$$\phi K'_{i,e} = \phi K_{i,e}\left(\frac{[A]_o}{K_A}\right) \tag{8.54}$$

Note that the magnitude of ϕ, or the ability of B to bind to EA', determines whether A^* is a noncompetitive or competitive inhibitor of the binding of B.

8.3.1.2 Inhibition by an Analog of Substrate B. Figure 8.7 shows an ordered mechanism with inhibition by B^*, an unreactive structural analog of substrate B. The rate equation for this mechanism is given below in Equation 8.55:

$$v_o = \frac{V_{\max}}{\dfrac{K_A}{[A]_o}\dfrac{K_B}{[B]_o} + \dfrac{K_B}{[B]_o}\left(1 + \dfrac{[B^*]_o}{K_{i,ea}}\right) + 1}. \tag{8.55}$$

This equation can be rearranged to isolate A, yielding

$$v_o = \frac{\dfrac{V_{\max}}{1 + \dfrac{K_B}{[B]_o} + \dfrac{K_B}{[B]_o}\dfrac{[B^*]_o}{K_{i,ea}}}}{K_A\left(\dfrac{\dfrac{K_B}{[B]_o}}{1 + \dfrac{K_B}{[B]_o} + \dfrac{K_B}{[B]_o}\dfrac{[B^*]_o}{K_{i,ea}}}\right) + [A]_o}. \tag{8.56}$$

Equations 8.57 and 8.58 are the expressions for $V_{A,obs}$ and $(V/K)_{A,obs}$ for this mechanism:

$$V_{A,obs} = \frac{V_{max}[B]_o}{K_B\left(1+\dfrac{[B^*]}{K_{i,ea}}\right)+[B]_o} \tag{8.57}$$

$$\left(\frac{V}{K}\right)_{A,obs} = \frac{V_{max}}{K_A}\frac{[B]_o}{K_B}. \tag{8.58}$$

We see from the above two equations, that B^* effects only V_{max}, so it is an uncompetitive inhibitor of the binding of substrate B.

Equation 8.55 can also be rearranged to isolate substrate B. This is expressed in Equation 8.59:

$$v_o = \frac{V_{max}[B]_o}{K_B\left(1-\dfrac{K_A}{[A]_o}+\dfrac{[B^*]_o}{K_{i,ea}}\right)+[B]_o}. \tag{8.59}$$

Equations 8.60 and 8.61 are the expressions for $V_{B,obs}$ and $(V/K)_{B,obs}$ for this mechanism:

$$\left(\frac{V}{K}\right)_{B,obs} = \frac{\dfrac{(V_{max}/K_B)[A]_o}{K_A+[A]_o}}{1+\dfrac{[B^*]_o}{K_{i,ea}\left(1+\dfrac{[A]_o}{K_A}\right)}} \tag{8.60}$$

$$V_{B,obs} = V_{max}. \tag{8.61}$$

From Equations 8.60 and 8.61, we see that only V_{max}/K_B is affected by B^*, meaning that it is a competitive inhibitor of the binding of B.

8.3.2 Substrate Analog Inhibition of Random Mechanisms

Given the symmetry of random mechanisms, our discussion of substrate analog inhibition need only consider inhibition by an analog of one of the two substrates, in this case substrate A.

Figure 8.7 depicts the inhibition of a random mechanism by substrate analog inhibitor A^*. The corresponding rate equation is given below in Equation 8.62:

$$v_o = \frac{V_{max}}{\dfrac{\alpha K_A}{[A]_o}\dfrac{K_B}{[B]_o}\left(1+\dfrac{[A^*]_o}{K_{i,e}}\right)+\dfrac{\alpha K_A}{[A]_o}\left(1+\dfrac{[A^*]_o}{\phi K_{i,e}}\right)+\dfrac{\alpha K_B}{[B]_o}+1}. \tag{8.62}$$

Rearrangement of this equation to isolate substrate A yields

$$v_o = \cfrac{\cfrac{V_{max}}{1 + \cfrac{\alpha K_B}{[B]_o}}[A]_o}{1 + \cfrac{\alpha K_A}{1 + \cfrac{\alpha K_B}{[B]_o}}\left\{\cfrac{K_B}{[B]_o}\left(1 + \cfrac{[A^*]_o}{K_{i,e}}\right) + \left(1 + \cfrac{[A^*]_o}{\phi K_{i,e}}\right)\right\} + [A]_o}. \tag{8.63}$$

$V_{A,obs}$ is independent of $[A^*]_o$, equaling $V_{max}/(1 + \alpha K_B/[B]_o)$, while $(V/K)_{A,obs}$ is dependent on $[A^*]_o$, equaling

$$\left(\frac{V}{K}\right)_{A,obs} = \cfrac{(V_{max}/\alpha K_A)[A]_o}{\cfrac{K_B}{[B]_o}\left(1 + \cfrac{[A^*]_o}{K_{i,e}}\right) + \left(1 + \cfrac{[A^*]_o}{\phi K_{i,e}}\right)}. \tag{8.64}$$

Equation 8.62 can also be rearranged to isolate substrate B:

$$v_o = \cfrac{\cfrac{V_{max}}{1 + \cfrac{\alpha K_A}{[A]_o}\left(1 + \cfrac{[A^*]_o}{\phi K_{i,e}}\right)}[B]_o}{\alpha K_B\left\{\cfrac{1 + \cfrac{K_A}{[A]_o}\left(1 + \cfrac{[A^*]_o}{K_{i,e}}\right)}{1 + \cfrac{\alpha K_A}{[A]_o}\left(1 + \cfrac{[A^*]_o}{\phi K_{i,e}}\right)}\right\} + [B]_o}. \tag{8.65}$$

Both $V_{A,obs}$ and $(V/K)_{A,obs}$ are dependent on $[A^*]_o$, according to the following equations:

$$V_{B,obs} = \cfrac{V_{max}}{1 + \cfrac{\alpha K_A}{[A]_o}\left(1 + \cfrac{[A^*]_o}{\phi K_{i,e}}\right)} \tag{8.66}$$

$$\left(\frac{V}{K}\right)_{B,obs} = \cfrac{(V_{max}/\alpha K_B)[B]_o}{1 + \cfrac{K_A}{[A]_o}\left(1 + \cfrac{[A^*]}{K_{i,e}}\right)}. \tag{8.67}$$

We see then that while A^* is competitive with substrate A, it is noncompetitive with B.

8.3.3 Substrate Analog Inhibition of Ping-Pong Mechanisms

Like the random mechanism we just discussed, ping-pong mechanisms have a symmetry that allows us to consider inhibition by an analog of only one of the two substrates.

Inhibition of a ping-pong mechanism by substrate analog A^* is shown in Figure 8.7, and its rate law is given in Equation 8.68:

$$v_o = \frac{V_{max}}{\dfrac{K_{m,A}}{[A]_o}\left(1 + \dfrac{[A^*]_o}{K_{i,e}}\right) + \dfrac{K_{m,B}}{[B]_o} + 1}. \tag{8.68}$$

Rearrangement of this equation to isolate substrate A yields

$$v_o = \frac{\dfrac{V_{max}}{1 + \dfrac{K_{m,B}}{[B]_o}}[A]_o}{K_{m,A}\left(\dfrac{1 + \dfrac{[A^*]_o}{K_{i,e}}}{1 + \dfrac{K_{m,B}}{[B]_o}}\right) + [A]_o}. \tag{8.69}$$

We see from Equation 8.69 that while $V_{A,obs}$ is independent of $[A^*]_o$, $(V/K)_{A,obs}$ is dependent on $[A^*]_o$:

$$\left(\frac{V}{K}\right)_{A,obs} = \frac{\dfrac{V_{max}/K_{m,A}}{1 + \dfrac{K_{m,B}}{[B]_o}}[A]_o}{1 + \dfrac{[A^*]_o}{K_{i,e}}}. \tag{8.70}$$

Isolating substrate B for inhibition of a ping-pong mechanism yields Equation 8.71:

$$v_o = \frac{\dfrac{V_{max}}{1 + \dfrac{K_{m,A}}{[A]_o}\left(1 + \dfrac{[A^*]_o}{K_{i,e}}\right)}[B]_o}{\dfrac{K_{m,B}}{1 + \dfrac{K_{m,A}}{[A]_o}\left(1 + \dfrac{[A^*]_o}{K_{i,e}}\right)} + [B]_o}. \tag{8.71}$$

While $(V/K)_{B,obs}$ can be seen to be independent of $[A^*]_o$, $V_{B,obs}$ is dependent on $[A^*]_o$ according to

$$V_{B,obs} = \frac{V_{max}}{1 + \dfrac{K_{m,A}}{[A]_o}\left(1 + \dfrac{[A^*]_o}{K_{i,e}}\right)}. \tag{8.72}$$

8.3.4 Summary of Inhibition Patterns and Utility in Distinguishing among Mechanisms

At the end of Chapter 7, I noted that the rate laws for rapid equilibrium random and steady-state ordered mechanisms have identical forms, making them indistinguishable by studies in which the initial velocity is determined as a function of both substrate concentrations. That is, simple bisubstrate analysis studies can distinguish between sequential and ping-pong mechanisms, but cannot distinguish between the two sequential mechanisms. However, from the discussion above, we can see that substrate analog inhibition studies can in fact can accomplish this. Table 8.1 summarizes the expected substrate analog inhibition mechanisms for the three standard, two-substrate mechanisms.

8.4 ANALYSIS OF SEQUENTIAL REACTIONS IN WHICH INHIBITOR BINDS TO ENZYME:PRODUCT COMPLEXES

Throughout this chapter, we have assumed rapid equilibrium for the sequential reactions we considered. In these cases, chemical transformation of $E:A:B$ into the complex of enzyme and product is rate-limiting, and is thus slower than both the binding of substrates and the release of products. If, on the other hand, product release is slow relative to chemistry, than complexes of enzyme and product will accumulate in the steady state and be available to bind inhibitor.

In situations where product release is rate-limiting, binding of inhibitor to an enzyme:product complex will decrease V_{max} but have no effect on V_{max}/K_m. Given this, we can predict the following changes in kinetic mechanism of inhibition when product release is rate-limiting.

Inhibitors that were competitive, will become noncompetitive:

$$v_o = \frac{V_{X,obs}[X]_o}{K_{X,obs}\left(1+\dfrac{[I]_o}{K_i}\right)+[X]_o} \Rightarrow v_o = \frac{\dfrac{V_{X,obs}}{1+\dfrac{[I]_o}{K_{ii}}}[X]_o}{K_{X,obs}\left(\dfrac{1+\dfrac{[I]_o}{K_i}}{1+\dfrac{[I]_o}{K_{ii}}}\right)+[X]_o}. \tag{8.73}$$

Inhibitors that are uncompetitive will remain uncompetitive, but now the $K_{i,app}$ on which $V_{X,obs}$ depends will be a combination of two inhibition constants:

$$v_o = \frac{\dfrac{V_{X,obs}}{1+\dfrac{[I]_o}{K_{ii}}}[X]_o}{K_{X,obs}+[X]_o} \Rightarrow v_o = \frac{\dfrac{V_{X,obs}}{\left(1+\dfrac{[I]_o}{K_{ii}}+\dfrac{[I]_o}{K_{iii}}\right)}[X]_o}{K_{X,obs}+[X]_o}. \tag{8.74}$$

TABLE 8.1. Inhibition Mechanisms for Substrate Analog Inhibitors

Mechanism	Substrate Analog	Mechanism versus Variable Substrate	
		A	B
Ordered	A*	C	C or N
Ordered	B*	U	C
Random	A*	C	N
Random	B*	N	C
Ping-Pong	A*	C	U
Ping-Pong	B*	U	C

TABLE 8.2. Substrate Analog Inhibition for Sequential Mechanisms in which Inhibitor Binds to Enzyme:Product Complexes

Mechanism		Substrate Analog	Mechanism versus Variable Substrate	
Substrate Addition	Product Release		A	B
Ordered	Ordered	A*	C	C or N
Ordered	Ordered	B*	U	N
Ordered	Random	A*	N	N
Ordered	Random	B*	U	N
Random	Ordered	A*	C	C or N
Random	Ordered	B*	N	N
Random	Random	A*	N	N
Random	Random	B*	N	N

Inhibitors that are noncompetitive remain noncompetitive, and obey the following general equation:

$$
v_o = \frac{\dfrac{V_{X,obs}}{1+\dfrac{[I]_o}{K_{ii}}}[X]_o}{K_{X,obs}\left(\dfrac{1+\dfrac{[I]_o}{K_i}}{1+\dfrac{[I]_o}{K_{ii}}}\right)+[X]_o} \Rightarrow v_o = \frac{\dfrac{V_{X,obs}}{\left(1+\dfrac{[I]_o}{K_{ii}}+\dfrac{[I]_o}{K_{iii}}\right)}[X]_o}{K_{X,obs}\left\{\dfrac{1+\dfrac{[I]_o}{K_i}}{\left(1+\dfrac{[I]_o}{K_{ii}}+\dfrac{[I]_o}{K_{iii}}\right)}\right\}+[X]}. \tag{8.75}
$$

These points are summarized in Table 8.2 where it is assumed that for ordered product release, the first product to leave is P_B, the product derived from substrate B, to generate $E:P_A$.

Rules for mechanism assignment that emerge from this table can be summarized as follows:

- Sequential mechanisms having random substrate addition cannot be confused with sequential mechanisms having ordered substrate addition. *Regardless of the order of product release, substrate analog inhibition studies will be able to distinguish sequential mechanisms with random substrate addition from ordered substrate addition.*
- Sequential mechanisms having ordered binding of substrates cannot be confused with sequential mechanisms having random binding of substrates.
- Sequential mechanisms having ordered substrate binding and random release of products can potentially be confused with sequential mechanisms having ordered substrate binding and ordered product release.

Comparison of the patterns of Tables 8.1 and 8.2 reveal no overlap, meaning that if an investigator observes a pattern shown in Table 8.2, he can conclude that product release is rate-limiting for the system under study.

8.4.1 Case Study: Inhibition of SIRT1 by Substrate Analogs to Establish Kinetic Mechanism

The sirtuins are a family of enzymes that catalyze the NAD^+-dependent deacetylation of ε-acetyl-Lys residues of proteins (Sauve et al. 2001; Sauve and Schramm 2003; Sauve et al. 2006). The chemical mechanism of sirtuin-catalyzed deacetylation is shown in Figure 8.8 and proceeds through an ADP-ribosyl-peptidyl imidate intermediate, formed by nucleophilic displacement of nicotinamide by the carbonyl oxygen of the acetyl group (Sauve et al. 2001; Sauve and Schramm 2003; Sauve et al. 2006). Collapse of this intermediate generates 1′-acetyl-ADP-ribose, liberating the deacetylated protein. Finally, the oxygen of the 2′-hydroxyl attacks the carbonyl carbon of the acetyl to form the third reaction product, 2′-O-acetyl-ADP-ribose.

Interest in these enzymes stems from the roles they are thought to play in human disease. Of particular interest is SIRT1, which has been implicated in a number of age-related diseases and biological functions involving cell survival, apoptosis, stress resistance, fat storage, insulin production, and glucose and lipid homeostasis (Baur 2010; Haigis and Sinclair 2010). Involvement in these diverse biologies is thought to occur through deacetylation of its many known in vivo protein substrates (Sauve et al. 2006; Guarente 2007).

Steady-state kinetic experiments of the sort described in Chapter 7 indicate that SIRT1 follows a sequential mechanism for deacetylation of Ac-Arg-His-Lys-**Lys**^{Ac}-Leu-Nle-Phe-NH_2 (p53-7mer), a peptide modeled after an acetylation site in the tumor suppressor protein p53, a natural substrate for SIRT1 (unpublished results of the author). Assuming a rapid equilibrium random mechanism, the data can be analyzed to yield the following parameter estimates: $K_{p53} = 1.4\,\mu M$, $K_{NAD} = 185\,\mu M$, $\alpha = 0.3$, and $k_c = 0.2\,s^{-1}$. As we learned in Chapter 7, the all-hyperbolic replots characteristic of

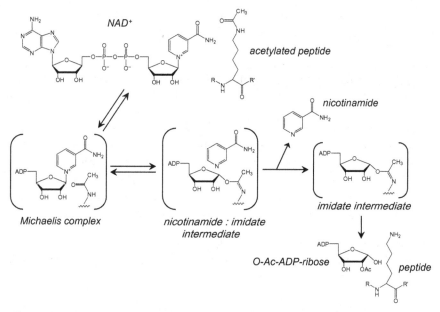

Figure 8.8. Chemical mechanism of NAD⁺-dependent sirtuin-catalyzed deacetylation.

random mechanisms are also seen with steady-state ordered mechanism. To distinguish between these two mechanisms, substrate analog inhibition studies were conducted. The NAD⁺ analog used was ADP-ribose (ADPR), and the p53-7mer analog was a trifluoroacetyl-Lys-containing peptide (TFA-peptide). Substitution of the acetyl group with a trifluoroacetyl group is known to produce a peptide that binds to SIRT1, but turns over extremely slowly (Smith and Denu 2007).

To determine the mechanism of inhibition, initial velocities were determined as a function of $[X]_o$, at several fixed concentrations of inhibitor, with $[Y]_o = K_{m,Y} = \alpha K_Y$ for all plots. Each dependence of v_o on $[X]_o$ was individually fit to the Michaelis–Menten equation to arrive at values of $V_{X,obs}$ and $(V/K)_{X,obs}$. Replots were then constructed of $V_{X,obs}$ and $(V/K)_{X,obs}$ versus $[I]_o$.

Figure 8.9 contains the replots for inhibition of SIRT1 by the TFA-peptide. In panels *A* and *B*, we see that while $V_{p53,obs}$ is independent of $[TFA\text{-}peptide]_o$, $(V/K)_{p53,obs}$ is dependent on $[TFA\text{-}peptide]_o$, adhering to a simple inhibition curve. This pattern indicates that the TFA-peptide is a competitive inhibitor of the interaction of the peptide substrate with SIRT1. In contrast, both $V_{NAD,obs}$ and $(V/K)_{NAD,obs}$ are dependent on the TFA-peptide, indicating a competitive mechanism of inhibition for NAD⁺.

Similar experiments were done with the NAD⁺ analog ADPR. The results of these experiments, together with those for the TFA-peptide, are summarized in Table 8.3, and indicate noncompetitive inhibition of both the p53-7mer and NAD⁺ by ADPR.

As the summaries of Tables 8.1 and 8.2 indicate, this pattern suggests a random mechanism with ADPR binding to an enzyme:product complex. The mechanism proposed is shown in Figure 8.10, in which ADPR binds to the complex of SIRT1 and

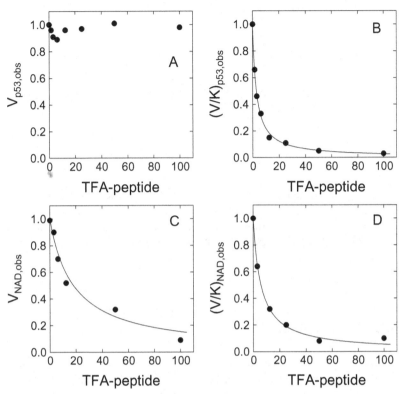

Figure 8.9. Replots for inhibition by substrate analog TFA-peptide of the SIRT1-catalyzed deacetylation of Ac-Arg-His-Lys-Lys^{Ac}-Leu-Nle-Phe-NH$_2$.

TABLE 8.3. Inhibition of SIRT1 by Substrate Analogs

Inhibitor	Substrate		Mechanism
	Varied	Fixed	
TFA-20mer	P53-7mer	NAD$^+$	C
TFA-20mer	NAD$^+$	p53-7mer	NC
ADPR	P53-7mer	NAD$^+$	NC
ADPR	NAD$^+$	p53-7mer	NC

deacetylated peptide product, requiring that release of the deacetylated product be at least partially rate-limiting. This mechanism is consistent with the chemical mechanism in which nicotinamide is the first substrate to depart from the active site. Note that random release of *O*-acetyl-ADP-ribose and deacetylated peptide would have given a pattern of inhibition in which TFA-peptide is a noncompetitive inhibitor of p53-7mer, since this inhibitor would be able to bind to the complex of SIRT1 and *O*-acetyl-ADP-ribose that would form in a random release mechanism.

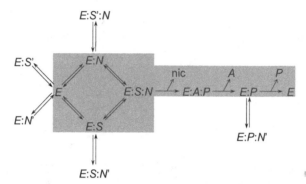

Figure 8.10. Mechanism of inhibition of SIRT1 by substrate analogs suggest random addition of substrates, and order release of products with deacetylated peptide departing last. The kinetic mechanism for substrate turnover is highlighted in the grey box. Abbreviations: S, p53-7mer; S′, TFA-peptide; N, NAD⁺; N′, ADPR; nic, nicotinamide; A, *O*-acetyl-ADDPR; and P, deacetylated peptide.

8.5   DRIVING SAR PROGRAMS FOR TWO-SUBSTRATE ENZYMATIC REACTIONS

Earlier in this chapter, I pointed out that determination of an inhibitor's kinetic mechanism provides information that could be used to optimize an inhibitor's binding affinity for a particular steady-state enzyme form. Admittedly, the idea of driving an SAR program with information that can only be gleaned from detailed mechanistic studies may seem beyond the scope of real-world drug discovery programs. However, I would like to suggest a workable research plan for a mechanism-driven SAR program:

1. Execute a high-throughput screen using an assay in which substrate concentrations are chosen to afford a balanced steady state (see Appendix C). Running the screen in this way will maximize the chances of identifying inhibitors of all mechanistic classes.
2. Determine the mechanism of inhibition of the most promising, validated hit(s). As usual in early-stage drug discovery programs, the criteria for choosing hits for further study will include potency, chemical/synthetic tractability, and structural novelty.
3. If the mechanistic studies reveal that one of the above hits binds to more than one steady-state form of the enzyme, two important decisions must be made. First, it must be decided if an SAR program should be initiated based on the hit. Second, if the inhibitor is to be pursued, the decision must be made as to enzyme form to which binding affinity of the inhibitor should be optimized. That is, if the inhibitor is found to bind to free enzyme, a complex of enzyme and substrate, and a complex of enzyme and product, it must be decided which of these binding modes will be optimized.

4. Finally, an assay must be devised with substrate concentrations and perhaps other conditions chosen to maximally increase the steady-state enzyme form chosen above. This assay will then be used to drive the SAR.

5. As the SAR matures and potency of inhibitors increase, mechanistic studies on one or more of the advanced compounds should be conducted to ensure that the desired binding mode has been optimized.

REFERENCES

Baur, J. A. (2010). "Resveratrol, sirtuins, and the promise of a DR mimetic." *Mech. Ageing Dev.* **131**: 261–269.

Guarente, L. (2007). "Sirtuins in aging and disease." *Cold Spring Harb. Symp. Quant. Biol.* **72**: 483–488.

Haigis, M. C. and D. Sinclair (2010). "Mammalian sirtuins: Biological insights and disease relevance." *Annu. Rev. Pathol.* **5**: 253–259.

Liu, M., et al. (2008). "Kinetic studies of Cdk5/p25 kinase: Phosphorylation of tau and complex inhibition by two prototype inhibitors." *Biochemistry* **47**: 8367–8377.

Roskoski, R. (2003). "STI-571: An anticancer protein-tyrosine kinase inhibitor." *Biochem. Biophys. Res. Commun.* **309**: 709–717.

Sauve, A. A., et al. (2001). "Chemistry of gene silencing: The mechanism of NAD-dependent deacetylation reactions." *Biochemistry* **40**: 15456–15463.

Sauve, A. A. and V. L. Schramm (2003). "Sir2 regulation by nicotinamide results from switching between base exchange and deacetylation chemistry." *Biochemistry* **42**: 9249–9256.

Sauve, A. A., et al. (2006). "The biochemistry of sirtuins." *Annu. Rev. Biochem.* **75**: 435–465.

Smith, B. C. and J. M. Denu (2007). "Mechanism-based inhibition of Sir2 deacetylases by thioacetyl-lysine peptine." *Biochemistry* **46**: 14478–14486.

9

ALLOSTERIC MODULATION OF ENZYME ACTIVITY

Perhaps the first observation of the allosteric modulation of an enzyme's catalytic activity was the demonstration by the Cori's together with Sidney Colowick that glycogen phosphorylase is activated by AMP (Cori et al. 1938). While a theoretical context was not yet available for their observation, they did understand that "the enzyme which forms the 1-ester from glycogen and inorganic phosphate is active only when *combined* with a nucleotide" (Cori et al. 1938, pp. 383–384; italics mine). That is, they understood the necessity for formation of a complex between enzyme and the effector molecule.

A theory to explain these observations was provided 25 years later by Monad, Changeux, and Jacob. This marked the introduction of enzyme allostery:

> In its most general form, the allosteric mechanism is defined by two statements:
> (1) No direct interaction of any kind need occur between the substrate of an allosteric protein and the regulatory metabolite which controls its activity. (2) The effect is entirely due to a reversible conformational alteration induced in the protein when it binds the specific effector. (Monad et al. 1963)

Kinetics of Enzyme Action: Essential Principles for Drug Hunters, First Edition. Ross L. Stein.
© 2011 John Wiley & Sons, Inc. Published 2011 by John Wiley & Sons, Inc.

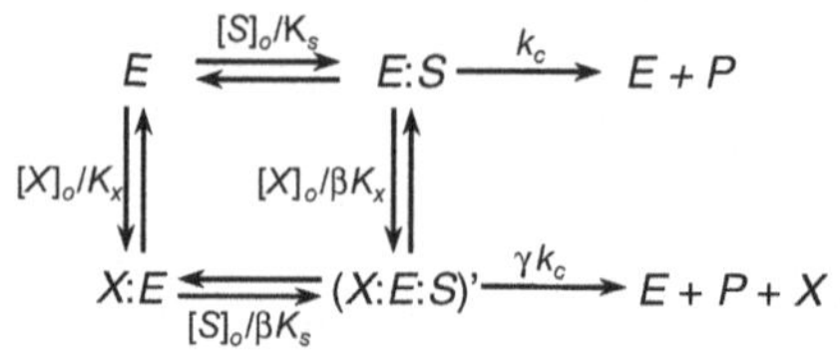

$$E \xrightleftharpoons{[S]_o/K_s} E{:}S \xrightarrow{k_c} E + P$$

$$[X]_o/K_x \updownarrow \qquad [X]_o/\beta K_x \updownarrow$$

$$X{:}E \xrightleftharpoons{[S]_o/\beta K_s} (X{:}E{:}S)' \xrightarrow{\gamma k_c} E + P + X$$

Figure 9.1. General mechanism for allosteric modulation of enzyme activity.

Since the initial discovery and description of enzyme allostery, the central role that this phenomena plays as a means of metabolic regulation has become increasingly clear. As any introductory biochemistry textbook will attest, hundreds of enzymes are subject to this mechanism of regulation. Also clear is the potential for drug discovery of the allosteric modulation of an enzyme's catalytic activity. Strategies have appeared both for the design and the discovery of allosteric modulators of kinases (Boboyevitch and Fairlie 2007; Lewis et al. 2008), proteases (Hauske et al. 2008), and enzymes in general (Hardy and Wells 2004). In a number of cases, these concepts have moved beyond theory into practice, with the introduction of allosteric modulators of enzyme activity into clinical use. For example, Nevirapine, which is used to treat HIV-1 infection, is a nonnucleoside reverse transcriptase inhibitor that works through an allosteric mechanism, and Gleevec, which is used to treat certain forms of cancer, is an allosteric inhibitor of Bcr-abl tyrosine kinase. In addition, several allosteric activators of glucokinase are in advanced preclinical development or are in clinical trials for the treatment of type II diabetes.

Descriptions of how such molecules modulate the catalytic activity of their target enzyme sometimes, but not always, require special kinetic treatments. For example, while Gleevec can be treated as a simple ATP-competitive inhibitor of Bcr-abl tyrosine kinase, activators of glucokinase require a much more sophisticated kinetic approach (Ralph et al. 2008). It is the purpose of this chapter to provide a description of enzyme allostery that can explain both the simple kinetic behavior of some modulators, as well as the more complex behavior of others.[1]

The kinetic formulations that we will develop in this chapter are based on the minimal mechanism for enzyme allostery shown in Figure 9.1. According to this mechanism, substrate S and modulator X can bind to different sites on enzyme E to form the catalytically competent ternary complex (XES)'. We will see in the last section of this chapter that this complex is designated (XES)', rather than XES, to signify that the former is produced as the result of a conformational isomerization of XES.

These kinetic discussions are prefaced by a description of the various mechanisms of enzyme modulation, in which allosteric and non-allosteric mechanisms of modulation are differentiated. From among the allosteric mechanisms, we identify those whose complex behavior require the kinetic formulation of the subsequent sections of this chapter.

---

[1] For a more comprehensive treatment of enzyme allostery, the reader should see Chapter 7 of Segel's book on enzyme kinetics, I. H. Segel, *Enzyme Kinetics* (New York: John Wiley & Sons, 1975).

9.1   MECHANISMS OF ENZYME MODULATION

Enzyme modulation can refer to either of two processes: inhibition or activation. Inhibition is any process in which modulator X is able to block the reaction pathway that produces product, while activation is any process in which X provides an alternate, more efficient pathway to product. As outlined below, these two broad categories of modulation can be broken down into specific mechanisms.

9.1.1   Inhibition: Blockade of Pathway to Product

- **Steric—X occupies active-site.** This is the mechanism of steric exclusivity that is implicit throughout this book. According to this mechanism, inhibition occurs when X occupies the same locus on the enzyme surface normally occupied by substrate.
- **Steric—X occupies allosteric site.** For enzymes requiring a bound effector molecule at an allosteric site for attainment of full catalytic activity, inhibition will be observed if X binds to the allosteric site and prevents occupancy of the site by the effector. This is not an allosteric mechanism, since when X binds to the allosteric site it does not induce an allosteric effect, but rather blocks one.
- **Allosteric—X occupies an allosteric site and triggers an activity-attenuating conformational change.** According to this mechanism, X cannot only bind to an allosteric site but, following binding, induces a conformational change[2] that causes substrate to bind with less affinity to the active site, or turnover more slowly to product, or both.
- **Allosteric, partial—allosteric inhibition with incomplete activity attenuation.** Here, the ternary complex of enzyme, substrate, and activator retains some fraction of the catalytic activity of the enzyme : substrate complex.

9.1.2   Activation: Provision of Alternate Pathway to Product

- **Chemical intervention—X diverts reaction intermediate.** X occupies a region of the active site that allows it to react with a reaction intermediate, with the liberation of free enzyme and an alternate reaction product.
- **Steric—X occupies inhibitor binding site.** According to this mechanism, X prevents binding of an antagonist by binding to the allosteric site to which the antagonist binds. Modulator X does not induce a conformation change of enzyme, and thus X behaves as a simple competitor of the antagonist.

---

[2] An alternate, but kinetically equivalent mechanism, states that rather than the conformational change following binding of X, that the conformation already exists as one of many the enzyme can adopt, and that X binds to this conformation and stabilizes it. Throughout this chapter, I will describe allosteric modulation in the context of the former mechanism.

- **Allosteric—X occupies allosteric site and triggers activity-enhancing conformational change.** X binds to an allosteric site and induces a conformational change that causes substrate to bind more strongly to the active site, or turnover more rapidly to product, or both.

9.2   KINETICS OF ALLOSTERIC MODULATION

The three allosteric mechanisms described above can all be explained in the context of the general mechanism of Figure 9.1. Derivation of the rate equation for this mechanism starts with the observation that the overall reaction rate can be described as the sum of the rates of the two pathways to product:

$$v_o = k_c[ES] + \gamma k_c[(XES)'].\tag{9.1}$$

As usual, we divide both sides by total enzyme, and obtain the expression of Equation 9.2:

$$\frac{v_o}{[E]_o} = \frac{k_c[ES]}{[E]+[XE]+[ES]+[(XES)']} + \frac{\gamma k_c[(XES)']}{[E]+[XE]+[ES]+[(XES)']}.\tag{9.2}$$

Division by the concentration of the appropriate catalytic species (i.e., $[ES]$ or $[(XES)']$) yields

$$\frac{v_o}{[E]_o} = \frac{k_c}{\dfrac{[E]}{[ES]}+\dfrac{[XE]}{[ES]}+1+\dfrac{[(XES)']}{[ES]}} + \frac{\gamma k_c}{\dfrac{[E]}{[(XES)']}+\dfrac{[XE]}{[(XES)']}+\dfrac{[ES]}{[(XES)']}+1}.\tag{9.3}$$

Substituting each ratio of enzyme species with a ratio of dissociation constants and $[S]_o$ or $[X]_o$, leads to Equation 9.4:

$$\frac{v_o}{[E]_o} = \frac{k_c}{\dfrac{K_s}{[S]_o}+\dfrac{K_s}{[S]_o}\dfrac{[X]_o}{K_x}+1+\dfrac{[X]_o}{\beta K_x}} + \frac{\gamma k_c}{\dfrac{K_x}{[X]_o}\dfrac{\beta K_s}{[S]_o}+\dfrac{\beta K_s}{[S]_o}+\dfrac{\beta K_x}{[X]_o}+1}\tag{9.4}$$

Collecting terms yields

$$\frac{v_o}{[E]_o} = \frac{k_c[S]_o}{K_s\left(1+\dfrac{[X]_o}{K_x}\right)+[S]_o\left(1+\dfrac{[X]_o}{\beta K_x}\right)} + \frac{\gamma k_c[S]_o}{\beta K_s\left(1+\dfrac{K_x}{[X]_o}\right)+[S]_o\left(1+\dfrac{\beta K_x}{[X]_o}\right)},\tag{9.5}$$

and finally,

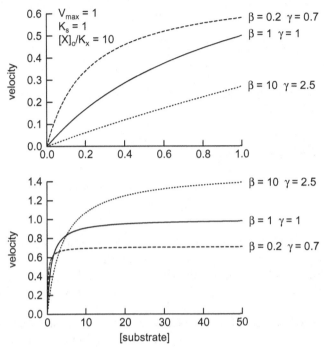

Figure 9.2. Simulations of dependencies of initial velocity on substrate concentration for the mechanism of Figure 9.1, according to Equation 9.6.

$$\frac{v_o}{[E]_o} = \frac{\dfrac{k_c}{1+\dfrac{[X]_o}{\beta K_x}}[S]_o}{K_s\left(\dfrac{1+\dfrac{[X]_o}{K_x}}{1+\dfrac{[X]_o}{\beta K_x}}\right)+[S]_o} + \frac{\dfrac{\gamma k_c}{1+\dfrac{\beta K_x}{[X]_o}}[S]_o}{\beta K_s\left(\dfrac{1+\dfrac{K_x}{[X]_o}}{1+\dfrac{\beta K_x}{[X]_o}}\right)+[S]_o}. \tag{9.6}$$

In Figure 9.2, I used the expression of Equation 9.6 to simulate dependencies of the initial velocity on substrate concentration, at the indicated values of α and β. In the upper panel of this figure, initial substrate concentrations are restricted to values less than K_s, while in the lower panel, $[S]_o \le 50K_s$.

Note how the mode of enzyme modulation by X is dependent on substrate concentration. When $\alpha = 0.2$ and $\beta = 0.7$, X behaves as an activator at low $[S]_o$, but behaves as an inhibitor at high $[S]_o$. Conversely, when $\alpha = 10$ and $\beta = 2.5$, X inhibits when $[S]_o$ is low, but activates when $[S]_o$ is high.

The origins of this bimodal behavior is better appreciated when Equation 9.6 is reexpressed with the condition $[X]_o \gg K_x$:

$$\frac{v_o}{[E]_o} = \frac{\gamma k_c [S]_o}{\beta K_s + [S]_o}.$$

(9.7)

At low substrate concentrations, Equation 9.7 becomes

$$\frac{v_o}{[E]_o} = \left(\frac{\gamma}{\beta}\right)\frac{k_c}{K_s}[S]_o$$

(9.8)

while at high substrate concentrations,

$$\frac{v_o}{[E]_o} = \gamma k_c.$$

(9.9)

We see that at low substrate concentrations the mode of modulation is governed by the ratio γ/β. If this ratio is greater than one, X will be seen to behave as an activator, but if γ/β is less than one, X will appear to be an inhibitor. In contrast, at high substrate concentrations, the mode of modulation is determined entirely by γ. If $\gamma > 1$, activation is observed, while if $\gamma < 1$, inhibition is observed.

In kinetic experiments where the mechanism of modulation is being determined, estimates of α and β can be obtained by generating a family of dependencies of v_o versus $[S]_o$ at several fixed values of $[X]_o$. These plots will all adhere to simple Michaelis–Menten kinetics with

$$\left(\frac{k_c}{K_m}\right)_{app} = \frac{k_c}{K_s}\left\{\frac{K_x + \dfrac{\gamma}{\beta}[X]_o}{K_x + [X]_o}\right\}$$

(9.10)

and

$$(k_c)_{app} = k_c\left\{\frac{\beta K_x + \gamma[X]_o}{\beta K_x + [X]_o}\right\}.$$

(9.11)

Equations 9.10 and 9.11 can then be used to analyze the two secondary plots that can be constructed, that is, $(k_c/K_m)_{app}$ versus $[X]_o$ and $(k_c)_{app}$ versus $[X]_o$. The dependence of $(k_c/K_m)_{app}$ on $[X]_o$ will be hyperbolic with a maximum of $(\gamma/\beta)(k_c/K_m)$ and half-maximal modulator concentration equal to K_x. Similarly, the dependence of $(k_c)_{app}$ on $[X]_o$ will be hyperbolic with a maximum of γk_c and half-maximal modulator concentration of βK_x. The estimate of γ comes directly from the secondary plot of $(k_c)_{app}$ versus $[X]_o$, while the estimate of β is obtained from the ratio of the half-maximal modulator concentrations from the two secondary plots.

There is, of course, another way to express the primary data, that is, as a family of dependencies of v_o on $[X]_o$, at various substrate concentrations. The rate equation for these curves is derived by first dividing v_x, the velocity in the presence of X, by v_o, the control velocity.

$$\frac{v_x}{v_o} = \frac{\dfrac{\dfrac{V_{max}}{1+\dfrac{[X]_o}{\beta K_x}}[S]_o}{K_s\left(\dfrac{1+\dfrac{[X]_o}{K_x}}{1+\dfrac{[X]_o}{\beta K_x}}\right)+[S]_o}}{\dfrac{V_{max}[S]_o}{K_s+[S]_o}} + \frac{\dfrac{\dfrac{\gamma V_{max}}{1+\dfrac{\beta K_x}{[X]_o}}[S]_o}{\beta K_s\left(\dfrac{1+\dfrac{K_x}{[X]_o}}{1+\dfrac{\beta K_x}{[X]_o}}\right)+[S]_o}}{\dfrac{V_{max}[S]_o}{K_s+[S]_o}} \tag{9.12}$$

The term $(K_s + [S]_o)/[S]_o$ can be factored out of the right side of Equation 9.12 to yield

$$\frac{v_x}{v_o} = \frac{K_s+[S]_o}{[S]_o}\left\{ \frac{\dfrac{1}{1+\dfrac{[X]_o}{K_x}}}{\dfrac{K_s}{[S]_o}\left(\dfrac{1+\dfrac{[X]_o}{\beta K_x}}{1+\dfrac{[X]_o}{K_x}}\right)+1} + \frac{\dfrac{\gamma}{1+\dfrac{\beta K_x}{[X]_o}}}{\dfrac{\beta K_s}{[S]_o}\left(\dfrac{1+\dfrac{K_x}{[X]_o}}{1+\dfrac{\beta K_x}{[X]_o}}\right)+1} \right\}. \tag{9.13}$$

Equation 9.13 can be rearranged to yield

$$\frac{v_x}{v_o} = \frac{K_s+[S]_o}{[S]_o}\left\{ \frac{1}{\dfrac{K_s}{[S]_o}\left(1+\dfrac{[X]_o}{K_x}\right)+\left(1+\dfrac{[X]_o}{\beta K_x}\right)} + \frac{\gamma}{\dfrac{\beta K_s}{[S]_o}\left(1+\dfrac{K_x}{[X]_o}\right)+\left(1+\dfrac{\beta K_x}{[X]_o}\right)} \right\} \tag{9.14}$$

and then,

$$\frac{v_x}{v_o} = \frac{1}{\dfrac{K_s}{K_s+[S]_o}\left(1+\dfrac{[X]_o}{K_x}\right)+\dfrac{[S]_o}{K_s+[S]_o}\left(1+\dfrac{[X]_o}{\beta K_x}\right)} + \frac{\gamma}{\dfrac{\alpha K_s}{K_s+[S]_o}\left(1+\dfrac{K_x}{[X]_o}\right)+\dfrac{[S]_o}{K_s+[S]_o}\left(1+\dfrac{\beta K_x}{[X]_o}\right)}. \tag{9.15}$$

If we then expand each denominator and collect terms, we arrive at

$$\frac{v_x}{v_o} = \frac{1}{1 + \dfrac{[X]_o}{K_x\left(1 + \dfrac{[S]_o}{K_s}\right)} + \dfrac{[X]_o}{\beta K_x\left(1 + \dfrac{K_z}{[X]_o}\right)}} + \frac{\gamma}{\dfrac{\beta K_s}{K_s + [S]_o} + \dfrac{[S]_o}{K_s + [S]_o} + \dfrac{\beta K_x}{[X]_o}} \qquad (9.16)$$

and finally

$$\frac{v_x}{v_o} = \frac{1}{1 + \dfrac{[X]_o}{\beta K_x\left(\dfrac{1 + \dfrac{K_s}{[S]_o}}{1 + \dfrac{\beta K_s}{[S]_o}}\right)}} + \frac{\gamma\left(\dfrac{1 + \dfrac{K_s}{[S]_o}}{1 + \dfrac{\beta K_s}{[S]_o}}\right)}{\beta K_x\left(\dfrac{1 + \dfrac{K_s}{[S]_o}}{1 + \dfrac{\beta K_s}{[S]_o}}\right)} \cdot \qquad (9.17)$$

Equation 9.17 can be simplified to Equation 9.18

$$\frac{v_x}{v_o} = \frac{1}{1 + \dfrac{[X]_o}{\beta K_{x,obs}}} + \frac{\gamma_{obs}}{1 + \dfrac{\beta K_{x,obs}}{[X]_o}} \qquad (9.18)$$

where

$$K_{x,obs} = K_x\left(\frac{1 + \dfrac{K_s}{[S]_o}}{1 + \dfrac{\beta K_s}{[S]_o}}\right) \qquad (9.19)$$

$$\gamma_{obs} = \gamma\left(\frac{1 + \dfrac{K_s}{[S]_o}}{1 + \dfrac{\beta K_s}{[S]_o}}\right) \cdot \qquad (9.20)$$

An interesting simplification of Equation 9.18 occurs when $\beta = 1$:

$$\frac{v_x}{v_o} = \frac{1}{1 + \dfrac{[X]_o}{K_x}} + \frac{\gamma}{1 + \dfrac{K_x}{[X]_o}} \cdot \qquad (9.21)$$

We see that under this special condition, the mode of modulation is independent of substrate concentration, and v_x/v_o simply titrates from 1 to γ, with dissociation constant K_x.

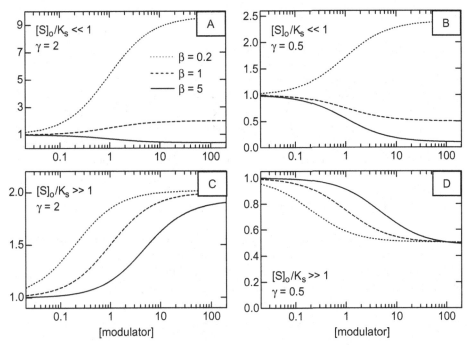

Figure 9.3. Simulations of dependencies of initial velocity on modulator concentration for the mechanism of Figure 9.1, according to Equation 9.17.

As can be seen from Equation 9.18, the shapes of modulator titration curves will depend on $[S]_o/K_s$, β, and γ. Figure 9.3 contains titration curve simulations in which these parameters have been systematically varied. Panels A and B correspond to cases in which $[S]_o/K_s \ll 1$, while panels C and D were drawn with $[S]_o/K_s \gg 1$. In all cases, $K_x = 1$.

To better understand the shapes of the curves in Figure 9.3, Equation 9.18 can be reexpressed under the two limits for $[S]_o/K_s$:

$$[S]_o/K_s \ll 1 \qquad \frac{v_x}{v_o} = \frac{1}{1+\dfrac{[X]_o}{K_x}} + \frac{\dfrac{\gamma}{\beta}}{1+\dfrac{K_x}{[X]_o}} \qquad (9.22)$$

$$[S]_o/K_s \gg 1 \qquad \frac{v_x}{v_o} = \frac{1}{1+\dfrac{[X]_o}{\beta K_x}} + \frac{\gamma}{1+\dfrac{\beta K_x}{[X]_o}} \qquad (9.23)$$

For panels A and B, where $[S]_o/K_s \ll 1$, Equation 9.22 tells us that the ratio γ/β entirely determines whether a modulator behaves as an activator or inhibitor. That is, the dependence of v_x/v_o on $[X]_o$ has a limiting value γ/β and a dissociation constant equal to K_x. In contrast, for panels C and D, where $[S]_o/K_s > 1$, Equation 9.23 tells us that γ determines the modality of modulation. The dependence of v_x/v_o on $[X]_o$ has a limiting value γ and a dissociation constant equal to βK_x.

In this section, we have seen from the simulations of Figures 9.2 and 9.3 that it is possible for an allosteric modulator to behave either as an activator or as an inhibitor, the mode of modulation being dependent on $[S]_o/K_s$, β, and γ. Thus, in characterizing enzyme modulators that are thought to work through allosteric mechanisms, it is imperative to conduct a thorough kinetic analysis, including an analysis of the dependence of velocity on both substrate and modulator concentration. Anything short of this can lead to inaccurate conclusions concerning the mode of modulation.

9.3   MEANING OF β AND γ

We saw in the first section of this chapter that for an allosteric modulator to exert its effect, it not only must bind to the enzyme but, upon binding, it must be able to trigger a conformational change that allows its potential as a modulator to be manifested. As we have seen, whether the mode of modulation is inhibition or activation depends on three factors: the ratio $[S]_o/K_s$, and the two system parameters β and γ. Providing an understanding of β and γ is the focus of this section.

Our discussion will take place in the context of Figure 9.4. This figure is an expanded version of Figure 9.1 that explicitly takes into account critical conformational changes. We see in Figure 9.4 that after enzyme binds modulator to form XE, a conformational change occurs that converts XE into (XE)′, which then binds substrate to form (XES)′. The thermodynamic equivalent pathway shown in this mechanism involves binding of S by enzyme to form ES, which then binds X to form XES. Finally, XES undergoes conformational isomerization to form (XES)′. Also in Figure 9.4, we see that product is formed via turnover of complexes ES, XES, and (XES)′, governed by first-order rate constants k_{es}, k_{xes}, and $k_{(XES)′}$, respectively.

According to this mechanism, modulation of binding affinity or activity occurs subsequent to the conformational isomerizations. This means that $K_{s,e} = K_{s,ex}$, $K_{x,e} = K_{x,es}$, and $k_{es} = k_{xes}$. The central, critical feature of this mechanism is the conformationally

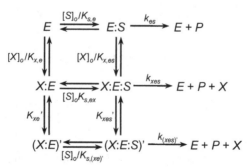

Figure 9.4. Expanded version of the mechanism of Figure 9.1 to explicitly include key conformational isomerizations.

altered ternary complex (XES)'. It is the nature of this complex, manifested in the magnitudes of $K_{s,(xe)'}$, K'_{xes}, and $k_{(XES)'}$, which determines whether the mode of modulation is complete inhibition, partial inhibition, or activation.

Our first step in understanding the meaning of β and γ is to undertake a kinetic comparison of the mechanisms of Figures 9.1 and 9.4. We will first consider γ.

9.3.1 The Meaning of γ

At saturating substrate concentrations, Equation 9.6 becomes

$$k_{c,obs} = \frac{k_c}{1 + \frac{[X]_o}{\beta K_x}} + \frac{\gamma k_c}{1 + \frac{\beta K_x}{[X]_o}} \tag{9.24}$$

From this we see that γ is the ratio of $k_{c,obs}$, in the presence of saturating concentrations of X, to $k_{c,obs}$, in the absence of X.

For the mechanism of Figure 9.4, we can derive the following expression for $k_{c,obs}$:

$$k_{c,obs} = \frac{k_{es}}{1 + \frac{[X]_o}{K_{x,es}} + \frac{[X]_o}{K_{x,es}} \frac{1}{K'_{xes}}} + \frac{k_{xes}}{\frac{K_{x,es}}{[X]_o} + 1 + \frac{1}{K'_{xes}}} + \frac{k_{(xes)'}}{\frac{K_{x,es}}{[X]_o} K'_{xes} + K'_{xes} + 1}. \tag{9.25}$$

In the presence of saturating concentrations of X, Equation 9.25 becomes

$$k_{c,obs} = \frac{k_{xes}}{1 + \frac{1}{K'_{xes}}} + \frac{k_{(xes)'}}{K'_{xes} + 1} \tag{9.26}$$

while in the absence of X, Equation 9.25 becomes

$$k_{c,obs} = k_{es}. \tag{9.27}$$

γ then is equal to

$$\gamma = \frac{k_{xes}}{k_{es}} \left(\frac{K'_{xes}}{K'_{xes} + 1} \right) + \frac{k_{(xes)'}}{k_{es}} \left(\frac{1}{K'_{xes} + 1} \right). \tag{9.28}$$

Since $k_{xes} = k_{ex}$,

$$\gamma = \frac{K'_{xes}}{K'_{xes} + 1} + \frac{k_{(xes)'}}{k_{es}} \left(\frac{1}{K'_{xes} + 1} \right), \tag{9.29}$$

which can be rearranged to produce

$$\gamma = \frac{K'_{xes} + \dfrac{k_{(xes)'}}{k_{es}}}{K'_{xes} + 1}. \tag{9.30}$$

We see from this equation that γ is not only a function of rate constants for the turnover of $E{:}S$ and $(X{:}E{:}S)'$, but also the equilibrium constant for conversion of $X{:}E{:}S$ and $(X{:}E{:}S)'$. If the equilibria favors formation of $(X{:}E{:}S)'$, then $\gamma = k_{(xes)'}/k_{es}$. On the other hand, if $X{:}E{:}S$ is favored, then $\gamma = 1$. We see then that γ reflects turnover of $(X{:}E{:}S)'$ only when the equilibria between $X{:}E{:}S$ and $(X{:}E{:}S)'$ favors the latter. Thus, the magnitude of k_{es} might be very different from that of $k_{(xes)'}$, but unless $K'_{xes} \ll 1$, this difference will not be manifested in the measurable steady state kinetic parameters for the system.

9.3.2 The Meaning of β

We can gain an understanding of the meaning of β if we set up equalities between the equivalent pathways of Figures 9.1 and 9.4. For the pathways that involve initial binding of X to E, we can equate the following:

$$K_x (\beta K_s) = K_{x,e} K'_{xe} K_{s,(xe)'}. \tag{9.31}$$

Since $K_x = K_{x,e}$,

$$\beta K_s = K'_{xe} K_{s,(xe)'}, \tag{9.32}$$

which can be rearranged to

$$\beta = K'_{xe}\left(\frac{K_{s,(xe)'}}{K_s}\right). \tag{9.33}$$

We see from Equation 9.33 that β is equal to the equilibrium constant for the conformational change that transforms $X{:}E$ into $(X{:}E{:}S)'$ multiplied by a term that expresses the relative magnitudes of the substrate dissociation constants for $X{:}E{:}S$ and $E{:}S$. If this ratio is close to unity, then β equals the equilibrium constant for the conformation change.

In similar fashion, we can analyze the pathway that involves initial binding of substrate. As before, an equality can be set up for the mechanisms of Figures 9.1 and 9.4:

$$K_s (\beta K_x) = K_{s,e} K_{x,e} K'_{xes}. \tag{9.34}$$

Since $K_s = K_{s,e}$,

$$\beta K_x = K_{x,e} K'_{xes}, \tag{9.35}$$

which can be rewritten as

$$\beta = K'_{xes} \left(\frac{K_{x,e}}{K_x} \right). \tag{9.36}$$

As before, we see that β is equal to the equilibrium constant for a critical conformational change multiplied by a term that express the relative magnitudes of two substrate dissociation constants. Again, if this ratio is close to unity, then β equals the equilibrium constant for the conformation change.

9.3.3 The Nature of $(X:E:S)'$ Determines the Mode and Extent of Allosteric Modulation

The analyses of the previous section reveals that

- β approximates the equilibrium constant for the conformational isomerizations that lead to the formation of $(X:E:S)'$.
- γ reflects the catalytic efficiency of $(X:E:S)'$, offset by a term that reflects the equilibrium between $X:E:S$ and $(X:E:S)'$.

Both β and γ are functions of the properties of $(X:E:S)'$ and thus, as I pointed out above, it is the nature of $(XES)'$ that determines the mode and extent of allosteric modulation.

A concept that will help us understand the nature of $(X:E:S)'$ is the idea that a receptor/ligand complex should not be construed simply as two molecular entities in association, but rather as a *single* molecular entity. It is more accurate to think of $X:E:S$, as a molecular unit, undergoing a conformational change to form $(X:E:S)'$, than it is to think of the enzyme-derived portion of $X:E:S$ undergoing a conformational change to form $X:E':S$. Once formed, the properties of $(X:E:S)'$ should be viewed as emergent properties of $(X:E:S)'$ as a molecular entity; properties that are not reducible to the sum of properties of modulator, substrate, and enzyme.[3]

From this discussion, we see that structural features of the substrate, as well of course as those of the modulator, will determine the properties of $(X:E:S)'$ and thus the mode and extent of modulation. This understanding of the nature $(XES)'$ has important implications for drug discovery programs trying to identify allosteric modulators

---

[3] The concept that a complex of ligand and receptor should be viewed as a single molecular entity is an ontological idea derived from systems theory and systems philosophy, as well as process philosophy. Readers with a philosophical bent are referred to the works of Ludwig von Bertalanffy, L. V. Bertalanffy, *General Systems Theory—Foundations, Development, Applications* (New York: George Braziller, 1968), and Ervin Laszlo E., Laszlo, *Introduction to Systems Philosophy—Toward a New Paradigm of Contemporary Thought* (New York: Harper & Row Publishers, 1972), on systems theory and systems philosophy, respectively, and a paper by the author that discusses enzymology in the context of process philosophy, R. L. Stein, "A Process Theory of Enzyme Catalytic Power—The Interplay of Science and Metaphysics," *Foundations of Chemistry* 8 (2006): 3–29.

of enzymes that have broad substrate specificities, such as kinases, proteases, and enzymes involved in posttranslational modification of proteins. Three predictions can be seen to follow:

1. The structure–activity relationship (SAR) for a series of allosteric modulators will be dependent, in part, on structural features of the substrate.
2. Whether a particular modulator behaves as an activator or an inhibitor will be dependent, in part, on structural features of the substrate.
3. Structural features of the transition state for turnover of (XES)′ will differ from those for turnover of XES.

These hypothesis-based predictions become expectations for drug discovery programs that are developing allosteric modulators for enzymes that work on multiple targets. It is clear that such programs must exercise caution in the interpretation of *in vitro* data based on reaction of the enzyme with a single substrate. Ideally, SAR data would be generated using the substrate that is most relevant to the disease or cellular process that has been targeted.

In the case studies below, we see that these predictions have been borne out.

9.4   CASE STUDIES: DEPENDENCE OF ALLOSTERIC MODULATION ON STRUCTURAL FEATURES OF THE SUBSTRATE

9.4.1   Prediction 1: Allosteric Activation of SIRT1

Sirtuins are enzymes that remove the acetyl moiety from ε-acetylated lysine residues of proteins (Sauve et al. 2006). The reaction these enzymes catalyze is NAD^+-dependent, generating nicotinamide and 2′-O-acetyl-ribose-ADP, along with the deacetylated Lys residue. Sirtuins are a family of enzymes with at least seven members (SIRT1–SIRT7) that all function as key regulators of cellular physiology. SIRT1 is of particular interest because of the central role it plays in many signaling cascades. In particular, it is a principal modulator of pathways that produce beneficial effects on glucose homeostasis and insulin sensitivity. Given this, it has been hypothesized that activators of SIRT1 might have therapeutic potential in type II diabetes, and perhaps other metabolic diseases (Milne et al. 2007). In 2003, the natural product resveratrol (Fig. 9.5) was reported to be an activator of this enzyme, and appeared to act by lowering K_m values for both NAD^+ and the Lys(Ac)-containing peptide Arg-His-Lys-Lys(Ac)-AMC (Howitz et al. 2003). Interestingly, subsequent studies revealed that activation is dependent on structural features of the substrate. Investigators found that resveratrol (100 μM) activated deacetylation of Arg-His-Lys-Lys(Ac)-AMC 5-fold, while only 1.4-fold for deacetylation of Lys-Gly-Gly-Ala-Lys(Ac)-AMC (Kaeberlein et al. 2005). No activation is seen if the AMC moiety is removed from the peptides.

Another study has also documented that activation of SIRT1 is dependent on substrate structure (Dai et al. 2010). In this study, a new series of SIRT1 activating compounds (STACs) are described that can activate the deacetylation of Ac-Arg-His-Lys-

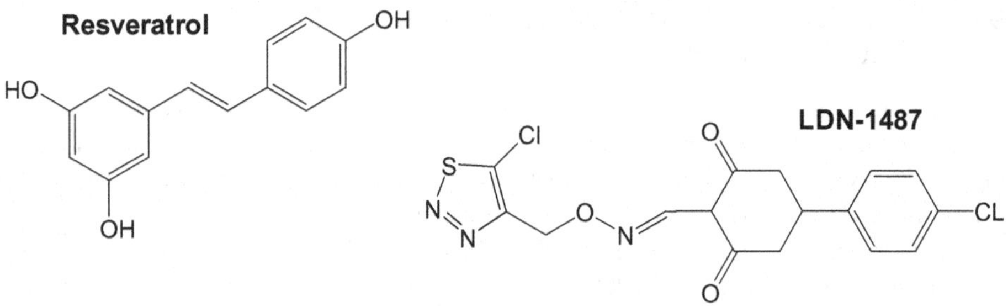

Figure 9.5. Structures of resveratrol, an activator of SIRT1, and LDN-1487, an activator of IDE.

Lys(Ac)-Xaa-NH$_2$ when Xaa is Trp or Phe, but not when Xaa is Ala. From these studies, it appears that SIRT1-catalyzed deacetylation is accelerated by STACs only when the peptide bears specific ring systems that were termed "activation cofactors." It is unclear how activation cofactors facilitate the activity of STACs. In broad strokes, this must occur by enhancing the binding of activator to enzyme:substrate complexes or by promoting the conformational change that produces (X:E:S)′, or both. Any mechanistic proposal for SIRT1 activation by STACs would have to include provisions that allow the activation cofactor to reside at a number of positions relative to the acetylated Lys residue of substrates. One can speculate that *in vivo* the activation cofactor might not reside on protein substrates but rather, may be presented to SIRT1 during interaction with accessory proteins with which SIRT1 interacts.

9.4.2 Prediction 1: Partial Inhibition of the Hydrolytic Activity of the Complex Tissue Factor:Factor VIIa by Peptides that Bind to a Factor VIIa Exosite

Initiation of the blood coagulation cascade occurs when tissue factor (TF), a nonvascular cell membrane protein, is exposed to circulating Factor VII (FVII), an inactive serine protease zymogen. Although blood and its constituents are not normally exposed to TF, upon vessel or tissue damage, FVII can bind to TF and become activated to FVIIa forming a TF:TVIIa complex. This complex binds and activates Factor IX (FIX) to FIXa, FX to FXa, and FVII to FVIIa, ultimately leading to the generation of thrombin.

In a 2001 publication, investigators at Genetech reported on a series of peptides, derived from phage display, that are apparently able to bind to an allosteric site on the TF:FVIIa complex and inhibit the hydrolysis of FX as well as the peptide *p*-nitroanilide substrate *N*-methylsulfonyl-(D)Phe-Gly-Arg-pNA (Dennis et al. 2001). One such peptide is A-183 and has the sequence EEWEVLCWTWETCER. The interaction of A-183 with TF:FVIIa brings about partial inhibition of FX hydrolysis with an IC$_{50}$ of 0.3 nM and a maximum extent of inhibition of 80%. While A-183 is also a partial inhibitor of the hydrolysis of *N*-methylsulfonyl-(D)Phe-Gly-Arg-pNA, the IC$_{50}$ for this system is 2 nM, and the full extent of inhibition is only 35%.

9.4.3 Predictions 1 and 2: Allosteric Modulation of Insulin-Degrading Enzyme (IDE)

IDE is another enzyme whose allosteric modulation is dependent on structural features of the substrate. This enzyme not only has a role in the catabolism of insulin, but is the principal enzyme involved in the degradation of amyloid fragments that play a causative role in Alzheimer's disease (AD) (Qiu and Folstein 2006).

Early studies on this enzyme revealed that its activity is allosterically modulated by ATP and other nucleotides. In 2005, Hersh and colleagues published a study in which this modulation was examined as function of substrate structure (Song et al. 2004). Some of the data from that study is summarized in the Table 9.1.

For ATP, we see activation of the hydrolysis of both the fluorogenic substrate and dynorphan, but no effect at all on the hydrolysis of insulin and $A\beta_{1\text{-}40}$. Also, note that the activation of dynorphan hydrolysis is 10 times that of the activation of hydrolysis of the fluorogenic substrate. CTP is seen to activate the hydrolysis of the fluorogenic substrate but inhibit the hydrolysis of $A\beta_{1\text{-}40}$. Finally, while triphosphate is similar to ATP in its activation and has similar effects on the hydrolysis of the fluorogenic substrate and dynorphan, it is an inhibitor of the hydrolysis of insulin and $A\beta_{1\text{-}40}$.

With its role in the catabolism of amyloid peptides, it was reasoned that activators of IDE might provide a treatment for AD since accelerated degradation of these peptides might inhibit plaque formation. A high-throughput screening campaign at Harvard's Laboratory for Drug Discovery in Neurodegeneration identified two classes of IDE activators (Cabrol et al. 2009). The structure of a representative from one of these classes is shown in Figure 9.5.

In studies to characterize this activation, assays involving two fluorogenic substrates were employed, one assay using 7-methoxycoumarin-3-acetyl-Gly-Gly-Phe-Leu-Arg-Lys-Val-Gln-Lys(2,4-dinitrophenyl)-NH₂ (1) and the other using 2-aminobenzoyl-Gly-Gly-Phe-Leu-Arg-Lys-His-Gln-2,4-dinitroanilide (2). Figure 9.6 is a plot of relative velocity versus [LDN-1487]. While solubility did not allow maximal activation to be reached, it is clear that the dose response curves are different for the two reactions, with the reaction involving substrate 1 showing appreciably more activation.

TABLE 9.1. Allosteric Modulation of IDE Is Substrate-Dependent

Modulator	$(k_c/K_m)_{\text{modulator}}/(k_c/K_m)_{\text{control}}$			
	Fluorogenic 9mer	Dynorphin B (1-9)	Insulin	$A\beta_{1\text{-}40}$
ATP	3.1	29	1.1	0.9
CTP	2.5	—	1.1	0.6
PPP$_i$	3.1	30	0.6	0.8

Modulator = 4 mM.

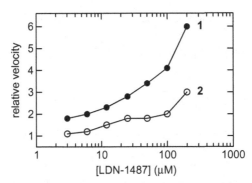

Figure 9.6. Dependence of relative velocity on the concentration of LDN-1487 for the IDE-catalyzed hydrolysis of peptide substrates **1** and **2**.

TABLE 9.2. Allosteric Modulation of Ribonucleotide Reductase Is Substrate-Dependent

Modulator	$v_{modulator}/v_{control}$		
	ATP	CTP	GTP
dGTP	5.3	0.2	0.33
dTTP	0.1	0.3	13

Substrate = 2 mM, modulator = 0.5 mM.

9.4.4 Predictions 1 and 2: Modulation of Ribonucleotide Reductase by Deoxyribonucleotides

The anaerobic ribonucleotide reductase from *Escherichia coli* catalyzes the reduction of ribonucleotides to deoxyribonucleotides. The activity this enzyme has toward ribonucleotides ATP, GTP, CTP, and UTP is allosterically modulated by the four deoxyribonucleotides dATP, dGTP, dCTP, and dTTP. Table 9.2 illustrates the mode of modulation by the deoxyribonucleotides is dependent on substrate structure (Eliasson et al. 1994).

Some of the data from that study is summarized in Table 9.2.

We see that while the modulator dGTP activates the reduction of ATP, it inhibits reduction of CTP and GTP. Similarly, while dTTP strongly activates reduction of GTP, it inhibits reduction of ATP and CTP.

9.4.5 Prediction 3: Allosteric Modulation of AMP Nucleosidase by ATP

AMP nucleosidase, which hydrolyzes the glycosidic bond of AMP to produce adenine and ribose, is allosterically activated by MgATP. Activation of AMP hydrolysis is between 10^2- and 10^3-fold depending on substrate concentration (Schramm 1974). Using kinetic isotope effects as their probe, Schramm and colleagues determined the transition state structures for the nucleosidase-catalyzed hydrolysis of AMP in the absence and presence of the allosteric activator MgATP (Mentch et al. 1987; Parkin and Schramm 1987) and found significant differences in transition state structure.

REFERENCES

Bertalanffy, L. V. (1968). *General Systems Theory—Foundations, Development, Applications.* New York, George Braziller.

Boboyevitch, M. A. and D. P. Fairlie (2007). "A new paradigm for protein kinase inhibition: Blocking phosphorylation without directly targeting ATP binding." *Drug Discov. Today* **12**: 622–633.

Cabrol, C., et al. (2009). "Small molecule activators of insulin degrading enzyme." *PLoS ONE* **4**: e5274.

Cori, G. T., et al. (1938). "The action of nucleotides in the disruptive phosphorylation of glycogen." *J. Biol. Chem.* **123**: 381–389.

Dai, H., et al. (2010). "SIRT1 activation by small molecules—Kinetic and biophysical evidence for direct interaction of enzyme and activator." *J. Biol. Chem.* **285**: 32695–32703.

Dennis, M. S., et al. (2001). "Selection and characterization of a new class of peptide exosite inhibitors of coagulation factor VIIa." *Biochemistry* **40**: 9513–9521.

Eliasson, R., et al. (1994). "Allosteric control of the substrate specificity of the anaerobic ribonucleotide reductase from *Escherichia coli*." *J. Biol. Chem.* **269**: 26052–26057.

Hardy, J. A. and J. A. Wells (2004). "Searching for new allosteric sites in enzymes." *Curr. Opin. Chem. Biol.* **14**: 706–715.

Hauske, P., et al. (2008). "Allosteric regulation of proteases." *Chembiochem* **9**: 2920–2928.

Howitz, K. T., et al. (2003). "Small molecule activators of sirtuins extend *Saccharomyces cerevisiae* lifespan." *Nature* **425**(6954): 191–196.

Kaeberlein, M., T., et al. (2005). "Substrate-specific activation of sirtuins by resveratrop." *J. Biol. Chem.* **280**: 17038–17045.

Laszlo, E. (1972). *Introduction to Systems Philosophy—Toward a New Paradigm of Contemporary Thought.* New York, Harper & Row Publishers.

Lewis, J. A., et al. (2008). "Allosteric modulation of kinases and GPCRs: Design principles and structural diversity." *Curr. Opin. Chem. Biol.* **12**: 269–280.

Mentch, F., et al. (1987). "Transition state structures for N-glycoside hydrolysis of AMP by acid adn by AMP nucleosidase in the presence and absence of allosteric activator." *Bichemistry* **26**: 921–930.

Milne, J. C., et al. (2007). "Small molecule activators of SIRT1 as therapeutics for the treatment of type 2 diabetes." *Nature* **450**(7170): 712–716.

Monad, J., et al. (1963). "Allosteric proteins and cellular control systems." *J. Mol. Biol.* **6**: 306–329.

Parkin, D. W. and V. L. Schramm (1987). "Catalytic and allosteric mechanism of AMP nucleosidase from primary, b-secondary, and multiple heavy atom kinetic isotope effects." *Biochemistry* **26**: 913–920.

Qiu, W. Q. and M. F. Folstein (2006). "Insulin, insulin-degrading enzyme and amyloid-beta peptide in Alzheimer's disease: Review and hypothesis." *Neurobiol. Aging* **27**: 190–198.

Ralph, E. C., et al. (2008). "Glucose modulation of glucokinase activation by small molecules." *Biochemistry* **47**: 5028–5036.

Sauve, A. A., et al. (2006). "The biochemistry of sirtuins." *Annu. Rev. Biochem.* **75**: 435–465.

Schramm, V. L. (1974). "Kinetic properties of allosteric adenosine monophosphate nucleosidase from *Azotobacter vinelandii*." *J. Biol. Chem.* **249**: 1729–1736.

Segel, I. H. (1975). *Enzyme Kinetics*. New York, John Wiley & Sons.

Song, E. S., et al. (2004). "ATP effects on insulin-degrading enzyme are mediated primarily through its triphosphate moiety." *J. Biol. Chem.* **279**: 54216–54220.

Stein, R. L. (2006). "A process theory of enzyme catalytic power—The interplay of science and metaphysics." *Found. Chem.* **8**: 3–29.

10

KINETICS-BASED PROBES OF MECHANISM

The kinetic methods described in previous chapters of this book answer specific questions regarding catalytic efficiency, binding affinity of substrates and modulators, and the steady-state enzyme forms to which these ligands bind. Taken together, these properties define an enzyme's kinetic behavior. However, kinetic behavior is not a primary property of an enzyme, but rather the manifestation of an enzyme's primary properties, which include active-site chemistry, the enzyme's means of stabilizing catalytic transition states, and the conformational isomerizations that are required for substrate turnover.

In this chapter, I introduce tools and methods that can be used to probe these and other enzymatic primary properties. In keeping with the rest of this book, these are all kinetics-based probes, and include pH, temperature, and viscosity dependencies, and kinetic isotope effects.

Kinetics of Enzyme Action: Essential Principles for Drug Hunters, First Edition. Ross L. Stein.
© 2011 John Wiley & Sons, Inc. Published 2011 by John Wiley & Sons, Inc.

10.1 pH DEPENDENCE OF ENZYMATIC REACTIONS

In this section, we discuss the experimental design, analysis, and interpretation of pH dependencies for enzymatic reactions. Before this discussion, however, it will be helpful to consider several general aspects of pH dependencies.

10.1.1 pH Dependencies of Enzymatic Reactions: General Considerations

10.1.1.1 Mechanistic Inadequacy of Plots of Initial Velocity versus pH. The pH dependence of an enzyme-catalyzed reaction can reveal a number of important mechanistic features, including pK_a values of ionizable amino acid residues that are important for catalysis, existence of intermediates, and alternate reaction pathways. To acquire this knowledge, the investigator must determine pH dependencies of the two limiting velocities V_{max} and $(V_{max}/K_m)[S]$, not merely the pH dependence of v_o at a single substrate concentration. The pH dependence of initial velocity reflects contributions from the pH dependencies of the two limiting velocities, and thus cannot readily be interpreted mechanistically.

10.1.1.2 Choice of Parameters Plotted in pH Dependencies. While pH dependencies of enzymatic reactions are often expressed as plots of $(k_c)_{obs}$ versus pH and $(k_c/K_m)_{obs}$ versus pH, this convention is not strictly correct. Changes in pH bring about changes in the distribution of possible ionized states of the enzyme and the substrate, not changes in rate constants. Rate constants are, by definition, insensitive to changes in pH because they reflect free energy differences between a specific reactant state and transition state. Ideally, the investigator would plot the pH dependencies of the two limiting velocities, $(V_{max})_{obs}$ and $\{(V_{max}/K_m)[S]_{standard-state}\}_{obs}$. Plotting the data in this way would emphasize the point that it is the ionization of enzyme and substrate that underlies pH dependencies of enzymatic reactions. But given that this is not common practice, the reader is advised to plot data as $(k_c)_{obs}$ versus pH and $(k_c/K_m)_{obs}$ versus pH. In the remainder of this section, pH dependencies will be expressed in this way.

10.1.1.3 pH Dependencies Reflect Reactant State Ionizations. The final general point to be made is that the pH dependence of a limiting rate constant reflects catalytically important ionizations of the reactant state species for the process governed by that rate constant.

- **pH dependence of k_c/K_m** not only reflects catalytically important ionizations of free enzyme, but also the substrate. This means that special care must be taken in the interpretation of pH dependencies of k_c/K_m when the substrate can ionize in the pH range of the study, often between 4 and 10.
- **pH dependence of k_c** reflects ionizations of those species that accumulate in the steady state at saturating concentrations of substrate. These species include the Michaelis complex and/or reaction intermediates that populate the steady

state. The specific moieties that undergo ionization may not only be those of enzyme amino acid residues, but also ionizations of bound substrate, or product if product release is rate-limiting.

10.1.2 Experimental Design

10.1.2.1 Separating the Effects of pH on Catalysis from those Causing Irreversible Changes in Enzyme Structure. Enzymatic activity is dependent on pH for two reasons: ionizations of amino acid side chains that are important for catalysis and the inherent instability of protein structure at extremes of pH, leading to denaturation or aggregation. Since our concern is with determining the effects of pH on catalysis, separating the effects of pH on enzyme stability from those on catalysis is the essential first step in any investigation of the pH dependence of an enzyme catalyzed-reaction.

To identify the pH values at which irreversible protein changes occur, an experiment is conducted in which initial velocities, at a single concentration of substrate, are determined over a broad pH range, say from 3–11 (see below for choice of buffers). The plot of v_o versus pH will be bell-shape, characterized by a pH, or pH plateau, at which maximum activity is observed, flanked by decreases in activity at pH values less than and greater than this optimal pH. With this information in hand, the investigator incubates the enzyme at pH values of low activity for the length of time needed to gather enough data for an accurate initial velocity estimation. These incubation mixtures are then adjusted to the pH where maximum activity is observed. Diminution of activity in these post-incubation/neutralization measurements that is greater than 20% of the control velocity in which the enzyme has never seen extreme pH, indicates an irreversible change in the enzyme structure, and eliminates that particular pH from those that can be used in studying the pH dependence of catalysis.

10.1.2.2 Choice of Buffers. The buffer salts of choice are the 12 zwitterionic buffer salts developed by Norman Good in the 1960s (Good et al. 1966). These include the familiar MES, HEPES, and CAPS buffer salts, and span the pK_a range from 6.2 to 10.4. A typical pH dependence might use the following buffers (pK_a values are in parentheses): formate (3.8), acetate (4.8), MES (6.2), PIPES (6.8), HEPES (7.5), HEPBS (8.3), CHES (9.3), and CAPS (10.4). With these buffers, studies can be conducted in which pH is varied from 3 to 11.

Buffers salts should be present at a concentrations between 10 mM and 50 mM, and final buffer solutions should contain any additives (e.g., DTT, EDTA) that were determined, during assay development, to be necessary for enzyme activity (see Section 2.5.1.2). Finally, the ionic strength of the buffer solutions should all be the same, and equal to a value determined to be optimal.

Buffer "crossover" experiments should be conducted to detect buffer salt-specific activity perturbation. For example, in a pH-dependence study in which a HEPBS buffer is used at pH 7.5 and a HEPBS buffer is used at pH 8.0, an experiment should be conducted in which enzyme activity is measured in HEPES buffer at pH 8.0. Identical rates

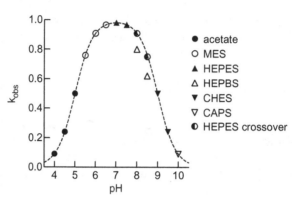

Figure 10.1. pH-rate profile illustrating buffer choices and the need for buffer crossover experiments.

in the two pH 8.0 buffers indicate the absence of buffer salt effects. However, if there is difference in reaction rate between the two solutions, further studies need to be conducted to understand these effects. Typically, this involves examining the entire pH dependence, including crossovers, with a view toward identifying outliers. This is illustrated in Figure 10.1.

In this figure, a bell-shaped dependence of $(k_c/K_m)_{obs}$ on pH is simulated. Data points at pH values of 8.0 and 8.5 were conducted in HEPBS buffer, with crossover using a HEPES buffer. The $(k_c/K_m)_{obs}$ values that were measured using the HEPBS buffer can be seen to be depressed relative to both the theoretical curve and the crossover values determined in HEPES, which conform to the theoretical line. In an actual experiment where this is observed, the entire data set, minus any values at pH 8.5 and 9.0, would be fit to the rate law for a bell-shaped pH dependence (see below) and a judgment is made concerning choice of which of the two buffers is more appropriate.

10.1.2.3 Data Collection. Initial velocities are determined as a function of substrate concentration over a range of pH, typically from 4–10 at intervals of 0.5 pH unit. Primary plots of $v_o/[E]_o$ versus $[S]_o$ are constructed and examined for any obvious departures from simple Michaelis–Menten kinetics. The data are then fit to the Michaelis–Menten equation to calculate estimates of pH-dependent values of k_c and k_c/K_m. Using these parameters, theoretical lines are drawn through the data, and the resulting plots are examined for systematic departure from the theoretical line. If simple kinetics are observed at all pH values, secondary plots are constructed: $k_{c,obs}$ versus pH and $(k_c/K_m)_{obs}$ versus pH. Analysis of secondary plots is described below in 10.1.3.

If non-Michaelian kinetics is observed, the data are analyzed using the rate equation that best describes the observed kinetic behavior (e.g., substrate inhibition or cooperativity). Secondary plots are then constructed that express the pH dependence of each of the model-specific parameters.

10.1.3 Analysis of pH Dependencies of Steady-State Kinetic Parameters

Analysis begins with an examination of secondary plots of $k_{c,obs}$ versus pH and $(k_c/K_m)_{obs}$ versus pH to identify the general shape (i.e., sigmoidal or bell-shaped) and the existence of any "fine structure" (see below). The three shapes that are typically observed, along with the rate law describing the pH dependencies, are as follows:

- **Sigmoidal with Activity Increasing with pH.** Achieving maximal activity is dependent on deprotonation of an acidic group (see Fig. 10.2A):

$$k_{obs} = \frac{k}{1 + \dfrac{[H^+]}{K_a}}. \tag{10.1}$$

- **Sigmoidal with Activity Decreasing with pH.** Achieving maximal activity is dependent on protonation of a basic group:

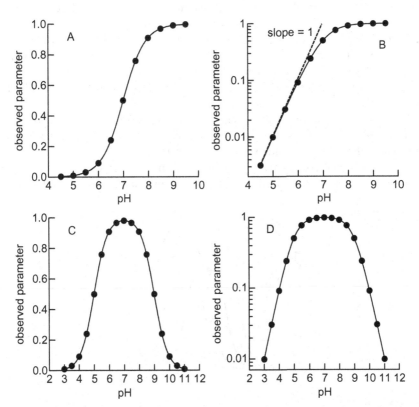

Figure 10.2. pH-rate profiles using linear-pH presentation (plots A and C) or log-pH presentation (plots B and D).

$$k_{obs} = \frac{k}{1 + \dfrac{K_a}{[H^+]}}. \tag{10.2}$$

- **Bell-Shaped.** Achieving maximal activity is dependent both on the deproton-ation of an acidic group and protonation of a basic group (see Fig. 10.2C):

$$k_{obs} = \frac{k}{1 + \dfrac{[H^+]}{K_{a,2}} + \dfrac{K_{a,1}}{[H^+]}}. \tag{10.3}$$

Note that these equations hold for both $k_{c,obs}$ and $(k_c/K_m)_{obs}$. Furthermore, the pH dependencies of $k_{c,obs}$ and $(k_c/K_m)_{obs}$ need not have the same shape.

There are two standard ways of plotting the dependence of k_{obs} on pH: k_{obs} versus pH or $\log(k_{obs})$ versus pH. In Figure 10.2A, B, are the two types of plots for a reaction in which activity increases as an acidic group is deprotonated. For simple situations, where there is no "fine structure" to the dependence (see below), the plot of $\log(k_{obs})$ versus pH is often preferred because the slope of the linear portion of the curve indicates the number of ionizable groups that contribute to the observed pH dependence. For the system simulated in Figure 10.2B, the linear portion of the curve at low pH has a slope of 1, indicating that full activity is attained when a single acidic group is deprotonated. The same holds true for the bell-shaped curves of Figure 10.2C, D. The legs of the bell the logarithmic plot of Figure 10.2D have slopes of 1 and −1, indicating that full activity is achieved when one group is deprotonated and another is protonated.

If a $\log(k_{obs})$ versus pH plot indicates that a single ionization is responsible for activity, the plot can be fit to one of the three rate equations above to calculate estimates of pK_a and k, the actual, pH-independent rate constant.

A few additional comments regarding bell-shaped pH dependencies are in order before moving on to the next section. Figure 10.3 shows simulations of bell-shaped pH

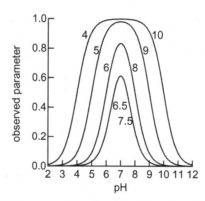

Figure 10.3. Bell-shaped pH-rate profiles illustrating how shape dependences on the difference between the two pK_a values that define the curve.

dependencies, in which, starting from the outermost curve, $pK_{a,1}$ and $pK_{a,2}$ values become increasingly close in magnitude. In these simulations, the maximal rate constant (i.e., k of Eq. 10.3) is 1.0. What is evident in this figure, is that only when these values differ by 4 or more pH units is the experimentally observed value of k_{limit} similar in magnitude to k. Estimates of $pK_{a,1}$ and $pK_{a,2}$ as well as k become increasingly less accurate as ΔpK_a decreases below 4, due to coupling of the three rate parameters in the rate expression of Equation 10.3.

10.1.4 Examples of pH Dependencies and Their Analyses and Interpretation

10.1.4.1 Building a Mechanism from pH Dependencies of kc,obs and (kc/Km)obs.

In 2003, a study was published with the goal of trying to understand the mechanistic origins of the substrate selectivity of α-chymotrypsin (Case and Stein 2003). For α-chymotrypsin and other serine proteases, catalytic efficiency of the enzyme increases as the length of the peptide substrate increases. In this study, three substrates were studied: Suc-Phe-pNA, Suc-Ala-Phe-pNA, and Suc-Ala-Ala-Pro-Phe-pNA (Suc, N-succinyl; pNA, p-nitroanilide), with k_c/K_m values of 29, 233, and 852,000 $M^{-1} s^{-1}$, respectively. One of the mechanistic probes used in this study was the pH dependence of catalysis. Here, we will look at the pH dependence of one of these substrates, Suc-Ala-Phe-pNA.

Figure 10.3 shows the pH dependencies of $k_{c,obs}$ and $(k_c/K_m)_{obs}$ for the α-chymotrypsin catalyzed hydrolysis of Suc-Ala-Phe-pNA. Note that while the pH dependence of $(k_c/K_m)_{obs}$ is bell-shaped with pK_a values of 6.45 ± 0.04 and 9.69 ± 0.04, and a pH-independent value of k_c/K_m equals to $233 \pm 5 M^{-1} s^{-1}$, the pH dependence of $k_{c,obs}$ is sigmoidal with a pK_a of 7.4 ± 0.1, and a pH-independent value of k_c equal to $0.91 \pm 0.03 s^{-1}$. The question before us now is how to construct an overall mechanism for this reaction using these results. We start with another result from this paper: Suc-Ala-Phe-pNA is hydrolyzed by α-chymotrypsin with k_c rate-limited by acylation (i.e., $k_{acylation} = 0.9 s^{-1} \ll k_{deacylation} = 42 s^{-1}$). This means that the pH dependence of $k_{c,obs}$ reflects ionization of the Michaelis complex and not the acyl–enzyme, and further that $K_m = K_s$.

In Figure 10.5A is the mechanism describing the pH dependence of $(k_c/K_m)_{obs}$ for the α-chymotrypsin catalyzed hydrolysis of Suc-Ala-Phe-pNA. In it, we see that the doubly protonated, inactive species EH_2 undergoes ionization with pK_a 6.5 to produce the active species EH, which can undergo ionization with pK_a 9.7 to produce inactive E. Given the pH dependence of $k_{c,obs}$, this mechanism is expanded to that shown in Figure 10.5B. In this mechanism, the Michaelis complex can exist as the inactive species $EH_2 : S$ or the active EH : S.

In this analysis, the rate constants calculated are pH-independent values and thus reflect the catalytic efficacy of the active ionized states of α-chymotrypsin. Given that the pK_a of the N-terminal blocking group succinyl is about 4.5, the pK_a values corresponded to catalytically important amino acid residues of the enzyme.

Based on a variety of biophysical studies, it has been possible to identify the imidazole moiety of the active site His-57 as the group with the pK_a around 6.5. Given its

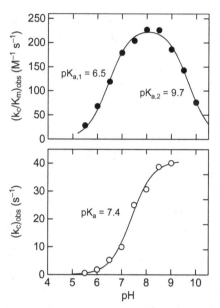

Figure 10.4. pH-rate profiles for the α-chymotrypsin-catalyzed hydrolysis of Suc-Ala-Phe-pNA.

A. Mechanism for the pH Dependence of k_c/K_m

$$EH_2$$
$$[H^+]_o/K_{a,E2} \Updownarrow$$
$$EH \xrightarrow{(k_c/K_m)[S]_o} EH + P$$
$$[H^+]_o/K_{a,E1} \Updownarrow$$
$$E$$

B. Combined Mechanism for the pH Dependencies of k_c/K_m, and k_c

$$EH_2 \underset{}{\overset{[S]_o/K_{s,2}}{\rightleftharpoons}} EH_2{:}S$$
$$[H^+]_o/K_{a,E2} \Updownarrow \qquad [H^+]_o/K_{a,E2S} \Updownarrow$$
$$EH \underset{}{\overset{[S]_o/K_{s,1}}{\rightleftharpoons}} EH{:}S \xrightarrow{k_c} EH + P$$
$$[H^+]_o/K_{a,E1} \Updownarrow$$
$$E$$

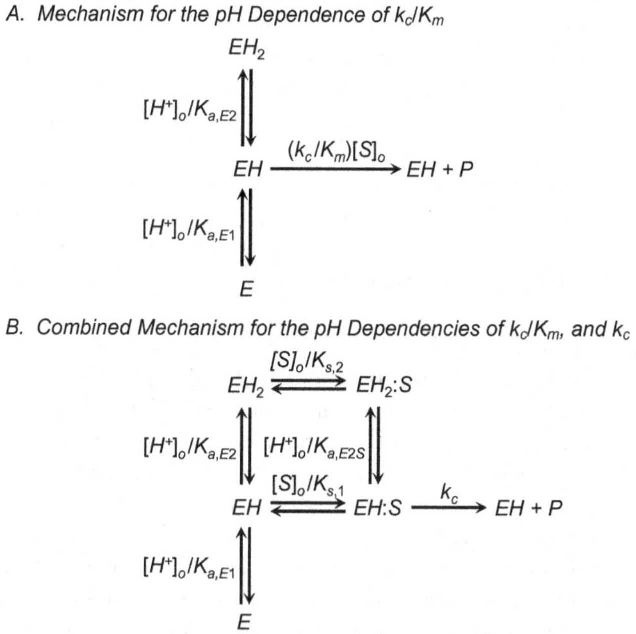

Figure 10.5. Kinetic mechanisms for the pH dependencies of the α-chymotrypsin-catalyzed hydrolysis of Suc-Ala-Phe-pNA. *Panel A:* Minimal mechanisms that explain the dependence of k_c/K_m on pH. *Panel B:* Overall mechanism that combines the pH dependencies of both k_c/K_m and k_c.

assumed role as a general-base catalyst, protonation of His-57 would be expected to generate an inactive form of enzyme. Likewise, the pK_a around 10 has been identified with disruption of an ion pair (Bender and Killgeffer 1973) between the amino group of Ile-16 and the carboxyl of Asp-194. This ion pair resides within the nonpolar interior of the protein upon activation of the zymogen and is critical for maintenance of the enzyme's catalytic activity.

10.1.4.2 Complex Mechanisms: The Existence of "Fine Structure" in pH Dependencies.

Penicillin-binding proteins (PBPs) are bacterial enzymes that catalyze two essential reactions in the biosynthesis of cell wall peptidoglycan from the glycopeptide precursor, Lipid II (1-3): transglycosylation, which polymerizes the disaccharides of Lipid II monomeric units into the backbone of the peptidoglycan, and transpeptidation, which crosslinks this backbone into a tight web-like structure. These two reactions are catalyzed by either low-molecular-weight, monofunctional PBPs or by certain high-molecular-weight PBPs that contain two independent catalytic domains for these activities.

PBPs derive their name from their facile interaction with penicillin and other β-lactam antibiotics (1-3). The transpeptidase domains of PBPs contain an active site serine that nucleophilically attacks the carbonyl carbon of the β-lactam ring of these antibiotics to form an acyl–enzyme that is stable and catalytically inactive. The chemistry of the transpeptidase active site has led to the development of structurally simple, thioester substrates, such as Bz-(D)Ala-^{S}Gly, which mimic the (D)Ala-(D)Ala bond that PBPs recognize in their natural substrate. Hydrolysis of Bz-(D)Ala-^{S}Gly is thought to proceed according to a mechanism involving the intermediacy of an acyl–enzyme.

In this section, we review a pH-dependence study that was conducted with PBP 2x, an enzyme that is essential for the viability of the pathogenic organism, *Streptococcus pneumoniae* (Thomas et al. 2001). In this study, the investigators determined the pH dependence of $(k_c/K_m)_{obs}$ for the hydrolysis of Bz-(D)Ala-^{S}Gly. This dependence is shown in Figure 10.6 and is bell-shaped with a shoulder on the alkaline side of the bell. The exis-

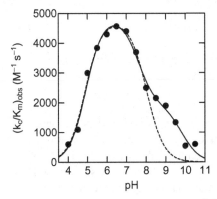

Figure 10.6. Dependence of k_c/K_m on pH for the PBPx-catalyzed hydrolysis of Bz-(D) Ala-^{S}Gly.

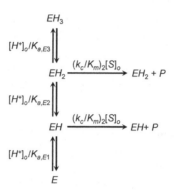

Figure 10.7. Minimal kinetic mechanism to explain the pH dependence of k_c/K_m on pH for the PBPx-catalyzed hydrolysis of of Bz-(D)Ala-^{5}Gly (see Fig. 10.6).

tence of a shoulder, and the necessity of including it in developing a mechanistic model for the pH dependence of this reaction, is highlighted by the dashed line, which was drawn using best-fit parameters to Equation 10.3: $k_c/K_m = 4.8 \pm 0.1\,\text{mM}^{-1}\text{s}^{-1}$, $pK_2 = 4.9 \pm 0.1$, $pK_1 = 8.0 \pm 0.1$. In this analysis, only data points from pH 4–8 were used.

The "upward" departure of observed values of k_c/K_m from the theoretical line based on a simple model incorporating two ionizable group, such as seen in Figure 10.5A, indicates that at these higher pH values an alternate route to products becomes available. Such a mechanism is shown in Figure 10.7, in which EH_2 and EH are catalytically active forms of PBP 2x and EH_3 and E are the catalytically inactive forms that predominate at extremes of acid and alkaline, respectively. The rate expression that describes the mechanism of 10.7 was derived using rapid equilibrium assumptions and is shown in Equation 10.4:

$$k_c/K_m = \frac{(k_c/K_m)_2}{\dfrac{[H^+]}{K_{a,3}} + 1 + \dfrac{K_{a,2}}{[H^+]} + \dfrac{K_{a,2}K_{a,1}}{[H^+]^2}} + \frac{(k_c/K_m)_1}{\dfrac{[H^+]^2}{K_{a,3}K_{a,2}} + \dfrac{[H^+]}{K_{a,2}} + 1 + \dfrac{K_{a,1}}{[H^+]}}. \quad (10.4)$$

Best fit parameters are: $(k_c/K_m)_2 = 4.9 \pm 0.2\,\text{mM}^{-1}\text{s}^{-1}$, $(k_c/K_m)_1 = 1.9 \pm 0.2\,\text{mM}^{-1}\text{s}^{-1}$, $pK_3 = 4.9 \pm 0.1$, $pK_2 = 7.6 \pm 0.2$, $pK_1 = 9.9 \pm 0.2$.

Protonation of an amino acid residue with pK_a of 5 and deprotonation of an amino acid residue with pK_a of 10 lead to completely inactive forms of the enzyme and likely are amino acids that play an active role in the chemistry of substrate hydrolysis. In contrast, deprotonation of a residue with pK_a of 7.6 leads to a form of PBP 2x that is still catalytically active, losing only about half its catalytic efficiency. While this residue is not critical in PBP 2x-catalyzed hydrolyses reactions, it plays a role of modulating the enzyme's activity. Given this role, the residue need not reside in the active site, but may in principle be anywhere on the enzyme surface, so long as its ionization leads to activity-altering conformational change. Together with data from a number of other mechanistic probes, a detailed mechanistic proposal has been offered by Thomas et al. (2001).

10.2 TEMPERATURE DEPENDENCE OF ENZYMATIC REACTIONS

The topics and organization of this section on the temperature dependence of enzyme-catalyzed reactions parallel those of the previous section. Here, we discuss the experimental design, analysis, and interpretation of temperature dependencies for enzymatic reactions, but first several general comments.

10.2.1 Temperature Dependencies of Enzymatic Reactions: General Considerations

10.2.1.1 Basic Interpretational Framework: Transition State Rate Theory. Transition state theory (see Chapter 3) provides the basis for analysis of the temperature dependence of enzymatic reactions. According to transition state theory, first-order rate constants for unimolecular processes have the following dependence on temperature, where $\Delta G^{\neq}$, κ, k_B, h, and R are the Gibbs free

$$k = \kappa \frac{k_B T}{h} \exp\left[-\frac{\Delta G^{\neq}}{RT} \right] \qquad (10.5)$$

energy of activation, the transmission coefficient (assumed here equal to 1), and the Boltzmann, Planck, and gas constants, respectively. For enzymatic reactions, k will be either k_c or $(k_c/K_m)[E]_{standard-state}$. Multiplication of k_c/K_m by a standard-state enzyme concentration allows k_c/K_m to be treated as a first-order rate constant, a requirement of Equation 10.5. $[E]_{standard-state}$ should be a value near the experimental enzyme concentration. Note that simply using k_c/K_m in the analyses implies that $[E]_{standard-state} = 1\,M$, and will lead to the calculation of unrealistically large values of $\Delta S^{\neq}$.

The expression of Equation 10.5 can be recast in the quasi-thermodynamic form of Equation 10.6,

$$k = \frac{k_B T}{h} \exp\left[-\left(\frac{\Delta H^{\neq}}{RT} - \frac{\Delta S^{\neq}}{R} \right) \right] \qquad (10.6)$$

where $\Delta H^{\neq}$ and $\Delta S^{\neq}$, are the enthalpy and entropy of activation, respectively. Finally, Equation 10.6 can be rearranged to Equation 10.7:

$$\ln\left[k\left(\frac{h}{k_B T} \right) \right] = -\frac{\Delta H^{\neq}}{RT} + \frac{\Delta S^{\neq}}{R}. \qquad (10.7)$$

Equation 10.7 predicts that a plot of $\ln[k(h/k_B T)]$ versus inverse temperature (i.e., an Eyring plot) will be linear slope and intercept equal to $\Delta H^{\neq}/R$ and $\Delta S^{\neq}/R$, respectively.

10.2.1.2 The Importance of pH in Temperature-Dependence Studies of Enzymatic Reactions. Temperature dependencies can reveal a number of mechanistically important features of enzyme-catalyzed reactions, including energies and enthalpies of activation, and the existence of intermediates and alternate reaction pathways. But to confidently draw mechanistic conclusion from these studies, it is crucial that pH-independent values of k_c and k_c/K_m be used in the analysis. If $k_{c,obs}$ or $(k_c/K_m)_{obs}$ are used in the analysis, the temperature dependence will not only reflect the temperature dependence of k_c or k_c/K_m, but also the temperature dependencies of pK_a values of catalytically important ionizations, rendering mechanistic conclusions ambiguous at best.

Determining pH-independent values of k_{obs} is not a problem, if the pH dependence of the reaction under study has a well-defined plateau. However, the situation is not so simple for bell-shaped pH dependencies where the two defining pK_a differ by less than 4 pH units, or sigmoidal pH dependencies in which rate constants on the plateau cannot be measured due to enzyme instability at the extremes of pH that define the plateau.

In cases where pH-independent values of k_{obs} cannot be directly measured, they may be calculated from the rate law that expresses the pH dependence of k_{obs}. For example, if the pH dependence of k_{obs} is bell-shaped, as expressed in Equation 10.3, then pH-independent k can be calculated at any pH according to

$$k = k_{obs}\left(1 + \frac{[H^+]}{K_{a,2}} + \frac{K_{a,1}}{[H^+]}\right). \tag{10.8}$$

What makes this method not as straightforward as it may first seem is that $pK_{a,1}$ and $pK_{a,2}$ are themselves temperature-dependent, as noted above. This of course requires the investigator to determine pH dependencies at each of the experimental temperatures, allowing calculation pH-independent rate constants at each temperature.

An alternate to this admittedly labor-intensive approach is to calculate pH dependencies at the two limiting temperatures and a temperature midway between them, say, 5, 25, and 45°C. If the defining pK_a values are relatively insensitive to temperature, the investigator can proceed with the method discussed above of calculating pH-independent rate constants at each temperature.

10.2.2 Experimental Design

10.2.2.1 Separating the Effects of Temperature on Catalysis from Those Causing Irreversible Changes in Enzyme Structure. Temperature not only affects rate constants for all the steps of an enzymatic reaction, but also the structure of the enzyme, with high temperature leading to denaturation or aggregation. In cases where lnk is linearly dependent on 1/T, the highest experimental temperature must be below temperatures at which the enzyme denatures and the plot solely reflects temperature effects on enzyme reaction steps, including substrate binding, active site chemistry, product release, and any reversible conformational changes. However, if the plot of lnk versus 1/T is curved, with rate constants at high temperature falling below those predicted based on the data at lower temperature, it is critical to determine if the curvature is due to a feature of reaction mechanism or to irreversible denaturation.

To make this assessment, an experiment is conducted in which enzyme is incubated at the high temperatures in question for the length of time required to make an initial velocity determination. The solution is the cooled to the standard assay temperature and the initial velocity is measured. If the initial velocity is less than 20% of the control velocity in which the enzyme has never seen extreme temperatures, an irreversible change in the enzyme structure has likely occurred. This eliminates that particular temperature from those that can be used in studying the temperature dependence of catalysis.

10.2.2.2   *Effect of Temperature on Buffer pH.* Changes in temperature will cause a change in the pK_a of the buffer salt, which leads to temperature-dependent changes in pH of reaction solutions. For the Good buffers, this effect is relatively small with an average $\Delta pK_a/\Delta C°$ value of about −0.010. In contrast is the temperature dependence of Tris ionization, for which $\Delta pK_a/\Delta C° = -0.031$. If the temperature range for temperature dependence covers 5–45°C, the pK_a of the buffer salt will have changed as much as −0.5 pH units even if a Good buffer was used in study. This means if the pH of a solution is 7.0 at 45°C, it will be 7.4 for a Good buffer and 8.2 for a Tris buffer. The impact this will have on the measured value of k_{obs} will be dictated where the two limiting pH values lie on the plot of pH versus k_{obs}. If these pH values lie on a plateau, this will result in little change in pH. On the other hand, if these pH values are near the pK_a of the enzymatic reaction, this change will have a significant effect on k_{obs}. To avoid these sorts of problems, it is recommended that Good buffers be used, and that separate buffer solutions be prepared at each of the experimental temperatures.

10.2.2.3   *Data Collection.* At least five experimental temperatures should be chosen that span the widest possible range, typically from 5°C to 40–60°C. At each of these temperatures, the dependence of initial velocity on substrate concentration is then determined, and is used to create Michaelis–Menten plots for estimation of k_c and k_c/K_m values. Finally, using these temperature-dependent parameters, Eyring plots are constructed. If the resulting Eyring plots appear as if they may be curved, kinetic parameters at additional temperatures should be measured to create Eyring plots with better resolution, so that the suspected curvature can be verified and then better analyzed.

10.2.3   Analysis and Interpretation of Temperature Dependencies of Steady-State Kinetic Parameters

10.2.3.1   *Linear Eyring Plots.* Linear Eyring plots present no problems in their analysis. Simple linear regression of the data to Equation 10.7 yields best fit values for the activation parameters ΔH^* and ΔS^*. Often, the calculated value of ΔS^* is multiplied by a reference temperature (e.g., 25°C = 298°K) so the entropy of activation is in units of kcal/mol, the same units as the enthalpy of activation.

It should be borne in mind that the calculation of ΔS^* requires an extrapolation to a point very far removed from the experimental data, an aspect of analysis that is seldom mentioned (Cornish-Bowden 2002). This is illustrated in Figure 10.8, where I replotted published data for the temperature dependence of the

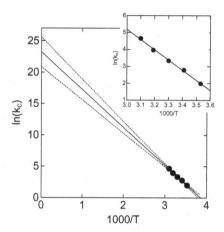

Figure 10.8. Eyring plot illustrating the error associated with the extrapolation to infinite temperature that is necessary to calculate $\Delta S^{\ddagger}$ values.

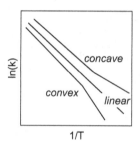

Figure 10.9. Classification of curved Eyring plots.

α-chymotrypsin-catalyzed hydrolysis of Suc-Ala-Phe-pNA (Case and Stein 2003). The inset is the data plotted in the conventional way, with the $1/T$ axis covering the range of the experimental temperatures. Visual inspection tells us that slope of the line, which is proportional to $\Delta H^{\ddagger}$, is well defined by these data, but does not tell us anything about y-intercept of the line and how well $\Delta S^{\ddagger}$ is defined. The main plot of Figure 10.8 illustrates that the y-intercept is not well defined at all. The dashed lines represent 95% confidence levels, meaning that $\ln(k_c)$ can have a value from about 21 to 26. This means that investigators should exercise caution in the interpretation $\Delta S^{\ddagger}$.

10.2.3.2 Nonlinear Eyring Plots.

Nonlinear Eyring plots for enzymatic reactions are not uncommon, and can be classified as concave or convex (see Fig. 10.9). While analyses of curved Eyring plots is not as straightforward as the analyses of linear plots, curved Eyring plots are much richer in mechanistic information.

Concave Eyring plots are typically caused by a reaction involving more than a single pathway to product. For examples, if Michaelis complex E:S can break down to products by two pathways governed by k_1 and k_2, the Eyring plot of k_c ($=k_1 + k_2$) will be concave unless the activation parameters are identical for the two parallel paths. Another cause of concave Eyring plots is a positive heat capacity for activation in a one step process.

Convex Eyring plots are more common than concave plots, and are generally thought to have one of at least four causes (Case and Stein 2003; Hengge and Stein 2004).

- For reactions proceeding through at least two sequential steps, convex plots will be observed if the rate-limiting step changes with temperature.
- For one step reactions, convex Eyring plots will occur if that step has a negative heat capacity of activation.
- Mechanisms involving fluctuations among enzyme conformers can generate nonlinear Eyring plots. These conformational isomers must exist as unstable intermediates on multiple, parallel reaction paths from the Michaelis complex to product, and these conformers must be in equilibrium with one another. If an equilibrium does not exist among the conformers, this mechanism reduces to one involving several independent paths to product and would generate concave Eyring plots. In essence, this third explanation represents a "pre-organization" mechanism in which the conformational change of the Michaelis complex might be viewed as adjusting the active site residues for optimal reaction with bound substrate.
- Convex Eyring plots can be caused by a temperature-sensitive coupling of an enzyme conformational change to the chemical step of catalysis; that is, the coupling of protein conformational isomerization to the reaction coordinate during the transformation of the Michaelis complex to product (Case and Stein 2003; Hengge and Stein 2004). This sort of coupling was discussed some decades back (Lumry and Biltonen 1969; Careri et al. 1979; Welch et al. 1982; Somogyi et al. 1984; Welch 1986), but was "rediscovered" in the early part of this century with the observation of the apparent coupling of protein motions to active site hydrogen tunneling in a number of hydrogen transfer reactions (Antoniou et al. 2002; Knapp and Klinman 2002).

10.2.4   Interpretation of Activation Parameters

Enzymatic reactions involve a number of temperature-dependent processes that can contribute to setting to magnitude of activation parameters: heavy atom rearrangements, conformational changes of substrate and enzyme, desolvation and establishment of hydrogen bonding networks, and changes in solvation of substrate and enzyme. It is essential to understand that each of these processes have both enthalpic and entropic elements. Because of this, interpreting experimentally determined values of $\Delta H^{\neq}$ and $\Delta S^{\neq}$ is problematic at best. Interpretation of the activation parameters for an enzymatic can only be done if those parameters can be put in some context, or if the range of mechanistic possibilities can be narrowed with independent data. To illustrate this point, we will consider the acylation of α-chymotrypsin.

TABLE 10.1. Activation Parameters for Acylation of α-Chymotrypsin[a]

Substrate	k_{acyl} (s^{-1})	$\Delta G^{\neq}$	$\Delta H^{\neq}$	$-T\Delta S^{\neq b}$
Suc-Phe-pNA	0.05	20	18	2
Suc-Ala-Phe-pNA	1.0	18	15	3
Suc-Ala-Ala-Pro-Phe-pNA	110	15	13	2

[a]Case and Stein (2003).
[b]T = 303 K.

10.2.4.1 Acylation of α-Chymotrypsin.

The acylation of α-chymotrypsin by Suc-Ala-Phe-pNA occurs with a rate constant of $1 s^{-1}$, and activation parameters: $\Delta G^{\neq} = 18 \, kcal \, mol^{-1}$, $\Delta H^{\neq} = 15 \, kcal \, mol^{-1}$, and $-T\Delta S^{\neq} = 3 \, kcal \, mol^{-1}$ (T = 303°K). In the absence of context, these parameters hold little mechanistic significance. However, activation parameters for acylation of α-chymotrypsin by a series of related substrates, including Suc-Ala-Phe-pNA (Table 10.1), reveals significant insights into the mechanism of this enzyme.

For α-CT and other serine proteases, the rate of reaction within the Michaelis complex to form the acyl–enzyme shows a dramatic dependence on peptide substrate chain length. In Table 10.1, we see that extending the substrate from Suc-Phe-pNA to Suc-Ala-Ala-Pro-Phe-pNA is accompanied by a 2300-fold increase in k_{acyl}, reflecting a $\Delta\Delta G^{\neq}$ of nearly $5 \, kcal \, mol^{-1}$. The temperature dependencies of k_{acyl} for reaction of three substrates reveal large values of $\Delta H^{\neq}$ that, overall, differ by about $5 \, kcal \, mol^{-1}$, but small and nearly identical values of $\Delta S^{\neq}$. Thus, the observed increase in catalytic efficiency that is observed when Suc-Phe-pNA is extended to Suc-Ala-Ala-Pro-Phe-pNA is entirely enthalpic in origin. Since identical chemical transformations occur for these substrates, the large observed enthalpy difference must reflect changes of the protein or protein:substrate complex that occur as the Michaelis complex is transformed into the activated complex for acylation.

One mechanism that can explain these results is catalysis by distortion. According to this mechanism, nucleophilic attack of the active site serine hydroxyl on the carbonyl carbon of the amide substrate can be facilitated if the scissile amide bond is twisted out its stable, planar conformation. This distortion disrupts the resonance stabilization of the amide and renders the carbonyl moiety of the amide more ester-like and more reactive toward nucleophilic attack. A serine protease can effect catalysis by distortion if it can support a mechanism in which reaction progress from the Michaelis complex to the activated complex for acylation includes the conformational isomerization of the active site to a geometry that can only accommodate a twisted scissile amide bond. This enzyme conformation is presumably one of a large ensemble of energetically similar conformations that are sampled by the enzyme and are stabilized, and thus "selected," by substrates that can engage in extended subsite interactions. We see then that extended substrates which can engage in subsite interactions react faster than shorter substrates for which these interactions are not available and, therefore, cannot select the most reactive conformers of the protease.

10.3 VISCOSITY DEPENDENCE OF ENZYMATIC REACTIONS

The topics and organization of this section on the viscosity dependence of enzyme-catalyzed reactions mirror those of the previous two sections. Here, we discuss the experimental design, analysis, and interpretation of viscosity dependencies for enzymatic reactions, but first several general comments.

10.3.1 Viscosity Dependencies of Enzymatic Reactions: Basic Interpretational Framework

Solvent viscosity can influence rates of enzyme-catalyzed reactions by two principle mechanisms:

1. Since molecular diffusion coefficients vary inversely with the viscosity of the medium, an increase in solvent viscosity will lead to decreases in rate constants for association and dissociation of enzyme with substrates, products, or other ligands (e.g., inhibitors). For this mechanism,

$$k_{obs} = \frac{k}{\eta_{rel}} \tag{10.9}$$

$$\frac{k}{k_{obs}} = \eta_{rel} \tag{10.10}$$

where η_{rel} is the relative viscosity, defined as the ratio of viscosity of the reaction solution in the presence of viscogen to the viscosity in the absence of viscogen.

2. Since solvent viscosity dampens structural fluctuations of proteins through frictional effects, increases in solvent viscosity will lead to decreases in reaction rates for processes that are coupled to enzyme structural fluctuations. For this mechanism,

$$k_{obs} = \frac{k}{\eta_{rel}^{\delta}} \tag{10.11}$$

$$\frac{k}{k_{obs}} = \eta_{rel}^{\delta} \tag{10.12}$$

where δ is a coupling constant that measures the sensitivity of a reaction to solvent viscosity. Theory predicts that δ can vary from 0 to 1 (Gavish 1980; Welch et al. 1982; Somogyi et al. 1984). δ will be 0 for reactions of enzymes whose active sites are largely uncoupled to solvent, while δ will be 1 for reactions of enzymes whose active sites are tightly coupled to solvent. Coupling of active sites and catalytic processes to solvent occurs through dynamic fluctuations of the enzyme.

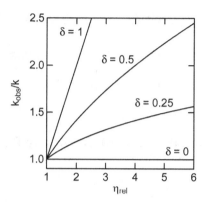

Figure 10.10. Dependence of relative velocity on solvent viscosity, at different values of δ, a coupling constant that measures the sensitivity of a reaction to solvent viscosity.

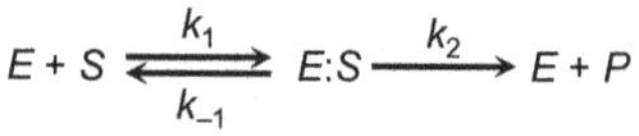

$$E + S \underset{k_{-1}}{\overset{k_1}{\rightleftharpoons}} E{:}S \xrightarrow{k_2} E + P$$

Figure 10.11. Steady-state mechanism for the enzymatic conversion of S into P.

When $\delta = 1$, the rate equations for the two mechanisms are of course identical. However, when $\delta < 1$, the two mechanisms can be differentiated, as illustrated in Figure 10.10. We see in this Figure that when $\delta = 1$, the dependence of k/k_{obs} on η_{rel} is linear, while at values of δ less than 1, this dependence is characterized by distinct curvature.

The above analysis holds for one-step reactions, or for multistep reactions in which one of the reaction steps is entirely rate-limiting. But, as we have seen in previous chapters, it is often the case in enzymatic reactions that the rate is not determined by a single reaction step. In such situations, observed viscosity dependencies can reflect contributions from both viscosity-sensitive and viscosity-insensitive reaction steps. To illustrate this, consider the reaction of Figure 10.11.

In this reaction, k_1 and k_{-1} will be sensitive to solvent viscosity due to diffusional effects on substrate binding and product release, and k_2, which is assumed to be entirely rate-limited by chemistry, may be sensitive to viscosity if active site chemistry is coupled to protein fluctuations. We will now consider the effect of viscosity on the process governed by k_c/K_m.

The rate law for the $(k_c/K_m)_{obs}$, abbreviated here as $k_{E,obs}$, is given in Equation 10.13,

$$k_{E,obs} = \frac{k_{1,obs}k_{2,obs}}{k_{-1,obs} + k_{2,obs}} \tag{10.13}$$

where

$$k_{1,obs} = \frac{k_1}{\eta_{rel}} \tag{10.14}$$

$$k_{-1,obs} = \frac{k_{-1}}{\eta_{rel}} \tag{10.15}$$

and

$$k_{2,obs} = \frac{k_2}{\eta_{rel}^{\delta}}. \tag{10.16}$$

Expressed as the ratio of k_E to $k_{E,obs}$, Equation 10.13 becomes

$$\frac{k_E}{k_{E,obs}} = \frac{k_1}{k_{1,obs}} \frac{k_2}{k_{2,obs}} \frac{k_{-1,obs} + k_{2,obs}}{k_{-1} + k_2} \tag{10.17}$$

$$\frac{k_E}{k_{E,obs}} = \eta_{rel} \eta_{rel}^{\delta} \frac{\eta_{rel}^{-1} + \eta_{rel}^{-\delta} \dfrac{k_2}{k_{-1}}}{1 + \dfrac{k_2}{k_{-1}}} \tag{10.18}$$

$$\frac{k_E}{k_{E,obs}} = \frac{\eta_{rel}^{\delta}}{1 + \dfrac{k_2}{k_{-1}}} + \frac{\eta_{rel} \dfrac{k_2}{k_{-1}}}{1 + \dfrac{k_2}{k_{-1}}}. \tag{10.19}$$

Equation 10.19 can be rearranged to

$$\frac{k_E}{k_{E,obs}} = \eta_{rel}^{\delta}(1 - f) + \eta_{rel} f \tag{10.20}$$

where

$$f = \frac{k_2}{k_{-1} + k_2} \tag{10.21}$$

f varies from 0 to 1, and reflects the extent to which k_E is rate-limited by substrate binding. For situations in which $f = 0$ (i.e., $k_{-1} \gg k_2$), substrate binding does not contribute to limiting k_E, while when $f = 1$ (i.e., $k_{-1} \ll k_2$), substrate binding is entirely rate-limiting. Figure 10.12 shows simulations according to Equation 10.20 for four values of f, at $\delta = 0$, 0.5, and 0.8. In these simulations, η_{rel} varies from 1 to 6, which is the upper limit for relative viscosities that can readily be attained experimentally (see below).

When $\delta = 0$, and the chemistry step governed by k_2 is not coupled to protein fluctuations, the dependencies of $k_E/k_{E,obs}$ on η_{rel} are linear with slopes equal to f, meaning that as rate limitation by substrate binding becomes less important, the k_E becomes increasingly less sensitive to viscosity. For $0 < \delta < 1$, coupling of active site chemistry

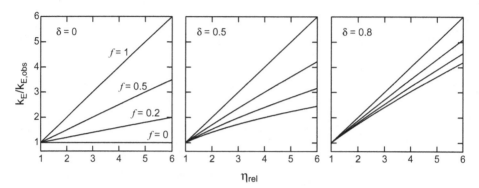

Figure 10.12. Simulations of the dependence of relative rate parameter $k_E/k_{E,obs}$ on viscosity as a function of four different f values at three $\delta = 0$, 0.5, and 0.8, where f is a partition factor (see text).

to protein fluctuations is a significant feature of the reaction mechanism. In these situations, both possible rate-limiting steps are sensitive to viscosity. When $f = 1$ and k_E is entirely rate-limited by substrate binding, the dependence of k_E on η_{rel} is linear with a slope of unity. At the other extreme where $f = 0$ and the k_E is entirely rate-limited by k_2, the dependence of k_E on η_{rel} exhibits curvature, reminiscent of the plots of Figure 10.10. At the two intermediate values of f, k_E is only partially rate-limited by k_2, and the curvature of the dependencies is less pronounced. We see then, that for $0 < \delta < 1$, relative rates at any η_{rel} are closer in magnitude to the relative rate at $f = 1$. Finally, when $\delta = 1$ (not shown), chemistry is fully coupled to protein fluctuations and dependencies of relative rate on η_{rel} are all linear with a slope of unity.

10.3.2 Experimental Design

The dependence of initial velocity on substrate concentration should be determined as a function of viscogen concentration. At each viscogen concentration, values of $k_{c,obs}$ and $(k_c/K_m)_{obs}$ should be calculated and plotted versus η_{rel}. If either of both of these parameters decrease with increasing concentration of a viscogen, it is important to determine if the observed dependence is related to increasing viscosity, or to some other phenomenon not related to viscosity, such as enzyme inhibition. To accomplish this, the investigator must conduct these experiments using two different viscogens, such as glycerol and sucrose. These viscogens are structurally distinct and, as illustrated in Figure 10.13, the dependence of η_{rel} on viscogen concentration is very different for two substances.

10.3.3 Interpretation of Data

The interpretation of viscosity dependencies is not at all straightforward. A linear dependence with a slope of unity can either arise from some diffusion controlled process

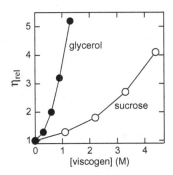

Figure 10.13. Dependence of viscosity on viscogen concentration, for viscogens glucose and sucrose.

(i.e., substrate binding for k_c/K_m, or product release for k_c) or from a mechanism with efficient coupling between active site chemistry and enzyme fluctuations. And while, in principle, dependencies that are less sensitive to viscosity can be analyzed for their departure from linearity, such departures may be too small relative to the error limits of k/k_{obs} values.

The only unambiguous way to interpret viscosity dependencies is in the context of independent experimental data, which addresses the issue of rate limitation. For example, if it can be shown through product inhibition studies or isotope effects (see below) that product release is rapid relative to chemistry, then the viscosity dependence of k_c can be interpreted to reflect coupling of chemistry to protein fluctuations.

Frequently, substrates of varying reactivities are used to allow interpretation of viscosity dependencies for k_c/K_m. The thought here is that if the reaction of poorly reactive substrates ($k_c/K_m < 10^5\,M^{-1}\,s^{-1}$) show no dependence on viscosity, while reaction of reactive substrates ($k_c/K_m > 10^6\,M^{-1}\,s^{-1}$) are viscosity-dependent, one may interpret the latter to reflect diffusional effects. However, one needs to exert caution here as well, for the subtle mechanistic changes that lead to variation in substrate reactivity may also lead to variation in dynamic coupling.

10.4   KINETIC ISOTOPE EFFECTS ON ENZYME-CATALYZED REACTIONS

A powerful tool for the study of enzyme mechanisms is the kinetic isotope effect, observed when rates of reactions are measured for molecules differing only in isotopic substitution. Isotope effects as probes of enzyme mechanism do not suffer from limitations inherent in many other mechanistic approaches. Since the potential energy surfaces for reactions differing only in the isotopic composition of reactants are identical (Born–Oppenheimer approximation), the mechanism and transition state structure may be probed without perturbing the reaction during the measurement itself.

In this section, we will consider isotope effects that arise from isotopic substitution of the substrate, as well as effects that result from isotopic substitution of the solvent.

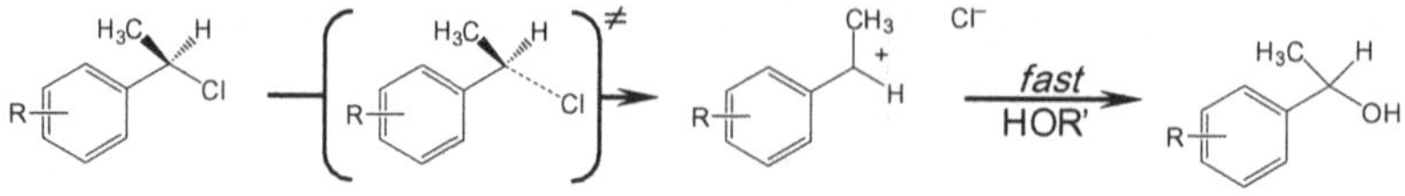

$$\text{I}^\circ \text{ Carbon KIE: } \frac{k_{C12}}{k_{C14}} = 1.020\text{--}1.060$$

$$\text{I}^\circ \text{ Chloride KIE: } \frac{k_{Cl35}}{k_{Cl37}} = 1.001\text{--}1.006$$

$$\text{II}^\circ \text{ }\alpha\text{D KIE: } \frac{k_{\alpha H}}{k_{\alpha D}} = 1.2\text{--}1.4$$

$$\text{II}^\circ \text{ }\beta\text{D KIE: } \frac{k_{\beta 3H}}{k_{\beta 3D}} = 1.00\text{--}1.20$$

Figure 10.14. Solvolysis of phenylethyl chlorides and the kinetic isotope that may be determined for this reaction.

The section ends with remarks concerning interpretation of isotope effects for enzymatic reactions.

10.4.1 Substrate Isotope Effects

Using solvolysis of phenethyl chloride as an example, Figure 10.14 illustrates the kinds of substrate isotope effects that can be measured for a reaction. Also included in this Figure are typical magnitudes for the various isotope effects for such a reaction.

When the isotopic substitution is to an atom that undergoes bond cleavage or formation, the observed effect is called a *primary isotope effect*. When the isotopic substitution is to an atom in the molecule that does not undergo bond cleavage or formation, the observed effect is called a *secondary kinetic effect*.

10.4.1.1 Primary Kinetic Isotope Effects. To understand the basis for primary isotope effects, we will consider the dissociation of the diatomic molecule, A–B, in which isotopic substitution is made in B (see Fig. 10.15). The zero-point energy is the lowest energy level for any molecule and, for A–B, may be approximated simply as $\frac{1}{2}hv$, where h is Planck's constant and v is vibrational frequency of the A–B bond. Since 99% of all molecules possess the zero-point energy level at room temperature, any energy difference between A–B and A–B' must result from differences in zero-point energies. Within the harmonic oscillator approximation for the vibration of the A–B bond, the difference in frequencies for this isotopic substitution is expressed as

$$\frac{v'}{v} = \sqrt{\frac{m_B(m_A + m_B)}{m_{B'}(m_A + m_B)}} = M \tag{10.22}$$

(where the m's are the atomic masses) and the difference in zero-point energies as

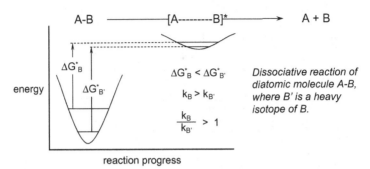

Figure 10.15. Origins of the primary isotope for dissociation of the diatomic molecule A-B.

$$E_{AB} - E_{AB'} = \frac{1}{2}hv(1-M).$$ (10.23)

We see that substitution with a heavy isotope leads to a lower zero-point energy.

These concepts as applied to the dissociation of A–B are illustrated in Figure 10.15. In the transition state for dissociation of A–B, the vibrational frequency approaches zero due to the near-zero bond order of the A–B in the transition state. This has the effect of causing E_{AB}–$E_{AB'}$ in the transition state to be very small. Since rate constants are dependent on the energy difference between a molecule in the ground state and transition state for a reaction, the overall effect is that $\Delta G_B^{\neq} < \Delta G_{B'}^{\neq}$, or $k_B > k_{B'}$. For more complex reactions, such as the solvolysis reaction of Figure 10.15, one need only consider the vibration of the bond being broken. Thus, in this reaction, a primary carbon effect is observed, because $\Delta G_{C12-Cl}^{\neq} < \Delta G_{C14-Cl}^{\neq}$.

10.4.1.2 Secondary Kinetic Isotope Effects.
While these effects can arise upon isotopic substitution of any atom within a molecule, the most common involve substitution of hydrogen, and these are the ones we will consider here. There are two types of secondary deuterium isotope effect: α-D and β-D isotope effects.

α-Deuterium isotope effects (expressed as the ratio of rate constants, k_H/k_D, and abbreviated as $^{\alpha D}k$) can be observed when a deuterium is substituted for one or more hydrogen atoms on a carbon atom undergoing reaction. α-D effects give information about out-of-plane bending vibrations of the C–H bond as the substrate proceeds from reactant state into the transition state. In the illustration of Figure 10.16, we see that as the C–X bond lengthens in the transition state, the reacting carbon becomes more planer. Thus, bending vibrations of the C–H bond are freer in the transition state than in the reactant state. As we saw in Figure 10.15, looser bonding leads to a shallower potential energy well. In these cases, $\Delta G_{C-H}^{\neq} < \Delta G_{C-D}^{\neq}$, and $^{\alpha D}k > 1$. These effects can range from 1.2 to 1.4.

β-Deuterium isotope effects ($^{\beta D}k$) can be observed when deuterium is substituted for one or more hydrogen atoms on a carbon that is bonded to an atom undergoing covalent bond changes in a reaction (see Fig. 10.16). βD isotope effects tell us about changes in hybridization at the β-carbon that occur as the substrate proceeds from the

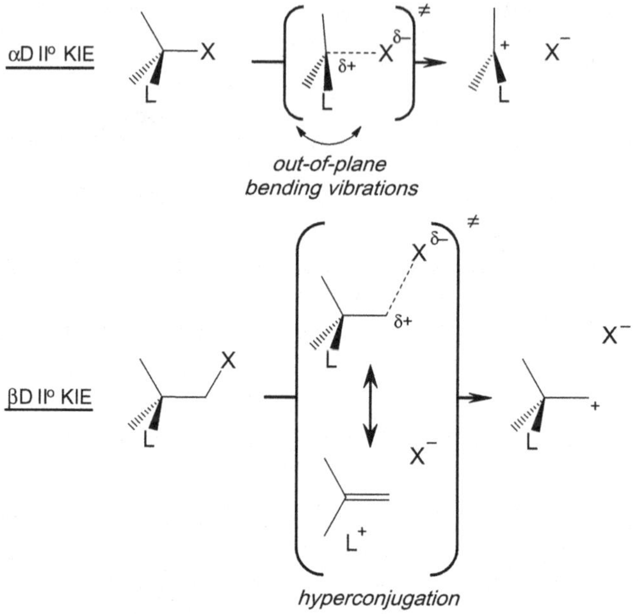

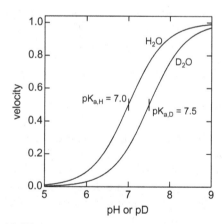

Figure 10.16. The two types of secondary deuterium isotope effects and their causes.

Figure 10.17. Dependence of reaction rate on pH and pD.

reactant state into the transition state. In proceeding to the transition state for the reaction of Figure 10.17, hyperconjugation from the βC–L bond is reduced, decreasing the electron density in the bonds and loosening them. As above, $\Delta G^{\neq}_{C-H} < \Delta G^{\neq}_{C-D}$, and $^{\alpha D}k > 1$. However, βD effects are smaller than αD effects and are typically no larger than about 6% per deuterium.

10.4.1.3 Measurement of Isotope Effects.

As the observed isotope effect decreases in magnitude, the more challenging its precise and accurate measurement becomes. Primary deuterium isotope effects, which can be as large as 7, or even larger if tunneling is involved, can easily be measured using standard techniques employed in the measurement of initial velocities. Secondary deuterium isotope effects can be measured directly if a continuous assay is available for the enzyme under study that allows initial velocities to be measured with standard deviations no larger than about 1%. If such an assay is not available, remote labeling techniques must be employed. For heavy atom isotope effects of C, O, N, S, and Cl, either remote labeling techniques or mass spectroscopy must be used. An excellent discussion of these measurement techniques can be found in Chapter 9 of a monograph by Cook and Cleland (Cook and Cleland 2007).

10.4.2 Solvent Deuterium Isotope Effects

Solvent isotope effects (SIE) are ratios of rate constants in H_2O and D_2O and reveal the transition state status of an enzyme's exchangeable hydrogenic sites (Quinn and Sutton 1991). Hydrogenic sites that produce a solvent isotope effect do so by forming so-called proton bridges with heteroatoms of the substrate, where the bridging proton can originate from either the substrate or the enzyme (Stein 2007). These sites include, but are not limited to, those that comprise an enzyme's catalytic machinery. Often, bridging protons of transition states are fully transferred. For example, during acylation of serine proteases, the proton residing on the hydroxyl of Ser[195] is completely transferred to N-1 of the imidazole of His[57] (α-chymotrypsin numbering) to form an imidazolium cation. However, full transfer of bridging protons does not always occur. Proton bridges that do not result in proton transfer can also be established in transition states. Again, for reactions of serine proteases, the proton of N-3 of His[57] forms a bridge to the carboxylate of Asp[102], but this proton is never fully transferred to Asp[102].

The magnitude of solvent deuterium isotope effects is of course dependent on the nature of the rate-limiting step and structural features of the transition state (Quinn and Sutton 1991). SIEs as large 4 can be observed for processes involving some form of general acid–base catalysis, as seen in reactions of serine proteases. Inverse effects of 0.5–0.8 are observed for reactions in which the rate-limiting step involves reaction of a sulfhydryl moiety, as in reactions of thiol proteases, or a metal-bound water molecule, as in reactions of metalloproteases. Solvent deuterium isotope effect have also been observed for conformational changes and product release (Zhou and Adams 1997; Vashishta et al. 2009).

10.4.2.1 Measurement of Solvent Isotope Effects. Care must be taken in the preparation of buffers for the determination of solvent isotope effects. First, there is a glass electrode effect, which, for solutions of D_2O, yields a pH-meter reading that is different from the actual pD of the solution, according to the relationship

$$pD = \text{meter reading} + 0.4. \tag{10.24}$$

Using this relationship, a series of pH/pD-matched solutions can be prepared and used to determine the dependencies of V_{max} and V_{max}/K_m on pH and pD for an enzymatic reaction. If the curves have well-defined plateaus and the data are precise enough, reliable estimates $^{D2O}(V_{max})$ and $^{D2O}(V_{max}/K_m)$ can be calculated directly from analysis of these curves.

Often it is impractical to determine full pL-rate profiles. In these cases, the investigator may choose to determine $^{D2O}k_c$ and $^{D2O}(k_c/K_m)$ at a single pH and pD equivalent, but not only must the investigator take into account the electrode effect but also the fact that the pK_a of active site amino acid residues are about 0.5 units higher in D_2O.

$$\Delta pK_a = pK_{a,D} - pK_{a,H} = 0.5 \tag{10.25}$$

Using Figure 10.17 as an example, if an investigator wants to conduct solvent isotope effects at the enzyme's pK_a, the H_2O solution would need to be buffered at a pH equal to the $pK_{a,H}$ of 7.0. The corresponding D_2O solution would need to have a pD equal to $pK_{a,HD}$, or pD = 7.5. This solution would be prepared such that the pH-meter reading equals pD−0.4, or 7.1. Taken together, this means that if a solution is to be buffered at the pD equivalent of a given pH, it would need to have a pH-meter reading that is 0.1 unit higher than the pH.

Estimates of the solvent isotope effects on k_c and k_c/K_m are obtained from initial velocity dependencies on substrate concentration for reactions conducted in appropriately buffered solutions of light and heavy water. $^{D2O}k_c$ and $^{D2O}(k_c/K_m)$ can then be calculated as the ratios of best-fit parameters from the two curves. Alternatively, if the same substrate concentrations are used for the reactions in H_2O and D_2O, $[S]_o$-dependent values of v_H/v_D can be calculated and then fit the expression of Equation 10.26, where $K_{m,H}$ is the K_m in light water

$$\frac{v_H}{v_D} = {}^{D2O}k_c \frac{[S]_o}{K_{m,H} + [S]_o} + {}^{D2O}(k_c/K_m) \frac{K_{m,H}}{K_{m,H} + [S]_o} \tag{10.26}$$

(Stein 1983; Stein 2002). This method has the advantage of giving direct estimates of the desired parameters.

10.4.2.2 Proton Inventory. The proton inventory is an extension of the solvent isotope effect, in which reaction velocities are measured in mixtures of isotopic waters. The shape of the dependence of velocity on mole fraction solvent deuterium, n_{D2O}, provides structural information about the transition state of the process under study (Venkatasubban and Schowen 1984; Stein 1985a,b; Stein et al. 1987), including the number of hydrogenic sites on the enzyme that form isotope effect-generating proton bridges in the transition state.

The starting point for the analysis of proton inventories is the Gross–Butler equation (Schowen 1978a; Venkatasubban and Schowen 1984):

$$k_n = k_0 \frac{\prod (1 - n + n\phi_i^T)}{\prod (1 - n + n\phi_j^R)} \tag{10.27}$$

where k_n is the rate constant at mole fraction solvent deuterium n, k_0 is the rate constant in pure H_2O (i.e., n = 0), and ϕ_i^T and ϕ_j^R are the isotopic fractionation factors for the ith and j[th] proton that is transferred in the transition state and reactant state, respectively.

Reactant state fractionation factors are often unity for enzyme-catalyzed reactions. The exceptions are enzymatic reactions that involve thiols or metal-bound water molecules in catalytic processes. For cases in which all $\phi_j^R = 1$, the Gross–Butler equation simplifies to:

$$k_n = k_0 \prod (1 - n + n\phi_i^T).$$ (10.28)

A final simplification is allowed by the fact that proton inventory data for enzymatic reactions is seldom of sufficient precision to allow estimation of the fractionation factors for individual protons. Values of ϕ_i are assumed to be approximately equal and Equation 10.28 can be recast as

$$k_n = k_0 (1 - n + n\phi^T)^i.$$ (10.29)

According to this expression, the shape of the proton inventory is dictated by the number active protons in the transition state (see Fig. 10.18). When i = 1, which corresponds to a single proton "in flight," the dependence of velocity on n_{D2O} is linear. Transition states in which i≥2 will produce concave or "bowl-shaped" proton inventories. The degree of downward curvature increases as the number of active proton increases.

In the limit of an infinite number of active protons (i.e., i≥3), Equation 10.29 takes on the form of Equation 10.30, where Φ^T is the product of all the individual fractionation factors, ϕ^T,

$$k_n = k_0 \left(\prod \phi^T \right)^n = k_0 (\Phi^T)^n = k_o \left(\frac{1}{SIE} \right)^n$$ (10.30)

which contribute to the isotope effect and is equal to the reciprocal of the observed solvent isotope effect.

The above analysis and the examples of Figure 10.18 assume a mechanism in which the chemical step is entirely rate-liming, reactant state fractionation factors are unity, and solvation effects accompanying substrate binding or product release are insignificant. For reactions where these assumptions do not hold, proton inventories with more complex shapes can arise. Analysis of such data has been discussed at length elsewhere (Venkatasubban and Schowen 1984; Stein 1985a,b; Stein et al. 1987; Quinn and Sutton 1991).

10.4.2.3 Interpretation of Isotope Effects for Enzymatic Reactions.
Kinetic isotope effects reveal structural features of a reaction's rate-limiting transition state. Often times, however, more than a single reaction step is rate-limiting. In such cases, the transition state is a "virtual transition state" (Schowen 1978b), a composite

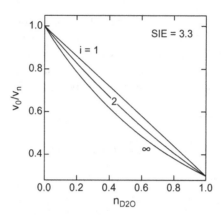

Figure 10.18. Proton inventories for a reaction with an overall solvent isotope effect 3.3. As the number of proton sites increases, so does the curvature.

reflecting structural features of the several real transition states, and the observed isotope effect is a weighted average of isotope effects for the steps that contribute to rate limitation (Stein 1981).

One hallmark of virtual transition states is that they can result in kinetic isotope effects of a magnitude that is smaller than predicted if chemistry were entirely rate-limiting. For example, in acyl-transfer reactions, βD kinetic isotope effects on the order of 4–6% per deuterium are expected due to loss of hyperconjugation that occurs as the acyl-moiety enters the transition state for its transfer, and the sp^2 hybridization of the carbonyl carbon becomes more sp^3-like. For such reactions, isotope effects less than 3% per deuterium can reflect either (1) a single transition state in which hybridization of the carbonyl carbon is largely sp^2 with only partial sp^3 character (i.e., either a very early or very late transition state), or (2) a virtual transition state comprising structural features of a trigonal transition state, with KIE of 6% per deuterium, and one or more transition states for isotope-insensitive reaction steps, such as conformation changes, substrate binding, or product release. To distinguish these alternatives, independent kinetic information is required that allows estimation of the extent to which the various steps contribute to rate limitation. With this data, the isotope effects on the various microscopic steps can be calculated (Stein 1981).

Another means to deconstruct a virtual transition state is to measure the isotope effect under various conditions that alter the reaction in some systematic manner. For example, $^{\alpha D}(k_c/K_m)$ for phosphorolysis of inosine catalyzed by purine nucleoside phosphorylase (PNPase) is dependent on pH as shown in Figure 10.19 (Stein and Cordes 1981). At neutral pH, where the enzyme is most active, an isotope effect of unity is observed. However, at extremes of pH, the full magnitude of the intrinsic isotope effect is expressed. These large αD isotope effects of 1.15–1.18 indicate a transition state in which the bending vibrations of the α-hydrogen are freer in the transition state than in the reactant state, and suggest the intermediacy of an oxocarbenium ion-like species, perhaps with some nucleophilic participation of the phosphate. At neutral pH, the isotope effect of unity reflects a virtual transition comprising the sp^2-hybridized transi-

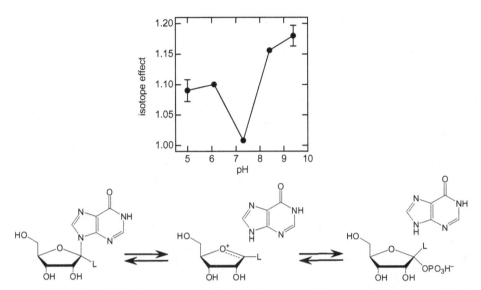

Figure 10.19. pH dependence of the α-deuterium isotope effect for purine nucleoside phosphorylase.

tion state of the chemical state and one or more nonchemical transition states, such as a conformational change or product release. At extremes of pH, the chemical step slows down and becomes rate-limiting, and the intrinsic isotope effect is expressed.

Altering substrate structure can also lead to changes in the structure of the virtual transition state. For example, solvent deuterium isotope effects were measured for the hydrolysis of series peptide substrates by human leukocyte elastase (Stein 1985b). The reactivity of these substrates toward HLE, as reflected in k_c/K_m values, ranged from $4.8 \times 10^3 \, M^{-1} \, s^{-1}$ to $6.8 \times 10^6 \, M^{-1} \, s^{-1}$. Significantly, $^{D2O}(k_c/K_m)$ decreased from 2.2 to 1.1 with this substrate-dependent increase in reactivity. In contrast, solvent isotope effects on k_c were very similar for the substrates with an average value of 2.8 ± 0.3. These results suggest that the observed solvent isotope effect reflects a virtual transition state comprising the transition state for acylation, with a large, normal isotope effect, and the transition state for an isotope-insensitive step, likely substrate binding. For the most reactive substrate, k_c/K_m is largely determined by substrate binding, with little contribution from acylation. As reactivity of the substrate decreases, the chemical step of acylation becomes slower and contributes more to rate limitation.

REFERENCES

Antoniou, D., et al. (2002). "Barrier passage and protein dynamics in enzymatically catalyzed reactions." *Eur. J. Biochem.* **269**: 3102–3112.

Bender, M. L. and J. V. Killgeffer (1973). "Chymotrypsins." *CRC Crit. Rev. Biochem.* **1**: 149–199.

Careri, G., et al. (1979). "Enzyme dynamics: The statistical physics approach." *Ann. Rev. Biophys. Bioeng.* **8**: 69–97.

Case, A. and R. L. Stein (2003). "Mechanistic origins of the substrate selectivity of serine proteases." *Biochemistry* **42**: 3335–3348.

Cook, P. F. and W. W. Cleland (2007). *Enzyme Kinetics and Mechanism.* London, Garland Science.

Cornish-Bowden, A. (2002). "Enthalpy-entropy compensation: A phantom phenomenon." *J. Biosci.* **27**: 121–126.

Gavish, B. (1980). "Position-dependent viscosity effects on rate constants." *Phys. Rev. Lett.* **44**: 1160.

Good, N. E., et al. (1966). "Hydrogen ion buffers for biological research." *Biochemistry* **5**: 467–477.

Hengge, A. C. and R. L. Stein (2004). "Role of protein conformational mobility in enzyme catalysis: Acylation of .alpha.-chymotrypsin by specific peptide substrates." *Biochemistry* **43**(3): 742–747.

Knapp, M. J. and J. P. Klinman (2002). "Environmentally coupled hydrogen tunneling—Linking catalysis to dynamics." *Eur. J. Biochem.* **269**: 3113–3121.

Lumry, R. and R. Biltonen (1969). "Thermodynamics and kinetic aspects of protein conformations in relation to physiological functions." In *Structure and Stability of Biological Macromolecules*, S. Timasheff and G. Fasman, eds. New York, Dekker.

Quinn, D. M. and L. D. Sutton (1991). Solvent isotope in enzymic reactions. In *Enzyme Mechanism from Isotope Effects*, P. F. Cook, ed. Boston, MA, CRC Press: 73–126.

Schowen, K. B. J. (1978a). Solvent hydrogen isotope effects. In *Transition States of Biochemical Processes*, R. D. Gandour and R. L. Schowen, eds. New York, Plenum Press: 225–284.

Schowen, R. L. (1978b). Catalytic power and transition-state stabilization. In *Transition States of Biochemical Processes*, R. D. Gandour and R. L. Schowen, eds. New York, Plenum Press: 77–114.

Somogyi, B., et al. (1984). "The dynamic basis of energy transduction in enzymes." *Biochim. Biophys. Acta.* **768**: 81–112.

Stein, R. L. (1981). "Analysis of kinetic isotope effects on complex reactions utilizing the concept of the virtual transition state." *J. Org. Chem.* **46**: 3328–3330.

Stein, R. L. (1983). "Catalysis by human leukocyte elastase. 1. Substrate structural dependence of rate-limiting protolytic catalysis and operation of the charge relay system." *J. Am. Chem. Soc.* **105**(15): 5111–5116.

Stein, R. L. (1985a). "Catalysis by human leukocyte elastase. 5. Structural features of the virtual transition state for acylation." *J. Am. Chem. Soc.* **107**(25): 7768–7769.

Stein, R. L. (1985b). "Catalysis by human leukocyte elastase. 4. Role of secondary-subsite interactions." *J. Am. Chem. Soc.* **107**: 5767–5775.

Stein, R. L. (2002). "Enzymatic hydrolysis of *p*-nitroacetanilide: Mechanistic studies of the aryl acylamidase from *Pseudomonas fluorescens*." *Biochemistry* **41**: 991–1000.

Stein, R. L. (2007). Proton transfer during catalysis by hydrolases. In *Hydrogen Transfer Reactions*, Vol. 4, J. T., Hynes, J. P. Klinman, H. H. Limbach, and R. L. Schowen, eds. Weinheim: Wiley-VCH Verlag: 1455–1472.

Stein, R. L. and E. H. Cordes (1981). "Kinetic a-deuterium isotope effects for *E. coli* purine nucleoside phosphorylase-catalyzed phosphorolysis of adenosine and inosine." *J. Biol. Chem.* **256**: 767–772.

Stein, R. L., et al. (1987). "Catalysis by human leukocyte elastase. 7. Proton inventory as a mechanistic probe." *Biochemistry* **26**: 1305–1314.

Thomas, B., et al. (2001). "Kinetic and mechanistic studies of penicillin-binding protein 2x from *Streptococcus pneumoniae*." *Biochemistry* **40**: 15811–15823.

Vashishta, A. K., et al. (2009). "Chemical mechanism of saccaropine reductase from *Saccharomyces cerevisiae*." *Biochemistry* **48**: 5899–5907.

Venkatasubban, K. S. and R. L. Schowen (1984). "The proton inventory technique." *CRC Crit. Rev. Biochem.* **17**: 1–44.

Welch, G. R. (1986). *The Fluctuating Enzyme*. New York, John Wiley & Sons.

Welch, G. R., et al. (1982). "The role of protein fluctuations in enzyme action: A review." *Prog. Biophys. Molec. Biol.* **39**: 109–146.

Zhou, J. and J. A. Adams (1997). "Is there a catalytic base in the active site of cAMP-dependent protein kinase?" *Biochemistry* **36**: 2977–2984.

APPENDIX A

BASIC PRINCIPLES OF CHEMICAL KINETICS

Chemical kinetics is the study of the time dependence of chemical reactions. The history of chemical kinetics is a distinct branch of chemistry that can be traced to the pioneering work of Ludwig Wilhelmy, who, in 1850, reported on the kinetics of the acid-catalyzed hydrolysis of sucrose to produce glucose and fructose (Wilhelmy 1850). In the quote below, Wilhelmy explains the goal of his studies.

> It is known that the action of acids on cane sugar converts it into fruit sugar. Since now with the aid of a polarization apparatus, readings of how far this change has proceeded can be made with great ease in an instant, it seemed to me to offer the possibility of finding the law of the process which we are discussing. (Wilhelmy 1850)

And, indeed, Wilhelmy did discover the law of this process. He found that the amount of sugar converted (dZ) in an element of time (dT) is proportional to the amount of sugar (Z) at time T and the amount of acid (S). He expressed this relationship as the differential equation shown below,

$$-\frac{dZ}{dT} = MZS \tag{A.1}$$

Kinetics of Enzyme Action: Essential Principles for Drug Hunters, First Edition. Ross L. Stein.
© 2011 John Wiley & Sons, Inc. Published 2011 by John Wiley & Sons, Inc.

where M was denoted a "velocity coefficient," which upon integration gave him the rate equation,

$$\log Z_o - \log Z = MST \tag{A.2}$$

where Z_o is the value of Z at the initiation of the reaction.

Just as important as this very specific discovery, Wilhelmy understood that the acid-catalyzed hydrolysis of cane sugar "is certainly only one member of a greater series of phenomena which all follow general laws of nature." He understood that chemical reactions proceed according to "the laws of the process," and further, through investigation of a reaction's time dependence, these laws can be discovered and expressed mathematically.

During the century and a half that has passed since Wilhelmy's discovery, chemical kinetics has matured into a sophisticated science that has become the principal tool of chemists interested in elucidating the mechanisms of chemical reactions. While there are many excellent treatments of chemical kinetics, such as *Kinetics and Mechanism* by Moore and Pearson (Moore and Pearson 1981), I nonetheless thought it of value to review some the basic principles here. My coverage is selective, focusing only on those areas that have relevance to this book's topic: the kinetics of enzyme action.

A.1   ONE-STEP, IRREVERSIBLE, UNIMOLECULAR REACTIONS

We start with the simplest of chemical reactions, a unimolecular reaction in which

$$A \xrightarrow{k_1} B$$

reactant A is transformed into product B.

The rate law for this reaction is given in Equation A.3, which tells us that the rate of disappearance of reactant A is equal to the rate of formation of product B,

$$\frac{-d[A]}{dt} = \frac{d[B]}{dt} = k_1[A] \tag{A.3}$$

both of which equal the concentration of A multiplied by the proportionality constant k_1. Since the reaction rate is proportional to the concentration of a single reactant raised to the first power, the reaction is called a first-order reaction. k_1 is a first-order rate constant with units of reciprocal time.

The differential form of this rate law appears in Equation A.4:

$$-d[A] = k_1[A]dt. \tag{A.4}$$

Integration of this equation yields

$$\int \frac{d[A]}{[A]} = -k_1 \int dt \tag{A.5}$$

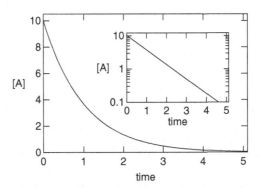

Figure A.1. Progress curve simulation of the first-order conversion of reactant A into product C. Inset is a plot of the same simulation with a logarithmic y-axis. $k_1 = 1$.

$$\ln[A] - C = -k_1 t. \tag{A.6}$$

The constant of integration C can be evaluated using the boundary condition that when t = 0,

$$\ln[A]_o - C = -k_1(0) \tag{A.7}$$

and thus

$$C = \ln[A]_o. \tag{A.8}$$

Substituting Equation A.8 into Equation A.6 leads to

$$\ln\left(\frac{[A]}{[A]_o}\right) = -k_1 t \tag{A.9}$$

or

$$[A] = [A]_o e^{-k_1 t}. \tag{A.10}$$

Equation A.10 predicts a simple exponential decay curve for the first-order transformation of A into B. This is illustrated in the simulation of Figure A.1. The inset of this figure shows the logarithmic form of the rate law as expressed in Equation A.9 where we see a linear dependence of log[A] on time with a slope of $k_1/2.3$

A.2 ONE-STEP, IRREVERSIBLE, BIMOLECULAR REACTIONS

In the mechanism shown below, one molecule of A reacts with one of B to produce

$$A + B \xrightarrow{k_1} C$$

one molecule of C. While the reaction is first order in both A and B, overall it is a second-order reaction. The rate expression for this mechanism is given in Equation A.11, where k_1 is

$$-\frac{d[A]}{dt} = -\frac{d[B]}{dt} = \frac{d[C]}{dt} = k_1[A][B] \qquad (A.11)$$

a second-order rate constant, with units of $M^{-1}s^{-1}$.

The final form that the rate law of Equation A.11 takes depends on the relative magnitudes of $[A]_o$ and $[B]_o$. While rate laws exist for the general case $[A]_o \sim [B]_o$, and the special case $[A]_o = [B]_o$, they are cumbersome and are difficult to apply in practice. Without question, the most useful and mathematically tractable case is the one in which the concentration of one of the reactants is much greater than concentration of the other.

When $[A]_o \ll [B]_o$, the concentration of the latter does not undergo a significant decrease during the course of the reaction. Thus, the product of k_1 and $[B]_o$ can be taken as a constant, allowing us to express Equation A.11 as

$$-\frac{d[A]}{dt} = k_1'[A] \qquad (A.12)$$

where k_1' is the pseudo-first order rate constant

$$k_1' = k_1[B]_o. \qquad (A.13)$$

Setting up a second-order reaction under pseudo-first-order conditions allows it to be treated with the analytical simplicity with which one treats a first-order reaction.

A.3   ONE-STEP, REVERSIBLE REACTIONS

For the reversible reaction shown below, the rate of change of concentration of either the

$$A \underset{k_{-1}}{\overset{k_1}{\rightleftharpoons}} C$$

reactant or the product is equal to the difference in the reaction rates in the two directions:

$$-\frac{d[A]}{dt} = \frac{d[C]}{dt} = k_1[A] - k_{-1}[C]. \qquad (A.14)$$

Derivation of this mechanism's rate law from Equation A.14 proceeds most conveniently in terms of the dependent variable x, where x is a molar quantity reflecting the extent to which A has been converted into C. Thus, the concentrations of A and B can be expressed as

$$[A] = [A]_o - x \tag{A.15}$$

$$[C] = [C]_o + x \tag{A.16}$$

and the rate of change of x as

$$\frac{dx}{dt} = k_1([A]_o - x) - k_{-1}([C]_o + x) \tag{A.17}$$

or

$$\frac{dx}{dt} = k_1[A]_o - k_{-1}[C]_o - (k_1 - k_{-1})x. \tag{A.18}$$

Integration of Equation A.18, after substituting x = 0 when t = 0, yields

$$\ln \frac{k_1[A]_o - k_{-1}[C]_o}{k_1[A]_o - k_{-1}[C]_o - (k_1 - +k_{-1})x} = (k_1 - +k_{-1})t. \tag{A.19}$$

Given sufficient time, the reaction comes to a state of equilibrium, and x attains a value of x_{eq}. Now, as t approaches infinity, so must the left side of Equation A.19, which requires the denominator of the left side of Equation A.19 to approach zero, that is,

$$(k_1[A]_o - k_{-1}[C]_o) - (k_1 + k_{-1})x_e = 0. \tag{A.20}$$

Equation A.20 can be rearranged to yield Equation A.21, an expression for x_{eq}, which when

$$x_{eq} = \frac{k_1[A]_o - k_{-1}[C]_o}{k_1 + k_{-1}} \tag{A.21}$$

substituted into Equation A.19 gives us

$$\ln \frac{x_{eq}}{x_{eq} - x} = (k_1 - +k_{-1})t \tag{A.22}$$

or,

$$\frac{x_{eq}}{x_{eq} - x} = e^{k_{eq}t} \tag{A.23}$$

where k_{eq} is the first-order rate constant for the approach of the reaction to equilibrium:

$$k_{eq} = k_1 + k_{-1}. \tag{A.24}$$

Experimentally, one will monitor the formation of C or disappearance of A with time. If the latter is the metric, analysis will proceed using a form of Equation A.23 in which x_{eq} and x are replaced by $([A]_0 - [A]_{eq})$ and $([A]_0 - [A])$, respectively:

$$\frac{[A]_o - [A]_{eq}}{([A]_o - [A]_{eq}) - ([A]_o - [A])} = e^{k_{eq}t}. \tag{A.25}$$

Upon rearrangement, Equation A.25 yields

$$\frac{[A] - [A]_{eq}}{[A]_0 - [A]_{eq}} = e^{-k_{eq}t} \tag{A.26}$$

and finally

$$[A] = ([A]_o - [A]_{eq}) e^{-k_{eq}t} + [A]_{eq}. \tag{A.27}$$

Analysis of a reaction progress curve for a reversible reaction according to Equation A.27 will yield estimates of $[A]_0$, $[A]_{eq}$, and k_{eq}. To estimate the individual values k_1 and k_{-1} that comprise k_{eq}, one first calculates their ratio, the equilibrium constant for the reaction, according to Equation A.28:

$$K_{eq} = \frac{k_1}{k_{-1}} = \frac{[C]_o + x_{eq}}{[A]_o - x_{eq}} \tag{A.28}$$

$$K_{eq} = \frac{[C]_o + ([A]_o - [A]_{eq})}{[A]_o - ([A]_o - [A]_{eq})} \tag{A.29}$$

or

$$K_{eq} = \frac{[C]_o + ([A]_o - [A]_{eq})}{[A]_{eq}}. \tag{A.30}$$

When $[C]_0 = 0$,

$$K_{eq} = \frac{[A]_o}{[A]_{eq}} - 1. \tag{A.31}$$

Knowledge of K_{eq} $(=k_1/k_{-1})$ and k_{eq} $(=k_1 + k_{-1})$, allows calculation of k_1 and k_{-1}.

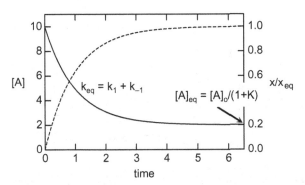

Figure A.2. Progress curve simulations of the reversible, first-order conversion of A into C. Reactant A can be seen to decrease in con centration, with a rate constant k_{eq}, to its equilibrium value. The dashed line is the time-dependence of the extent of reaction (see text). $k_1 = k_{-1} = 1$.

A reversible reaction of the sort we have been considering appears in Figure A.2. We see that as the reaction progresses, the concentration of reactant A decreases from $[A]_o$ to a nonzero, equilibrium value of $[A]_{eq}$. Also illustrated in the figure is the time course for the extent of reaction, which was drawn using a rearranged form of Equation A.23:

$$\frac{x}{x_{eq}} = 1 - e^{-k_{eq}t}. \tag{A.32}$$

The extent of reaction for a reversible reaction increases from zero to one, with first-order rate constant k_{eq}.

A.4 TWO-STEP, IRREVERSIBLE REACTIONS

In one step reactions, there is an equivalence between the rate of reactant disappearance and production formation. This is not the case for multistep reactions, where these two rates can be very different. For reactions that proceed through an intermediate, the investigator needs to specify which species is being monitored and then to consider the often complex progress curve that corresponds to it.

Consider the reaction below, in which the formation of C from A proceeds through intermediate B.

$$A \xrightarrow{k_1} B \xrightarrow{k_2} C$$

The rate of disappearance of A is defined quite simply in Equation A.33 as equal to the concentration

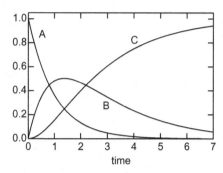

Figure A.3. Progress curve simulations of the first-order sequential conversion of reactant A into intermediate B into product C. $k_1 = k_2 = 1$.

$$-\frac{d[A]}{dt} = k_1[A] \tag{A.33}$$

of A multiplied by first-order rate constant k_1. Integration of Equation A.33 yields the expression of Equation A.34 for the exponential disappearance of A from its initial concentration to zero. This is

$$[A] = [A]_o e^{-k_1 t} \tag{A.34}$$

illustrated in Figure A.3.

The rate of concentration change of B equals the rate of its formation from A, via k_1, minus the rate of its transformation to C, via k_2:

$$\frac{d[B]}{dt} = k_1[A] - k_2[B]. \tag{A.35}$$

Substituting the expression of Equation A.34 for [A] yields Equation A.36, which upon integration

$$\frac{d[B]}{dt} = k_1[A]_o e^{-k_1 t} - k_2[B] \tag{A.36}$$

gives us Equation A.37. This equation predicts that the concentration will rise exponentially,

$$[B] = [A]_o \frac{k_1}{k_2 - k_1} \left(e^{-k_1 t} - e^{-k_2 t} \right) \tag{A.37}$$

with rate constant k_1, to a maximum concentration and then decrease to zero, with rate constant k_2 (Fig. A.3).

The concentration of C can be calculated from the simple stoichiometry of the reaction,

$$[C] = [A]_o - ([A] + [B]) \tag{A.38}$$

which becomes

$$[C] = [A]_o \left[1 + \frac{1}{k_1 - k_2} \left(k_2 e^{-k_1 t} - k_1 e^{-k_2 t} \right) \right]. \tag{A.39}$$

As shown in Figure A.3, the production of C is characterized by an initial lag phase, followed by an exponential rise.

A.5 TWO-STEP REACTION, WITH REVERSIBLE FIRST STEP

Next we consider the reaction shown below, in which the formation of B from A is

$$A \underset{k_{-1}}{\overset{k_1}{\rightleftarrows}} B \overset{k_2}{\longrightarrow} C$$

a reversible process. Despite this seemingly modest change in mechanism, the exact solutions for the time dependencies of the concentrations of A, B, and C are of such complexity as to make them of little use (Moore and Pearson 1981). There are two practical ways to approach this mechanism: the steady-state approximation and simulations.

The steady-state approximation assumes that the concentration of intermediate species does not change during the course of the reaction. This derivation starts by first stating the three differential equations for the concentrations of A, B, and C.

$$\frac{d[A]}{dt} = -k_1[A] + k_{-1}[B] \tag{A.40}$$

$$\frac{d[B]}{dt} = -(k_{-1} + k_2)[B] + k_1[A] \tag{A.41}$$

$$\frac{d[C]}{dt} = k_2[B]. \tag{A.42}$$

According to the steady-state approximation,

$$\frac{d[B]}{dt} = 0. \tag{A.43}$$

Given this assumption, Equation A.41 can be rearranged to produce

$$[B]_{ss} = \frac{k_1}{k_{-1} + k_2}[A].$$

(A.44)

Substituting this into Equation A.42 yields

$$\frac{d[C]}{dt} = \frac{k_1 k_2}{k_{-1} + k_2}[A]$$

(A.45)

and

$$-\frac{d[A]}{dt} = -k_1 \left\{ \frac{(k_{-1} + k_1) + k_2}{k_{-1} + k_1} \right\}[A].$$

(A.46)

If $(k_{-1} + k_1) \ll k_2$, which is a condition for the steady-state approximation holding (Moore and Pearson 1981, p. 314), then

$$\frac{d[C]}{dt} = -\frac{d[A]}{dt} = \frac{k_1 k_2}{k_{-1} + k_2}[A].$$

(A.47)

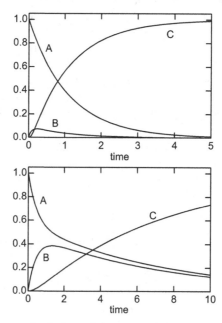

Figure A.4. Progress curve simulations of the sequential conversion of A into B into C, with a reversible A to B reaction. Upper panel: $k_1 = k_{-1} = 1$, $k_2 = 10$. Lower panel: $k_1 = k_{-1} = 1$, $k_2 = 0.3$.

We see from the expression of A.47, that the appearance of C and the disappearance of A will be first-order processes governed by

$$k_{obs} = \frac{k_1 k_2}{k_{-1} + k_2}. \tag{A.48}$$

In Figure A.4 are simulations of this reaction in which the condition $(k_{-1} + k_1) \ll k_2$ is met (upper panel) and not met (lower panel). In the upper panel, the concentration of B is very low throughout the course of the reaction and both the disappearance of A and appearance of C have the appearance of simple first-order reactions. In the lower panel, however, concentration of B rises to an appreciable extent, the disappearance of A is clearly biphasic, and the appearance of A is characterized by an appreciable lag phase.

Simulations of the sort shown in Figure A.4 can be used to model complex reactions and then to estimate rate parameters from experimental data, by matching curves to data by manually adjusting parameters by the investigator. Alternatively, many of the programs used to simulate reaction progress curves can also perform nonlinear least-squares analysis to estimate parameter values. The simulation program used throughout this book is Berkley Madonna (http://www.berkeleymadonna.com/), and can simulate reaction progress curves as well as fit data to models.

REFERENCES

Moore, J. W. and R. G. Pearson (1981). *Kinetics and Mechanism.* New York, John Wiley & Sons.
Wilhelmy, L. (1850). "The law by which the action of acid on cane sugar occurs." *Ann. Phys. Chem.* **81**: 413–433.

APPENDIX B

TRANSITION STATE THEORY AND ENZYMOLOGY: ENZYME CATALYTIC POWER AND INHIBITOR DESIGN

In Chapter 3, I presented an overview of transition state theory and how it is used to construct free energy diagrams for enzymatic reactions. In this appendix, I discuss two other important areas in which transition state theory has had an impact: hypotheses for the catalytic power of enzymes and design of inhibitors.

B.1 CATALYTIC POWER OF ENZYMES

Enzymes are extraordinary catalysts, able to accelerate the rates of their reactions by factors that exceed 10^9. Attempts to explain the catalytic power of enzymes date back to at least 1894 when Emil Fischer proposed his famous "lock-and-key" model. Fischer had observed that the enzyme invertase, which he had isolated from brewer's yeast, hydrolyzed α-glucosides but not β-glucosides, while the related enzyme emulsin hydrolyzed β- but not α-glucosides. Reflecting on these observations, Fischer speculated that ". . . enzyme and glucoside have to fit to each other like lock and key in order to exert a chemical effect on each other." (Fischer 1894). This was a profound insight that accounted not only for the stereoselectivity that these enzymes possess, but also led to a construal of enzyme action in which a snug and intimate fit between enzyme and

Kinetics of Enzyme Action: Essential Principles for Drug Hunters, First Edition. Ross L. Stein.
© 2011 John Wiley & Sons, Inc. Published 2011 by John Wiley & Sons, Inc.

substrate is a necessary requirement for catalysis. This idea, which still has its utility today, held sway during the first-half of the twentieth century until Linus Pauling proposed that enzyme catalysis results not because the substrate is bound tightly by enzyme, but rather because the activated complex of the reaction is bound tightly (Pauling 1946, 1948).

> From the standpoint of molecular structure and the quantum mechanical theory of chemical reaction, the only reasonable picture of the catalytic activity of enzymes is that which involves an active region of the surface of the enzyme which is closely complementary in structure not to the substrate itself, in its normal configuration, but rather to the substrate molecule in a strained configuration, corresponding to the "activated complex" for the reaction catalyzed by the enzyme. The substrate molecule is attracted to the enzyme, and caused by the forces of attraction to assume the strained state which favors the chemical reaction—that is, the activation energy of the reaction is decreased to such an extent as to cause the reaction to proceed at an appreciable greater rate than it would in the absence of enzyme. (Pauling 1946)

Here, Pauling is telling us that enzymes are able to effect catalysis because they are able to bring to bear stabilizing interactions in the transition state that are absent in the ground state (Pauling 1946, 1948).

Twenty-five years later, these thoughts were formalized by Richard Wolfenden (Wolfenden 1972; Miller and Wolfenden 2002) as shown in Figure B.1.[1] In this scheme, chemical transformation of substrate into product is seen to occur both through an uncatalyzed and enzyme-catalyzed reaction. In the uncatalyzed reaction, transition state theory tells us that substrate in its reactant state and the transition state for the reaction exist in state of pseudo-equilibrium governed by constant $K_u^{\neq}$, where

$$K_u^{\neq} = \frac{[S]}{[S^{\neq}]} \tag{B.1}$$

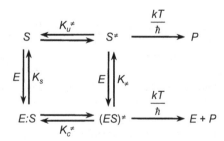

Figure B.1. Transition state theory and the origins of enzyme catalytic power.

---

[1] The simple thermodynamic relationship that Wolfenden proposed holds for catalysts of all types and may have been first described in 1921 by Michael Polanyi. Wolfenden (Miller and Wolfenden, 2002) provides an analysis of Michael Polanyi's work in this area.

and that the velocity of the uncatalyzed reaction can be expressed as

$$v_u = k_u[S] \tag{B.2}$$

where

$$k_u = \left(\frac{kT}{\hbar}\right)K_u^{\ddagger}. \tag{B.3}$$

The enzyme-catalyzed reaction proceeds by initial formation of an ES complex, which then enters the transition state for the enzyme-catalyzed reaction. The pseudo-equilibrium constant for the this process is given by

$$K_c^{\ddagger} = \frac{[ES]}{[(ES)^*]}. \tag{B.4}$$

This now lets us write the rate expression for the enzymatic reaction as

$$v_c = k_c[ES] \tag{B.5}$$

where

$$k_c = \left(\frac{kT}{\hbar}\right)K_c^{\ddagger}. \tag{B.6}$$

The ratio of Equations B.3 and B.6 is

$$\frac{K_c^{\ddagger}}{K_u^{\ddagger}} = \frac{k_c}{k_u}. \tag{B.7}$$

Since $K_c^{\ddagger}$ and $K_u^{\ddagger}$ are parallel arms of a thermodynamic box, Equation B.7 can be rewritten as

$$\frac{K_S}{K_{\ddagger}} = \frac{k_c}{k_u} = \chi \tag{B.8}$$

where $K_{\ddagger}$ is the dissociation constant for the complex of enzyme and substrate in the transition state of the reaction and χ is the catalytic efficiency. We see from Equation B.8 that an enzyme's catalytic efficiency originates in its ability to more tightly bind substrate in its transition state configuration than substrate in its reactant state configuration.

To continue our assessment of the catalytic power of enzymes, we need to express the catalytic efficiency in thermodynamic terms:

$$\chi = \frac{\dfrac{kT}{\hbar}\exp\left(-\Delta G_c^{\neq} / RT\right)}{\dfrac{kT}{\hbar}\exp\left(-\Delta G_u^{\neq} / RT\right)} \tag{B.9}$$

$$RT\left(\ln \chi\right) = \Delta G_u^{\neq} - \Delta G_c^{\neq} = \Delta G_{cat}^{\neq}. \tag{B.10}$$

$\Delta G_{cat}^{\neq}$ is the difference in activation energies between the catalyzed and uncatalyzed reactions, and thus reflects the difference in energies between the two reactions.

To understand the molecular origins of $\Delta G_{cat}^{\neq}$, which is to understand enzyme catalytic power, requires detailed structures for the following three species: (1) substrate and the enzyme in their reactant states, (2) activated complex of the uncatalyzed reaction, and (3) activated complex of the catalyzed reaction. If sufficiently detailed, these structures can serve as a basis for energetics calculations that will allow a dissection of the $\Delta G_{cat}^{\neq}$ value into specific interactions between enzyme and substrate. Such a strategy for investigating enzyme catalytic power was suggested by Schowen and coworkers over 30 years ago (Maggiora and Schowen 1977; Schowen 1978). Schowen explains that:

> The free energy liberated by the binding of the standard reaction transition state to the enzyme is calculable from experimental data, being the difference between the free energies of activation for the catalyzed and uncatalyzed reactions, and is thus called the *empirical free energy of binding of the transition state* [i.e., $\Delta G_{cat}^{\neq}$]. This empirical binding energy can usefully be separated conceptually into two additive terms: *the free energy of distortion* for distorting the free enzyme and the standard-reaction transition state into the forms they will have in the enzymic transition state ("poised structures") and *the vertical free energy of binding* for bringing the poised structures together from their standard-state distributions and forming the enzymic transition state by their interaction. This suggests a program of investigation of enzymic catalytic power involving (1) determination of standard-reaction and enzymic transition-state structures, (2) investigation of the energy requirements for distortion to the poised structures, and (3) investigation of the free-energy associated with combination of the poised structures. (Schowen 1978, p. 112)

As a specific example to illustrate these points, consider the hydrolysis of p-nitroacetanilide (see Fig. B.2). The hydrolysis of anilides and other structurally simple amides have been useful model systems in biochemistry and have served as reaction surrogates for protein hydrolysis, which for both technical and theoretical reasons represents a much more complex process to study. Both the nonenzymatic hydrolysis (Stein et al. 1984) and enzymatic hydrolysis by aryl acylamidase (Stein 2002) of p-nitroanilide has been the subject of thorough investigation.

The aryl acylamidase-catalyzed hydrolysis of p-nitroacetanilide occurs with a k_c/K_m value of $3 \times 10^5\,M^{-1}\,s^{-1}$ (Stein 2002). This rate constant reflects the energy difference between the reactant state of enzyme and substrate free in solution and the rate-limiting transition state for acylation, which likely corresponds to expulsion of the p-nitroanilide, leaving group from the tetrahedral intermediate to produce the acyl-enzyme. To assess

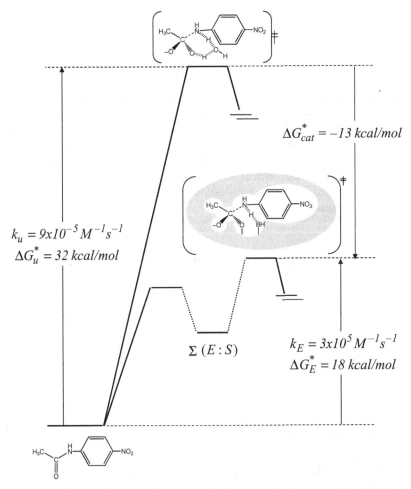

Figure B.2. Free-energy diagrams for the hydroxide- and enzyme-catalyzed hydrolysis of *p*-nitroacetanilide.

the catalytic enhancement that this enzyme brings about, we need to compare this reaction and rate constant to a nonenzymatic hydrolytic reaction of *p*-nitroacetanilide. An appropriate standard is the alkaline hydrolysis of *p*-nitroacetanilide at low hydroxide concentration. This reaction is governed by the second order-rate constant k_u which equals $9 \times 10^{-5} M^{-1} s^{-1}$ (Stein et al. 1984). k_u reflects the energy difference between the reactant state of hydroxide and substrate and the rate-limiting transition of decomposition of the tetrahedral intermediate (Stein et al. 1984). The clear chemical analogy between the reactions governed by k_E and k_u allows mechanistic interpretation of the ratio, k_E/k_u. This catalytic enhancement ratio of 3×10^9 corresponds to a $\Delta G^{\neq}_{cat}$ of $13 \, kcal \, mol^{-1}$ and reflects the stabilizing interactions that exist in the activated complex of the enzymatic reaction but do not exist in the hydroxide reaction.

At this point, our information concerning the origin of $\Delta G_{cat}^{\neq}$ is rudimentary and pertains chiefly to the substrate-derived portion of the activated complex for the uncatalyzed and catalyzed reactions. Experimental results from kinetic isotope effect studies (Stein et al. 1984) suggest that the activated complex of the uncatalyzed reaction has substantial tetrahedral character but less relative to the enzymatic reaction (Stein 2002). In addition, we also know that in both the enzymatic and standard reaction, *p*-nitroaniline expulsion from the tetrahedral intermediate is assisted by processes which donate a proton to the amide nitrogen. What is, of course, missing from this analysis is any structural information about either the enzyme in the activated complex or solvent reorganization. The former is essential for the assessment of the *free energy of distortion*, while the latter for the *vertical free energy of binding*. This information can, in principle, be obtained from a combination of kinetic and computational studies (Maggiora and Schowen 1977).

B.2   TRANSITION STATE ANALOG INHIBITION

The following simple rearrangements of Equation B.8 shown below led Wolfenden to predict that it

$$K_{\neq} = \frac{K_s}{\chi} \tag{B.11}$$

$$K_{\neq,assoc} = \chi K_{s,assoc} \tag{B.12}$$

might be possible to design new types of enzyme inhibitors with extraordinary potencies (Wolfenden 1969). Note in Equation B.12 equilibrium constants are expressed as association constants.

Equation B.12 tells us that the affinity with which an enzyme binds its catalytic transition state is equal to its affinity toward substrate multiplied by a factor equal to the enzyme's catalytic efficiency. Wolfenden reasoned that "analogs approaching the structure of the transition state for a chemical reaction should often show a high and unique affinity for the enzyme responsible for that reaction" (Wolfenden 1972). We see then that the K_i value for the "perfect" transition state analog inhibitor should approach $K_{\neq}$.

To design such inhibitors, the investigator must have insight into structural features of the substrate-derived portion of the catalytic transition state. This insight typically comes from a number of sources, including:

- Structures of substrates, intermediates, and products of the reaction.
- Structural features of transition states of pertinent nonenzymatic reactions.
- Correlation of k_c and/or k_c/K_m with systematic structural variation of the substrates or active-site residues.
- Kinetic isotope effects.

Based on these sources and other sources of mechanistic information, investigators have attempted to design transition state analog inhibitors for a large number of enzymes. As of 1995, there were reports of transition state analogs for at least 132 enzymes (Radzicka and Wolfenden 1995). However, examination of this and more recent literature reveals that claims of transition state analogy may not always be justified. Frequently, such claims are based on the observation of a compound's extreme inhibitor potency, or the mechanistic information used to design the inhibitor. However, potency is not in itself a sound justification for the claim of transition state analogy. And if the mechanistic information did not originate from direct observation of transition state structure, then this information might not lead to compounds that accurately reflect structural features of the catalytic transition state.

Two principle justifications exist for the claim that an inhibitor is a transition analog. One justification is that the analog was designed using structural details of the catalytic transition state derived from kinetic isotope effects (Schramm 2007). A case in point is the design of transition state analog inhibitors for purine nucleoside phosphorylase (PNPase). PNPase catalyzes the phosphorolysis of purine nucleosides to yield ribose-1-phosphate and purine base. For this reaction, it is estimated, based on a χ value of about 10^{13}, that the complex of the chemical transition state with the enzyme has a dissociation constant of 3×10^{-18} M, which is of course the limiting value for a perfect transition state analog of PNPase.

Based on α-deuterium kinetic isotope effects, it has been known for some time that reactions catalyzed by PNPase proceed through an S_N1 rather than an S_N2 reaction, with an oxocarbenium-like transition state, as shown in Figure B.3 (Stein et al. 1978;

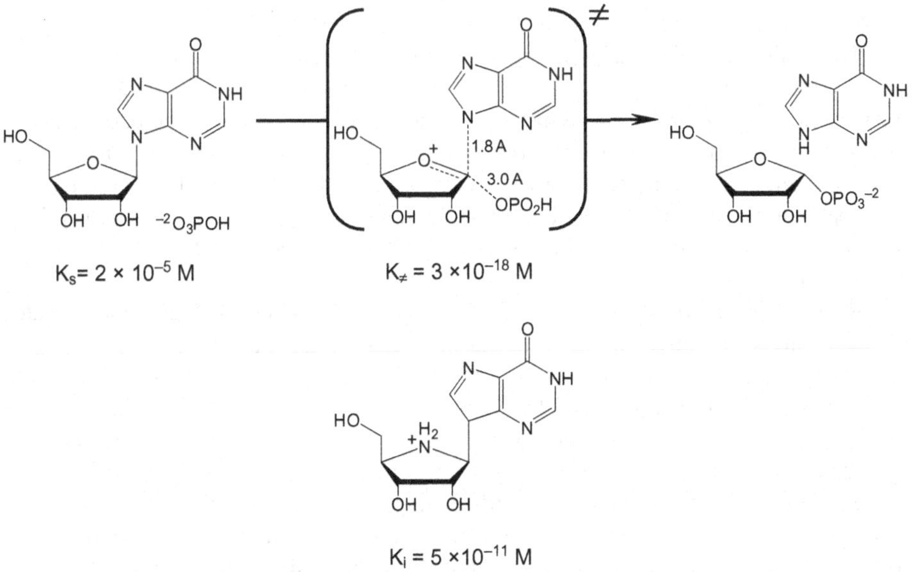

Figure B.3. Transition state analog inhibition of purine nucleoside phosphorylase by Immucillin H.

Stein and Cordes 1981). Schramm and coworkers extended this work to include isotope effects at other positions in nucleoside substrates and determined a detailed picture of PNPase's catalytic transition state (Kline and Schramm 1993). Based on these studies, Schramm was able to design a transition state analog inhibitor with a K_i value of 5 10^{-11} M (Miles et al. 1998). This compound, Immucillin-H, is currently in clinical trials for a number of T-cell-related diseases.

Another justification for the claim that an inhibitor is a transition analog relies on a thermodynamic argument. We know, from the above discussion, that the association constant for the perfect transition state analog inhibitor can be expressed as

$$K_{i,\text{assoc}} = \frac{\chi}{K_s} \tag{B.13}$$

or

$$K_i = \left(\frac{k_c}{K_s}\right) k_u = k_E k_u. \tag{B.14}$$

Taking the natural logarithm of each side yields

$$\ln\left(K_{i,\text{asoc}}\right) = \ln\left(k_E\right) + \ln\left(k_u\right) \tag{B.15}$$

which is equivalent to the linear free energy relationship of Equation B.16.

$$\Delta G_{i,\text{assoc}} = \Delta G_E^{\ddagger} + \Delta G_u^{\ddagger} \tag{B.16}$$

This tells us that for a homologous series of transition state analog inhibitors, $\Delta G_{i,\text{assoc}}$ should be a linear function of $\Delta G_E^{\ddagger}$ for the corresponding substrates, if $\Delta G_u^{\ddagger}$ is the same for all substrates. The slope of this linear dependence should be one, meaning that a structural change in the substrate that gives rise to a certain $\Delta\Delta G_E^{\ddagger}$, relative to some reference substrate, should produce $\Delta\Delta G_{i,\text{assoc}}$, for the corresponding inhibitors, of identical magnitude. To illustrate these concepts, we will consider the inhibition of the metalloprotease thermolysin by a series of peptide phosphoramidates (Bartlett and Marlowe 1983).

Reactions of thermolysin are thought to involve the attack of a Zn^{++}-bound water molecule on the carbonyl carbon of the substrate to form a tetrahedral intermediate (see Fig. B.4), which then collapse to amine and carboxylate products. The transition state for these reactions is characterized by partially tetrahedral carbonyl carbon with its two oxygens liganded to the active site Zn^{++} (Fig. B.4). χ for thermolysin-catalyzed peptide hydrolysis is about 10^{13}.

Bartlett reasoned that phosphoramidates might be able to displace the water molecule of the active site zinc form a bidentate complex, with the metal atom, which resemble the catalytic transition state. To test this hypothesis, Bartlett prepared a series of six peptide substrates of general structure Z-Gly-Leu-R (hydrolysis at the Gly-Leu bond) and their corresponding phosphoramidates.

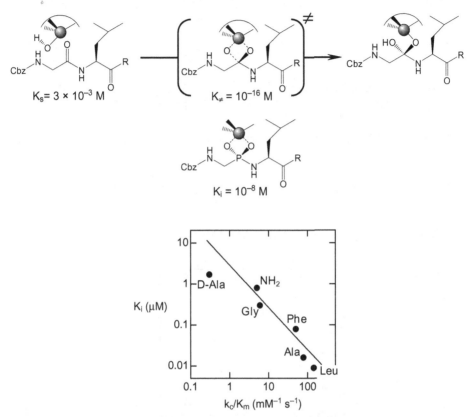

Figure B.4. Transition state analog inhibition of thermolysin by peptide phosphoramidates. Analysis performed with R = Leu.

The dependence of $\log(K_i)$ on $\log(k_c/K_m)$ for the five most potent inhibitors is linear with a slope of −1.2 (see Fig. B.4) and supports the claim that these peptide phosphoramidates are indeed transition state analog inhibitors of thermolysin. It is unclear why Z-NHCH$_2$PO$_2^-$-Leu-D-Ala falls significantly below the correlation line calculated for the other five inhibitors.

Another example is the inhibition of human leukocyte elastase (HLE) by peptide trifluoromethyl ketones (Stein et al. 1987). HLE-catalyzed reactions involve the attack of the oxygen of the active site serine on the carbonyl carbon of the substrate to form a tetrahedral intermediate, which then collapses with expulsion of the leaving group to form an acyl-enzyme. Hydrolysis of this species to liberate free enzyme completes the reaction. Carbonyl species that can react with the active site serine will form tetrahedral species that might bear resemblance to the transition states that lead to and then away from the tetrahedral intermediate.

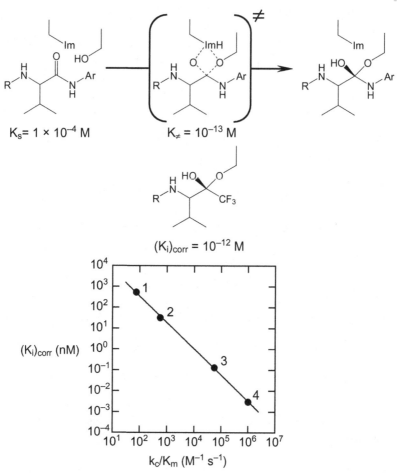

Figure B.5. Transition state analog inhibition of human leukocyte elastase by peptide trifluoromethyl ketones. 1, MeOSuc-Lys(Z)-Ala-Pro-; 2, MeOSuc-Ala-Pro-; 3, MeOSuc-Pro-; 4, MeOSuc-. Analysis performed with 1.

This hypothesis was tested for a series of peptide p-nitroanilide substrates, of general structure MeOSuc-R-Val-pNA (MeOSuc; methoxysuccinyl) and the corresponding trifluoromethyl ketones. Since the latter exist in aqueous solution as hydrates ($K_{assoc} \sim 100$), experimental inhibitor dissociation constants must be divided by a factor of 100 to yield $K_{i,corr}$ values that accurately reflect the dissociation of the hemiketal (see Fig. B.5) to the free ketone. The dependence of $\log(K_{i,corr})$ on $\log(k_c/K_m)$ is shown Figure B.4, and is seen to be linear with a slope of -1.2. This linear correlation supports the notion that these inhibitors form active site adducts that are analogs of the catalytic transition.

REFERENCES

Bartlett, P. A. and C. K. Marlowe (1983). "Phosphonamidates as transition-state analogue inhibitors of thermolysin." *Biochemistry* **22**(20): 4618–4624.

Fischer, E. (1894). "Einfluss der Configuration aur de Wirkung Derenzyme." *Berichte Deutschen Chemischen Gesellschaft* **27**: 2985–2993.

Kline, P. C. and V. L. Schramm (1993). "Purine nucleoside phosphorylase. Catalytic mechanism and transition-state analysis of the arsenolysis reaction." *Biochemistry* **32**(48): 13212–13219.

Maggiora, G. M. and R. L. Schowen (1977). The interplay of theory and experiment in bioorganic chemistry—Three case histories. In *Bioorganic Chemistry*, Vol. 1, E. E. van Tamelen, ed. New York, Academic Press.

Miles, R. W., et al. (1998). "One-third-the-sites transition-state inhibitors for purine nucleoside phosphorylase." *Biochemistry* **37**(24): 8615–8621.

Miller, B. G. and R. Wolfenden (2002). "Catalytic proficiency: The unusual case of OMP decarboxylase." *Annu. Rev. Biochem.* **71**: 847–885.

Pauling, L. (1946). "Molecular architecture and biological reactions." *Chem. Eng. News* **24**: 1375–1377.

Pauling, L. (1948). "The nature of forces between large molecules of biological interest." *Proc. R Inst. GB* **34**: 181–187.

Radzicka, A. and R. Wolfenden (1995). "Transition state and multisubstrate analogs." *Methods Enzymol.* **249**: 284–312.

Schowen, R. L. (1978). Catalytic power and transition state stabilization. In *Transition States of Biochemical Processes*, R. D. Gandour and R. L. Schowen, eds. New York, Plenum Press: 77–144.

Schramm, V. L. (2007). "Enzymatic transition state theory and transition state analogue design." *J. Biol. Chem.* **282**(39): 28297–28300.

Stein, R. L. (2002). "Enzymatic hydrolysis of p-nitroacetanilide: Mechanistic studies of the aryl acylamidase from pseudomonas fluorescens." *Biochemistry* **41**: 991–1000.

Stein, R. L. and E. H. Cordes (1981). "Kinetic α-deuterium isotope effects for *E. coli* purine nucleoside phosphorylase catalyzed phosphorolysis of adenosine and inosine." *J. Biol. Chem.* **256**: 767–772.

Stein, R. L., et al. (1984). "Transition-state structural features for anilide hydrolysis from β-deuterium isotope effects." *J. Am. Chem. Soc.* **106**: 1457–1461.

Stein, R. L., et al. (1978). "A Kinetic α-deuterium isotope effect for the binding of purine nucleosides to calf spleen purine nucleoside phosphorylase. Evidence for catalysis by distortion." *J. Am. Chem. Soc.* **100**: 6249–6251.

Stein, R. L., et al. (1987). "Mechanism of slow-binding inhibition of human leukocyte elastase by trifluormethyl ketones." *Biochemistry* **26**: 2682–2689.

Wolfenden, R. (1969). "Transition state ananogues for enzyme catalysis." *Nature* **223**: 704–705.

Wolfenden, R. (1972a). "Analog approaches to the structure of the transition state in enzyme reactions." *Acc. Chem. Res.* **5**: 10–18.

APPENDIX C

SELECTING SUBSTRATE CONCENTRATIONS FOR HIGH-THROUGHPUT SCREENS

In the design of an enzyme assay to be used in a high-throughput screen, a critical step is the selection of substrate concentrations. The concentrations of substrates establish the distribution of steady-state enzyme forms, and thus which enzyme forms are available to bind test compounds. For example, in the case of a simple one-substrate enzyme, there are only two steady-state enzyme forms to consider, E and E:S. If $[S]_o$ is set at a value much greater than K_m, ES will prevail in the steady-state and be available to bind inhibitors. Thus, an assay in which $[S]_o \gg K_m$ will be sensitive to uncompetitive inhibitors, with only the most potent competitive inhibitors being identified.

Enzyme assays will be equally sensitive to competitive, noncompetitive, and uncompetitive inhibitors if substrate concentrations are selected that produce what I call a "balanced steady state," that is, a situation in which all steady-state enzyme forms

Kinetics of Enzyme Action: Essential Principles for Drug Hunters, First Edition. Ross L. Stein.
© 2011 John Wiley & Sons, Inc. Published 2011 by John Wiley & Sons, Inc.

exist at equal concentrations. In the case of the one-substrate enzyme discussed above, the investigator would merely select a substrate concentration equal to K_m to achieve a balanced steady, since at this concentration $[E] = [E:S]$. However, for enzymes whose mechanisms involve more than a single steady-state intermediate or use two or more substrates, the substrate concentrations needed to achieve a balanced steady state or to established a particular distribution of steady-state enzyme forms cannot easily be ascertained by simple inspection of the mechanism.

In this appendix, I develop the mathematical methodology that allows these concentrations to be calculated. We start with one-substrate reactions, to illustrate the method, and then move on to increasing complex systems.

C.1   BALANCING THE STEADY STATE FOR ONE-SUBSTRATE REACTIONS

Figure C.1 illustrates the method for calculating the substrate concentration to achieve a balanced steady state for a one-substrate enzymatic reaction. The starting point in the derivation of the distribution equations is the statement that to achieve a balanced steady-state, f_E, the fraction of total enzyme that exists as E, must equal f_{EA}, the fraction of total enzyme that exists as EA. For more complex mechanisms with greater than two steady-state enzyme forms, we will see that to achieve a balanced steady state, the concentrations of all these forms must be equal.

In Figure C.1, the derivations of the distribution equations for f_E and f_{EA} develop in parallel, starting with the definition: $f_X = [X]/[E]_o$, where X is a steady-state enzyme form, E or EA in this case. Derivation of the distribution equation for each fraction is complete with f_X being shown to equal a quotient of a numerator, reflecting the ratio of $[X]$ to $[E]$, divided by a common denominator equal to the sum of all numerators. Finally, we can see that in the case of a one-substrate reaction, the condition $f_E = f_{EA}$ is met when $[A]_o = K_A$.

C.2   BALANCING THE STEADY STATE FOR TWO-SUBSTRATE, RAPID EQUILIBRIUM-ORDERED ENZYMATIC REACTIONS

Rapid equilibrium-ordered mechanisms have three steady-state enzyme forms: E, EA, and EAB. Thus, to achieve a balanced steady state, substrate concentrations must be chosen so that $f_E = f_{EA} = f_{EAB}$. Derivations of the three distribution equations proceed in precisely the same manner as above, except for the inclusion of one extra step (see Figure C.2). For the sake of clarity, Equation C.13 was inserted to explicitly show how the $[EAB]/[E]$ term in the numerator and denominator of Equation C.12 becomes $[A]_o[B]_o/K_A K_B$ in the denominator of Equation C.14. This "deconstruction" of an $[X]/[E]$ term into two (or more) fractions, each representing consecutive enzyme forms, will become an even more important tool for the steady-state mechanism we consider below.

$$E + A \underset{K_A}{\rightleftharpoons} EA \xrightarrow{k_c} E + P$$

The fraction of total enzyme as E and $E{:}A$ must be identical

$$f_E = f_{EA} \qquad\qquad\qquad\qquad \text{C.1}$$

$$f_E = \frac{[E]}{[E]_o} \qquad\qquad f_{EA} = \frac{[EA]}{[E]_o} \qquad\qquad \text{C.2}$$

$$f_E = \frac{[E]}{[E]+[EA]} \qquad\qquad f_{EA} = \frac{[EA]}{[E]+[EA]} \qquad\qquad \text{C.3}$$

$$f_E = \frac{1}{1+\dfrac{[EA]}{[E]}} \qquad\qquad f_{EA} = \frac{\dfrac{[EA]}{[E]}}{1+\dfrac{[EA]}{[E]}} \qquad\qquad \text{C.4}$$

$$f_E = \frac{1}{1+\dfrac{[A]_o}{K_A}} \qquad\qquad f_{EA} = \frac{\dfrac{[A]_o}{K_A}}{1+\dfrac{[A]_o}{K_A}} \qquad\qquad \text{C.5}$$

$$D = 1 + \frac{[A]_o}{K_A} \qquad\qquad \text{C.6}$$

$$\frac{1}{D} = \frac{\dfrac{[A]_o}{K_A}}{D} \qquad\qquad \text{C.7}$$

$$[A]_o = K_A \qquad\qquad \text{C.8}$$

Figure C.1. Derivation of distribution equations that allow calculation of substrate concentrations to achieve a balanced steady state for a one-substrate enzymatic reaction.

As above, each f_X can be shown to equal a quotient of a numerator, reflecting the ratio of [X] to [E], divided by a common denominator equal to the sum of all numerators. From inspection of Equation C.16, we see that $f_E = f_{EA} = f_{EAB}$ when $[A]_o = K_A$ and $[B]_o = K_B$.

Before moving on to the next section, it is instructive to note the relationship between the denominator of distribution equations and the denominator of rate equations. In derivation of the former, all steady-state enzyme forms are divided by free enzyme, while in the derivation of the latter, all steady-state enzyme forms are divided by substrate-bound enzyme, that is, the form that gives rise to product. This means that if the denominator of the final distribution equation is multiplied through by the reciprocal of the final term of the distribution equation, the denominator of the rate law is produced. So, in the case of an ordered mechansim,

$$E \underset{\longleftarrow}{\overset{[A]/K_A}{\rightleftharpoons}} EA \underset{\longleftarrow}{\overset{[B]/K_B}{\rightleftharpoons}} EAB \overset{k_c}{\longrightarrow} E + P$$

The fraction of total enzyme as E, EA, and EAB must be identical

$$f_E = f_{EA} = f_{EAB} \qquad \text{C.9}$$

$$f_E = \frac{[E]}{[E]_o} \qquad f_{EA} = \frac{[EA]}{[E]_o} \qquad f_{EAB} = \frac{[EAB]}{[E]_o} \qquad \text{C.10}$$

$$f_E = \frac{[E]}{[E] + [EA] + [EAB]} \qquad f_{EA} = \frac{[EA]}{[E] + [EA] + [EAB]} \qquad f_{EAB} = \frac{[EAB]}{[E] + [EA] + [EAB]} \qquad \text{C.11}$$

$$f_E = \frac{1}{1 + \dfrac{[EA]}{[E]} + \dfrac{[EAB]}{[E]}} \qquad f_{EA} = \frac{\dfrac{[EA]}{[E]}}{1 + \dfrac{[EA]}{[E]} + \dfrac{[EAB]}{[E]}} \qquad f_{EAB} = \frac{\dfrac{[EAB]}{[E]}}{1 + \dfrac{[EA]}{[E]} + \dfrac{[EAB]}{[E]}} \qquad \text{C.12}$$

$$f_E = \frac{1}{1 + \dfrac{[EA]}{[E]} + \dfrac{[EA]}{[E]}\dfrac{[E]}{[EAB]}} \qquad f_{EA} = \frac{\dfrac{[EA]}{[E]}}{1 + \dfrac{[EA]}{[E]} + \dfrac{[EA]}{[E]}\dfrac{[E]}{[EAB]}} \qquad f_{EAB} = \frac{\dfrac{[EAB]}{[E]}\dfrac{[EA]}{[E]}}{1 + \dfrac{[EA]}{[E]} + \dfrac{[EA]}{[E]}\dfrac{[EAB]}{[EA]}} \qquad \text{C.13}$$

$$f_E = \frac{1}{1 + \dfrac{[A]_o}{K_A} + \dfrac{[A]_o}{K_A}\dfrac{[B]_o}{K_B}} \qquad f_{EA} = \frac{\dfrac{[A]_o}{K_A}}{1 + \dfrac{[A]_o}{K_A} + \dfrac{[A]_o}{K_A}\dfrac{[B]_o}{K_B}} \qquad f_{EAB} = \frac{\dfrac{[A]_o}{K_A}\dfrac{[B]_o}{K_B}}{1 + \dfrac{[A]_o}{K_A} + \dfrac{[A]_o}{K_A}\dfrac{[B]_o}{K_B}} \qquad \text{C.14}$$

$$D = 1 + \frac{[A]_o}{K_A} + \frac{[A]_o}{K_A}\frac{[B]_o}{K_B} \qquad \text{C.15}$$

$$\frac{1}{D} = \frac{\dfrac{[A]_o}{K_A}}{D} = \frac{\dfrac{[A]_o}{K_A}\dfrac{[B]_o}{K_B}}{D} \qquad \text{C.16}$$

$$[A]_o = K_A \qquad [B]_o = K_B \qquad \text{C.17}$$

Figure C.2. Derivation of distribution equations that allow calculation of substrate concentrations to achieve a balanced steady-state for a two-substrate enzymatic reaction that proceeds through a rapid-equilibrium ordered mechanism.

$$\left\{ 1 + \frac{[A]_o}{K_A} + \frac{[A]_o}{K_A}\frac{[B]_o}{K_B} \right\} \left(\frac{[A]_o}{K_A}\frac{[B]_o}{K_B} \right)$$

equals

$$\frac{K_A}{[A]_o}\frac{K_A}{[A]_o} + \frac{K_B}{[B]_o} + 1$$

which is the denominator of the rate law for a rapid equilibrium-ordered mechanism (see Eq. 7.12). This relationship can be used to check the accuracy of distribution equations derived for complex mechanisms, or to quickly derive distribution equations by simply multiplying the denominator of the rate equation by the reciprocal of its first term.

C.3 BALANCING THE STEADY STATE FOR TWO-SUBSTRATE, RAPID EQUILIBRIUM RANDOM ENZYMATIC REACTIONS

In a rapid equilibrium random mechanism, there are four steady-state enzyme forms, so to achieve a balanced steady state, $f_E = f_{EA} = f_{EB} = f_{EAB}$ (see Fig. C.3). The derivation of the four distribution equations proceeds as before, arriving at Equation C.20. Inspection of this equation reveals that if $\alpha = 1$, a balanced steady state is achieved when $[A]_o = K_A$ and $[B]_o = K_B$. However, if $\alpha \neq 1$, there are no combination of substrate concentrations that can be chosen such that $f_E = f_{EA} = f_{EB} = f_{EAB}$. These concepts are illustrated in Figure C.4.

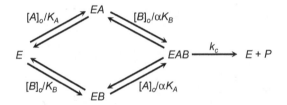

The fraction of total enzyme as E, EA, EB, and EAB must be identical

$$f_E = f_{EA} = f_{EB} = f_{EAB} \qquad \text{C.18}$$

$$\frac{1}{D} = \frac{\dfrac{[A]_o}{K_A}}{D} = \frac{\dfrac{[B]_o}{K_B}}{D} = \frac{\dfrac{1}{\alpha}\dfrac{[A]_o}{K_A}\dfrac{[B]_o}{K_B}}{D} \qquad \text{C.19}$$

$$D = 1 + \frac{[A]_o}{K_A} + \frac{[B]_o}{K_B} + \frac{1}{\alpha}\frac{[A]_o}{K_A}\frac{[B]_o}{K_B} \qquad \text{C.20}$$

If, only if, $\alpha = 1$

$$[A]_o = K_A \qquad [B]_o = K_B \qquad \text{C.21}$$

Figure C.3. Distribution equations that allow calculation of substrate concentrations to achieve a balanced steady-state for a two-substrate enzymatic reaction that proceeds through a rapid-equilibrium random mechanism.

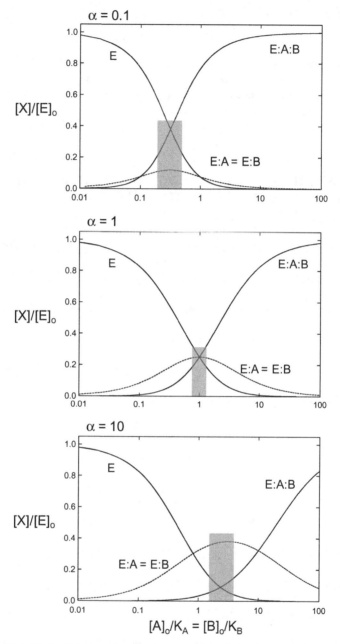

Figure C.4. Plots illustrating the substrate concentration-dependence of steady-state enzyme forms, for a rapid equilibrium random mechanism where $\alpha=0.1$, 1, 10. In these plots, $[A]_o/K_A$ and $[B]_o/K_B$ are equal to one another, and vary simultaneously from 0.01 to 100. In each plot, the shaded area reflects the $[A]_o/K_A$ and $[B]_o/K_B$ ratio that generates a distribution of steady-state enzyme forms that most closely achieve a balanced state. Because of the symmetry of random mechanisms and the fact that in these plots were constructed with $[A]_o/K_A=[B]_o/K_B$, $[EA]=[EB]$. Finally, note that only when $\alpha=1$ is balanced steady-state achieved, that is, $f_E=f_{EA}=f_{EB}=f_{EAB}=0.25$, and occurs with $[A]_o/K_A=[B]_o/K_B=1$.

$$E \underset{}{\overset{[A]/K_A}{\rightleftharpoons}} EA \overset{k_a}{\longrightarrow} E' \overset{k_b}{\longrightarrow} E + P$$

The fraction of total enzyme as E, EA, and E' must be identical

$$f_E = f_{EA} = f_{E'} \tag{C.22}$$

$$f_E = \frac{[E]}{[E]_o} \qquad f_{EA} = \frac{[EA]}{[E]_o} \qquad f_{E'} = \frac{[E']}{[E]_o} \tag{C.23}$$

$$f_E = \frac{[E]}{[E]+[EA]+[E']} \qquad f_{EA} = \frac{[EA]}{[E]+[EA]+[E']} \qquad f_{E'} = \frac{[E']}{[E]+[EA]+[E']} \tag{C.24}$$

$$f_E = \frac{1}{1+\dfrac{[EA]}{[E]}+\dfrac{[E']}{[E]}} \qquad f_{EA} = \frac{\dfrac{[EA]}{[E]}}{1+\dfrac{[EA]}{[E]}+\dfrac{[E']}{[E]}} \qquad f_{E'} = \frac{\dfrac{[E']}{[E]}}{1+\dfrac{[EA]}{[E]}+\dfrac{[E']}{[E]}} \tag{C.25}$$

$$f_E = \frac{1}{1+\dfrac{[EA]}{[E]}+\dfrac{[EA]}{[E]}\dfrac{[E']}{[EA]}} \qquad f_{EA} = \frac{\dfrac{[EA]}{[E]}}{1+\dfrac{[EA]}{[E]}+\dfrac{[EA]}{[E]}\dfrac{[E']}{[EA]}} \qquad f_{E'} = \frac{\dfrac{[EA]}{[E]}\dfrac{[E']}{[EA]}}{1+\dfrac{[EA]}{[E]}+\dfrac{[EA]}{[E]}\dfrac{[E']}{[EA]}} \tag{C.26}$$

$$f_E = \frac{1}{1+\dfrac{[A]_o}{K_A}+\dfrac{[A]_o}{K_A}\dfrac{[E']}{[EA]}} \qquad f_{EA} = \frac{\dfrac{[A]_o}{K_A}}{1+\dfrac{[A]_o}{K_A}+\dfrac{[A]_o}{K_A}\dfrac{[E']}{[EA]}} \qquad f_{E'} = \frac{\dfrac{[A]_o}{K_A}\dfrac{[E']}{[EA]}}{1+\dfrac{[A]_o}{K_A}+\dfrac{[A]_o}{K_A}\dfrac{[E']}{[EA]}} \tag{C.27}$$

In the steady state,

$$k_a[EA] = k_b[E'] \tag{C.28}$$

$$\frac{[E']}{[EA]} = \frac{k_a}{k_b} \tag{C.29}$$

$$f_E = \frac{1}{1+\dfrac{[A]_o}{K_A}+\dfrac{[A]_o}{K_A}\dfrac{k_a}{k_b}} \qquad f_{EA} = \frac{\dfrac{[A]_o}{K_A}}{1+\dfrac{[A]_o}{K_A}+\dfrac{[A]_o}{K_A}\dfrac{k_a}{k_b}} \qquad f_{E'} = \frac{\dfrac{[A]_o}{K_A}\dfrac{k_a}{k_b}}{1+\dfrac{[A]_o}{K_A}+\dfrac{[A]_o}{K_A}\dfrac{k_a}{k_b}} \tag{C.30}$$

$$\frac{1}{D} = \frac{\dfrac{[A]_o}{K_A}}{D} = \frac{\dfrac{[A]_o}{K_A}\dfrac{k_a}{k_b}}{D} \tag{C.31}$$

$$D = 1+\frac{[A]_o}{K_A}+\frac{[A]_o}{K_A}\frac{k_a}{k_b} \tag{C.32}$$

Figure C.5. Distribution equations that allow calculation of substrate concentrations to achieve a balanced steady-state for a one-substrate enzymatic reaction that proceeds through a mechanism with a two steady-state intermediates.

A number of points related to the choice of substrate concentrations for high-throughput screens come out of inspection of Figure C.4. All too often, substrate concentrations are chosen with no knowledge of an enzyme's kinetic mechanism. Frequently, substrate concentrations equal to K_m are chosen, assuming that this choice of substrate concentration will achieve something approaching a balanced steady state. Recall that K_m for substrate X is defined as the apparent substrate dissociation concentration measured at saturating concentrations of substrate Y, so in the context of the random mechanism of Figure C.3, $K_{m,A} = \alpha K_A$ and $K_{m,B} = \alpha K_B$. For example, assuming a mechanism in which $\alpha = 0.1$, $K_{m,X} = 0.1 K_X$. If an assay is run with $[X]_o = K_{m,X} = 0.1 K_X$, this means that $[X]_o/K_X = 0.1$. Examination of the plot of Figure C.3 where $\alpha = 0.1$ reveals that $[E]/[E]_o \sim 0.75$, and $[EA]/[E]_o = [EB]/[E]_o = [EAB]/[E]_o \sim 0.08$. We see that under conditions where the investigator expects a balanced steady state, free enzyme predominates over the other three species. Nothing approaching the hoped-for balanced steady state is achieved.

This illustrates the necessity of determining the kinetic mechanism of an enzyme before assay design for a screen is initiated.

C.4 BALANCING THE STEADY STATE FOR NONEQUILIBRIUM ENZYMATIC REACTIONS INVOLVING A SECOND STEADY-STATE INTERMEDIATE

The general reaction to be considered in this section is shown in Figure C.5, and includes an intermediate E' that forms irreversibly after the Michaelis complex. The distribution equations for this mechanism differs from that of the previous three derivations in a significant way. Unlike the steady-state concentration of enzyme forms of the mechanisms of Figures C.1–C.3 that are entirely dependent on substrate concentrations, the concentrations of EA and E' are, in part, independent of $[A]_o$, but instead have a strong dependence on the relative magnitudes of k_a and k_b. For example, if $k_a \gg k_b$, $[EA]/[E']$ approaches zero, regardless of $[A]_o$.

This has an important consequence for trying to balance the steady state for this mechanism. Inspection of Equations C.31 and C.32 reveal that a balanced steady state can be achieved with $[A]_o = K_A$, only under the very special condition that $k_a/k_b = 1$. For any other value of k_a/k_b, there is no concentration of substrate that will produce a balanced steady state. The only thing that can be achieved with adjustment of $[A]_o$ is to produce a steady-state situation where the free enzyme concentration is equal to the sum of the concentrations of the two intermediates. This can readily be demonstrated, as shown below.

$$f_E = f_{EA} + f_{E'} \tag{C.33}$$

$$\frac{1}{D} = \frac{\dfrac{[A]_o}{K_A}}{D} + \frac{\dfrac{[A]_o}{K_A}\dfrac{k_a}{k_b}}{D} \tag{C.34}$$

$$1 = \frac{[A]_o}{K_A}\left(1 + \frac{k_a}{k_b}\right) \tag{C.35}$$

$$1 = \frac{[A]_o}{K_{m,A}} \tag{C.36}$$

$$[A]_o = K_{m,A}. \tag{C.37}$$

We see that when the initial substrate concentration is set to a concentration equal to $K_{m,A}$, [E] will equal the sum of [EA] and [E'].

C.5 BALANCING THE STEADY STATE FOR TWO-SUBSTRATE, PING-PONG ENZYMATIC REACTIONS

In Figure C.6, I have derived the denominator of the four fraction terms of the distribution equation for the the ping-pong mechanism. Equation C.40 is an expansion of Equation C.39 in which each ratio reflects two adjacent species in the mechanism. In Equation C.41, the terms $[A]_o/K_A$ and $[B]_o/K_B$ are substituted for [EA]/[E] and [E'B]/[E'], respectively. The next step of the derivation, in which we move from Equation C.41–C.44, follows from the steady-state condition that the initial velocity for ping-pong mechanisms is equal to either $k_a[EA]$ or $k_b[E'B]$ (see Section 7.5.1 for the derivation of the rate law for the ping-pong mechanism). Equation C.46 results from substation $K_B/[B]_o$ for [E']/[E'B] in Equation C.44.

A balanced steady state can only be achieved in the special situation where $k_a/k_b = 1$, in which case substrate concentrations would need to be set at their respective dissociation constants. If it is not the case that $k_a/k_b = 1$, the simplest way to determine substrate concentrations that give the closest approximation to a balanced steady state is through simulations using the rate law and parameter estimates. This is illustrated in the example below.

Case Study: Balancing the Steady State for Tissue Transglutaminase. Recall in Chapter 4 (Section 4.4.2), I introduced tissue transglutaminase (TGase) as a therapeutic target of some interest. TGase catalyzes the formation of isopeptide bonds between the amide carbonyl of Gln residues and the ε-amine of Lys residues, thereby crosslinking proteins *in vivo*. *In vitro*, a small peptide and amino acid derivatives can serve the roles of carbonyl acceptor and amine donor.

In 2005, Case et al. reported the development of a TGase assay suitable for high-throughput screens (Case et al. 2005). One of the several steps involved in the development of this assay was the use of simulations to determine substrate concentrations that would produce a balanced steady state. To conduct these simulations, the investigators first determined the kinetic mechanism and rate constants for the TGase-catalyzed transamidation of Gln residues N-methyl-casein (NMC) by the fluorescent lysine derivative αN-Boc-Lys-NH-CH$_2$-CH$_2$-NH-dansyl (KXD). The mechanism they elucidated, together with rate constants, are shown in the inset Figure C.7. We have the information that is necessary to calculate values of [NMC]$_o$ and [KXD]$_o$ that are required to produce a balanced steady state.

$$E \underset{[A]/K_A}{\rightleftharpoons} EA \xrightarrow{k_a} E' \underset{[B]/K_B}{\rightleftharpoons} E'B \xrightarrow{k_b} E + P$$

The fraction of total enzyme as E, EA, E', and $E'B$ must be identical

$$f_E = f_{EA} = f_{E'} = f_{E'B} \qquad\qquad \text{C.38}$$

$$D = 1 + \frac{[EA]}{[E]} + \frac{[E']}{[E]} + \frac{[E'B]}{[E]} \qquad\qquad \text{C.39}$$

$$D = 1 + \frac{[EA]}{[E]} + \frac{[EA]}{[E]}\frac{[E']}{[EA]} + \frac{[EA]}{[E]}\frac{[E']}{[EA]}\frac{[E'B]}{[E']} \qquad\qquad \text{C.40}$$

$$D = 1 + \frac{[A]_0}{K_A} + \frac{[A]_0}{K_A}\frac{[E']}{[EA]} + \frac{[A]_0}{K_A}\frac{[E']}{[EA]}\frac{[B]_0}{K_B} \qquad\qquad \text{C.41}$$

$$v_0 = k_a[EA] = k_b[E'B] \qquad\qquad \text{C.42}$$

$$[EA] = \frac{k_b}{k_a}[E'B] \qquad\qquad \text{C.43}$$

$$D = 1 + \frac{[A]_0}{K_A} + \frac{[A]_0}{K_A}\left(\frac{[E']}{[E'B]}\frac{k_a}{k_b}\right) + \frac{[A]_0}{K_A}\left(\frac{[E']}{[E'B]}\frac{k_a}{k_b}\right)\frac{[B]_0}{K_B} \qquad\qquad \text{C.44}$$

$$D = 1 + \frac{[A]_0}{K_A} + \frac{[A]_0}{K_A}\left(\frac{K_B}{[B]_0}\frac{k_a}{k_b}\right) + \frac{[A]_0}{K_A}\left(\frac{K_B}{[B]_0}\frac{k_a}{k_b}\right)\frac{[B]_0}{K_B} \qquad\qquad \text{C.45}$$

$$D = 1 + \frac{[A]_0}{K_A} + \frac{[A]_0}{K_A}\frac{K_B}{[B]_0}\frac{k_a}{k_b} + \frac{[A]_0}{K_A}\frac{k_a}{k_b} \qquad\qquad \text{C.46}$$

Figure C.6. Distribution equations that allow calculation of substrate concentrations to achieve a balanced steady state for a two-substrate enzymatic reaction that proceeds through a ping-pong mechanism.

According to this mechanism, a Gln residue of NMC reacts with TGase to form acyl-enzyme, E-acyl, which either hydrolyzes to S′, an NMC molecule with one additional Glu residue, or reacts with the primary amine of KXD to form S-N, a covalently modified form of NMC with KXD incorporated. Note that E-acyl:N, the encounter of complex of E-acyl and KXD, is stable enough to accumulate in the steady state

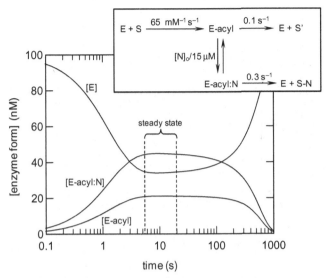

Figure C.7. Simulated reaction progress curves for TGase-catalyzed transamidation of NMC by KXD. [E], [E-acyl], and [E-acyl:N] were calculated over time for the mechanism summarized in the inset of this figure. In this simulation, initial concentrations were set at: $[E]_o = 0.1\,\mu M$, $[S]_o = 8\,\mu M$, and $[N]_o = 32\,\mu M$.

($K_N = 15\,\mu M$), the Michaelis complex of TGase and NMC is not, meaning that only three steady-state species need to be considered in simulations of this mechanism: E, E-acyl, and E-acyl.

Using the mechanism and rate constants that were determined, the investigators simulated time-dependent concentrations of E, E-acyl, and E-acyl for reactions at a variety of initial concentrations of NMC and KXD. An example is shown in Figure C.7, where we can see that the steady-state is from about 5 s to 11 s. In this time domain, $[E]_{ss} = 35\,\mu M$, $[E\text{-acyl}]_{ss} = 21\,\mu M$, and $[E\text{-acyl:N}]_{ss} = 42\,\mu M$, which clearly reflects an "unbalanced" steady-state situation. Simulations such as these were calculated for a range of initial S and N concentrations. The results of some of these are summarized in Figure C.8 where we have plotted $[E]_{ss}$, $[E\text{-acyl}]_{ss}$, and $[E\text{-acyl:N}]_{ss}$ as a function of $[KXD]_0$ at two constant concentrations of $[NMC]_0$, 4 and 8 μM. From these plots it is clear that a balanced steady state is attained when $[NMC]_0 = 8\,\mu M$ and $[KXD]_0 = 16\,\mu M$.

The investigators conclude that when the assay is run at these initial concentrations of NMC and KXD, it should be sensitive to compounds that bind not only to free enzyme (i.e., competitive inhibitors) but also to compounds that bind to the acyl-enzyme and the form of the acyl-enzyme in which KXD is bound (i.e., mixed inhibitors). Thus, this assay should allow the identification of a wide range of inhibitor types.

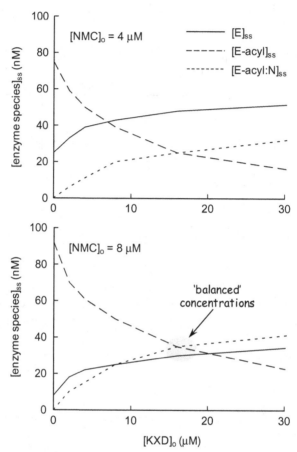

Figure C.8. Steady-state concentrations of E, E-acyl, and E-acyl:N as a function of substrate concentrations. Based on the mechanism of Figure C.7 and $[E]_{total} = 100\,nM$, simulations were performed as a function of $[NMC]_o$ and $[KXD]_o$ to determine steady-state levels of the principal enzyme species.

REFERENCE

Case, A., et al. (2005). "Development of a mechanism-based assay for tissue transglutaminase—Results of a high-throughput screen and discovery of inhibitors." *Anal. Biochem.* **338**: 237–244.

INDEX

A-183, 213

$A\beta_{1-40}$, 214

Ac-Arg-His-Lys-**Lys**^{Ac}-Leu-Nle-Phe-NH$_2$, 193, 195

Ac-Arg-His-Lys-Lys(Ac)-Xaa-NH$_2$, 212–213

ACES buffer, 33–34

Acetate, 221

1'-Acetyl-ADP-ribose, 193

2'-O-Acetyl-ADP-ribose, 193, 195, 212

Acetylases, 65

ε-Acetylated lysine residues, 212

Acetylcholinesterase, inhibition of, 110

ε-Acetyl-Lys residues, 193

3-Acetylpyridine adenine dinucleotide, 185

Acetylthiocholine, 110

Acetyltransferases, 111

Activated complex, 10, 42, 42n1, 264, 266

Activation, enzyme modulation by means of, 201–202

Activation cofactors, 213

Activation parameters, interpretation of, 233–234

Active center theory, 74

Active site, 6–7
  steric inhibition and, 201

Active site amino acids, 10

Active-site titrant, use of tight-binding inhibitor as, 119–121

Ac-Trp p-nitrophenyl ester, 49

Acylation
  of α-chymotrypsin, 49, 234
  esters and, 49
  of serine proteases, 243

Acyl-enzymes, 84

Acyl-transferases, 109

Addition compound theory of enzyme action, 7

Adenine, xanthine oxidase and, 8

5'-Adenylyl imidodiphosphate, 185

ADP (adenosine diphosphate)-ribose, 194, 195

D-Alanyl-D-alanine dipeptide, 111–112

Albumen, 34

Alchemy, 2

Alkaline hydrolysis of p-nitroacetanilide, 267

Allosteric activation, 202
  of SIRT1, 212–213

Allosteric inhibition, 201

Allosteric modulation of enzyme activity, 199–216
  case studies
    allosteric activation of SIRT1, 212–213
    allosteric modulation of AMP nucleosidase by ATP, 215–216
    allosteric modulation of insulin-degrading enzyme, 214–215
    modulation of ribonucleotide reductase by deoxyribonucleotides, 215
    partial inhibition of hydrolytic activity of complex tissue factor : factor VIIa, 213
  kinetics of, 202–208
  meaning of β and γ, 208–212
  mechanisms, 200, 201–202
    activation, 201–202
    inhibition, 201
  nature of $(X:E:S)'$ determines mode and extent of, 211–212

Kinetics of Enzyme Action: Essential Principles for Drug Hunters, First Edition. Ross L. Stein.
© 2011 John Wiley & Sons, Inc. Published 2011 by John Wiley & Sons, Inc.

Allosteric site
  allosteric activation and, 202
  steric inhibition and, 201
Allostery, 11
α-D (deuterium) isotope effects, 10, 241–242
  inhibitors as transition analogs and, 269–270
  pH dependence of, 247
αN-Boc-Lys-NH-CH$_2$-CH$_2$-NH-dansyl
  (KXD), 132, 283–286
Alzheimer's disease, 110, 158, 179, 214
Alzheimer's disease-associated kinases, 159
American Chemical Society, 13
Amidases, 109
Amides, acylation rate constants, 49
2-Aminobenzoyl-Gly-Gly-Phe-Leu-Arg-Lys-
  His-Gln-2,4-dinitroanilide, 214
N4-(6-Aminopyrimidin-4-yl)-sulfanilamide
  (APS), 99, 100, 101, 102
AMP nucleosidase, modulation by ATP,
  215–216
*Annals of the New York Academy of
  Sciences*, 8
Antimetabolites, 8
APS (N4-(6-aminopyrimidin-4-yl)-
  sulfanilamide), 99, 100, 101, 102
  inhibition of cdk5/p25 kinase by, 179–181
Arg-His-Lys-Lys(Ac)-AMC, 212
Armstrong, E. F., 8
Arrhenius, Svannte, 42
Aryl acylamidase-catalyzed hydrolysis of
  p-nitroacetanilide, 266–268
Asp-194, 227
Assay development, 32–35
  assay buffer solution, 33–34
  choice of method, 34
  implementation of method, 34–35
  selection of substrate, 32–33
Assays, troubleshooting, 34–35
Association constants, 95
ATP (adenosine triphosphate), 214
  modulation of AMP nucleosidase by, 215–216
Avogadro, Amedeo, 4
Axes, of free energy diagrams, 44–45

Bacon, Francis, 2
Balanced steady state, 275–276
  for nonequilibrium reactions involving
    second steady state intermediate, 281,
    282–283

  for one-substrate reactions, 276, 277
  for two-substrate, ping-pong reactions,
    283–286
  for two-substrate, rapid equilibrium-
    ordered reactions, 276–279
  for two substrate, rapid equilibrium
    random reactions, 279–282
Bardsley, William, 53
Barendrecht, H. F., 73
Bcr-abl tyrosine kinase, 200
Bell-shape of pH-rate profiles, 223, 224–225
Berkley Madonna, 261
Berman, H. A., 110
Berzelius, Jöns Jakob, 2, 7
β, meaning of, 208–209, 210–211
β-D (deuterium) isotope effects, 241–242,
  246
$\beta K_{i,assoc}$, 109–110
β-lactam antibiotics, 227
Binding isotherm, 56, 121
  Langmuir, 76
  parameters describing, 76
Biochemistry, interplay with organic
  chemistry, 2–4
Bisubstrate analysis studies, 191
Boc-Gly-Arg-Arg-AMC, hydrolysis of, 57, 58
Bodenstein, Max, 25
Boltzmann constant, 43, 229
Boyle, Robert, 2
Brown, Adrian, 6, 8, 20–21, 73
Buchner, Eduard, 4–5
Buffer, choice of, 221–222
Buffer crossover experiments, 221–222
Buffer pH, effect of temperature on, 231
Buffer salts, 33
Buffer solution, assay, 33–34
Bz-(D)Ala-^{S}Gly, 227–228

Calpain, 158
Cancers, ubiquitin C-terminal hydrolase-L1
  and, 83
CAPS, 221
Carbonyl addition reactions, 45
Catalysis, enzyme, 7–8
  activity of enzyme in, 19–21
  by distortion, 234
  pH effects on, 221
  power of enzymes in, 11, 263–268
  temperature effects on, 230–231

Catalytic efficiency, in thermodynamic terms, 265–266
Catalytic mechanism, 11, 13
cdk5/p25-catalyzed phosphorylation of tau, 158–160, 161
  inhibition by APS of, 179–181
cdk5/p25 kinase, 99, 100, 101, 102
Changeux, J.-P., 199
Charge relay system, 10
Chemical intervention, activation and, 201
Chemical kinetics, 5–7
  defined, 251
  principles of, 251–261
    one-step, irreversible, bimolecular reactions, 253–254
    one-step, irreversible, unimolecular reactions, 252–253
    one-step, reversible reactions, 254–257
    two-step, irreversible reactions, 257–259
    two-step reaction, with reversible first step, 259–261
Chemical mechanism, 11, 12
Chemistry, 2
  interplay between organic and biochemistry, 2–4
CHES buffer, 33, 221
Chymostatin, 122
Chymotrypsin, 10, 122
α-Chymotrypsin, acylation of, 49, 234
α-Chymotrypsin-catalyzed hydrolysis, of Suc-Ala-Phe-pNA, 225–226, 232
α-Chymotrypsin numbering, 243
^{13}C isotope effect, 10
Cleland, W.W., 9
Cognex, 110
Colowick, Sidney, 199
Commitment to catalysis factors, 9
Competitive inhibition, 90, 92, 93, 94, 97, 99, 102, 103, 104
  estimating inhibitor potency in, 116–117
  mechanism for, 126
Competitive inhibitors, 108
Completeness of inhibition, 75, 76
Complex tissue factor:factor VIIa, partial inhibition of hydrolytic activity of, 213
Concave Eyring plots, 232, 233
Conformational change
  allosteric activation and, 202

allosteric inhibition and, 201
allosteric modulation and, 208
Conformational transitions
  cooperativity or, 64–65
  substrate concentration-dependent, 63–64, 65
Conservation of enzyme equation, 154
Conservation of matter equations, 23
Continuous enzyme assays, 27–28
  data analysis in, 37
Control velocity, in enzyme inhibition, 76
Convex Eyring plots, 232, 233
Coombs, H. I., 73
Coombs, Herbert, 8
Cooperativity, 58–62
  conformational transitions or, 64–65
  for homodimeric enzyme, 60
  negative, 54, 59, 63, 64, 160
  positive, 54, 59, 62–63, 105, 160
Cori, Carl, 199
Cori, Gerty, 199
COT kinase, 111
Coupled enzyme assays, 29–32, 36
CTP, 214
Curve fitting software, 13
Cyclin-dependent kinase, 159
Cys residue, 84
Cysteine proteases, 109

Dakin, Henry, 6
Data analysis, 14, 37–40
Data collection, 35–37, 131–134, 222, 231
Deacetylases, 65, 111
Dehydrogenases, substrate analog inhibitor of, 185
$\Delta G\neq$ for overall reaction, 45–46
Deoxyribonucleotides, modulation of ribonucleotide reductase by, 215
Diabetes type II, 200, 212
Diastase, 3
Diatomic molecule A-B, primary isotope for dissociation of, 241
Dihydrotestosterone, 122
N-(2,4-Dinitrophenyl)-Arg-Pro-Lys-Leu-Nva-TrpNH$_2$, 28
Discontinuous enzyme assays, 28–29
  data analysis, 37–38
  direct measurement, 28
  measurement after separation, 29

Dissociation constants, 22
  conversion into association constants, 95
  for enzyme : inhibitor complexes, 89
  in nonprocessive reactions, 69–70
  in ordered mechanism, 176–177
  in slow-binding inhibition, 129
Distribution equations
  balanced steady state for one-substrate
      reaction with two steady-state
      intermediates, 281, 282–283
  balanced steady state for two-substrate,
      rapid equilibrium random reaction,
      278
  balanced steady state for two-substrate
      reaction with ping-pong
      mechanism, 284
Dixon, M., 8, 73
Drug discovery, 1
  enzyme allostery and, 200
  enzyme inhibition and, 8–9
    kinetic mechanism of, 89–90, 141,
        142
  high-throughput screening and, 142
  inhibition of acetylcholinesterase and, 110
  inhibition of two-substrate enzymatic
      reactions and, 170
  inhibitor potency and, 115–116
DTT, 34
Dynamics mechanism, 11, 13

E, 276–279
EA, 276–279
EAB, 276–279
E-acyl, 284–286
EDTA, 34
Eg51-367, thiazole inhibitors of monomeric,
    103–105, 106
Elastase, 247
  inhibition by MeOSuc-Val-Pro-Val-CF_3,
      134, 135
Empirical free energy of binding of the
    transition state, 266
Emulsin, 5, 263
Endothermic formation of enzyme : substrate
    complex, 46–47, 48
Enthalpy component, of formation of
    enzyme : substrate complex, 47
Entropy component, of formation of
    enzyme : substrate complex, 47–48

Enzyme allostery. *See* Allosteric modulation
    of enzyme activity
Enzyme assays, 26–32
  continuous, 27–28
  coupled, 29–32
  discontinuous, 28–29
    direct measurement, 28
    measurement after separation, 29
Enzyme catalysis, 7–8
Enzyme-catalyzed reactions, quantitative
    study of, 4–9
Enzyme hysteresis, 35
Enzyme inhibition. *See* Inhibition;
    Inhibitors
Enzyme : inhibitor complexes, 91, 92, 94, 97,
    106, 126, 129, 174, 178, 181
  dissociation constants for, 89
Enzyme : inhibitor species, multiple,
    sequentially formed, 93–95
Enzyme kinetics, 1–2
Enzyme kinetic theory, 9
Enzyme : product complexes, sequential
    reactions in which inhibitor binds to,
    191–196
Enzymes
  catalytic power of, 263–268
  with multiple intermediates, 51–53
  structural investigations of, 10–11
  that catalyze reactions of macromolecular
      substrates, 111
Enzyme structure
  separating effects of pH from those
      causing irreversible changes in, 221
  separating effects of temperature from
      those causing irreversible changes in,
      230–231
Enzyme : substrate (E : S) complex, 21–22,
    265
  kinetic consequences of, 46–50
    kinetic partitioning of the E : S complex,
        48–49
    meaning of steady-state kinetic
        parameters, 49–50
    relationship between initial velocity and
        steady-state kinetic parameters, 49
    summary, 49–50
    thermodynamics of saturation kinetics,
        46–48
  in steady state, 26

Enzymology
  contemporary, 9–11
  goals of, 11–13
  history of, 2–11
$[E]_o$, 118, 119, 120, 121, 124
Equilibrium constants, 93
  for overall reaction, 46
Equilibrium dialysis, 65
ES, 208
Esterases, 109
Esters, acylation and, 49
Exothermic formation of enzyme : substrate
    complex, 46–47
Eyring, Henry, 10, 42
Eyring plots, 231
  linear, 231–232
  nonlinear, 232–233

Factor IX, 213
Factor VII, 213
Factor VIIa exosite, 213
Falk, George, 6
Finasteride, inhibition of 5α-reductase by,
    122–123
Fine structure, in pH dependencies, 227–228
First-order rate constants, 20, 22–24, 42, 44,
    45, 60, 123, 131, 208, 229, 252,
    256–258
First-order rate law, 20
First-order reaction, 252, 253
Fischer, Emil, 5–6, 263
FK-506, 118–119
FKBP, 118–119
Formate, 221
Free energy changes, in formation of
    enzyme : substrate complex, 46–48
Free energy diagrams, xi, 44–46
  axes and construction of, 44–45
  $\Delta G \neq$ for overall reaction, 45–46
  equilibrium constant for overall reaction, 46
  for hydroxide- and enzyme-catalyzed
    hydrolysis of p-nitroacetanilide, 267
  inhibition, 94, 95
  for ordered mechanisms, 145, 165, 166
  for ping-pong mechanisms, 156, 157, 167
  for random mechanisms, 146, 163, 164
Free energy difference, 49–50
  in sequential mechanisms, 144
Free energy of distortion, 266, 268

γ, meaning of, 208–210, 211
γboroGlu, 136–137
γGlu-AMC, 136–137
γGTase-catalyzed hydrolysis of γGlu-AMC,
    136–137
Gas constant, 229
General Systems Theory (von Bertalanffy),
    211n3
Genetech, 213
Gibbs free energy, 45, 229
Gleevac, 200
Gln residues
  of NMC, 132
  of Z-Pro-Gln-Gly-Trp, 133
Global fit, 100–104, 160
  visual judgment of goodness-of-fit, 101,
    103–104
Global fitting, 158
Glucokinase, 200
α-Glucosides, 5, 263
β-Glucosides, 5, 263
Glutamate dehydrogenase, 133
γ-Glutamyl transpeptidase, 163
Glycerol, 34
Glycogen phosphorylase, 199
Glycohydrolases, 109, 111
Gly-Ome, 163
Good, Norman, 33, 221
Good buffer, 231
Gross-Butler equation, 244–245
GTP (guanosine-5'-triphosphate), inhibition
    of tissue transglutaminase by, 84–86
Guanine, xanthine oxidase and, 8

Haldane, J. B. S., 25
Harvard University, Laboratory for Drug
    Discovery in Neurodegeneration, 83,
    84, 214
Heavy atom isotope effect, 10
HEPBS buffer, 221
HEPES buffer, 33, 34, 221–222
Hexoses, effect on hydrolysis, 8
High-performance liquid chromatography, 29
High-throughput screening (HTS), 142, 159
  selecting substrate concentrations for,
    275–286
    balancing steady state for nonequilibrium
      reactions involving steady-state
      intermediate, 281, 282–283

High-throughput screening (HTS) (*cont'd*)
  balancing steady state for one-substrate
    reactions, 276, 277, 281
  balancing steady state for two-substrate,
    ping-pong reactions, 283–286
  balancing steady state for two-substrate,
    rapid equilibrium ordered reactions,
    276–279
  balancing steady state for two-substrate,
    rapid equilibrium random reactions,
    279–282
Hill coefficient, 76
His-57, 227
HLE. *See* Human leukocyte elastase (HLE)
Hofmeister, Franz, 4
Homodimeric enzyme, cooperativity for, 60
HTS. *See* High-throughput screening (HTS)
Human leukocyte elastase (HLE)
  hydrolysis of series peptide substrates by,
    247
  inactivation of, 138–139
  inhibition of by trifluoromethyl ketones,
    271–272
Hunter, F. Edmund, 9
Hydrogenic sites, 243
Hydrogen transfer reactions, 11
Hydrolases, 65, 109
Hydrolysis
  of Boc-Gly-Arg-Arg-AMC, 57, 58
  of Bx-(D)Ala-SGly, 227–228
  effect of hexoses on, 8
  of γGlu-AMC, 136–137
  of α-glucosides, 5
  of *p*-nitroacetanilide, 266–268
  of series peptide substrates, 247
  of substance P, 29, 30
  of Suc-Ala-Phe-pNA, 225–226, 232
  of sucrose, 5, 20–21, 251–252
  of Z-Glu(γ-AMC)-Gly, 85
Hyperbolic dependencies, 149, 151, 155, 163,
    164, 167, 168

IC_{50} values
  inhibitor potency and, 75, 77
  for irreversible inhibitors, 138
IDE. *See* Insulin-degrading enzyme (IDE)
I/E stoichiometry greater than one, 80–81
Ile-16, 227
Immucillin H, 269, 270

Inhibition, 8–9, 73–86. *See also* Irreversible
    inhibition; Slow-binding inhibition;
    Substrate inhibition; Tight-binding
    inhibition
  case studies, 83–86
    inhibition of tissue transglutaminase by
      GTP, 84–86
    inhibition of ubiquitin C-terminal
      hydrylase-L1 by LDN-91946, 83–84
  competitive, 90, 92, 93, 94, 97, 99, 102,
    103, 104
  completeness of, 75
  enzyme modulation by means of, 201
  kinetic mechanism of, 12, 90–93, 99, 142
  kinetic mechanism of, in one-substrate
    enzymatic reactions, 89–114
  basic mechanisms and rate expressions,
    90–93
  depletion of substrate, 111–114
  drug discovery and, 89–90
  inhibitor binding site and, 93
  initial velocity data analysis, 96–108
  meaning of K_i for, 93–96
  one-substrate, two-intermediate
    reactions, 108–110
  kinetic mechanism of, in two-substrate
    enzymatic reactions, 169–197
  conceptual understanding of, 170–173
  drug discovery and, 170
  elucidation of, 173–175
  ordered mechanisms, 175–177
    substrate analog inhibition of, 185–188
  ping-pong mechanisms, 181–184
    substrate analog inhibition, 189–190
  random mechanisms, 177–181
    substrate analog inhibition, 188–189
  SAR programs, 196–197
  sequential reactions in which inhibitor
    binds to enzyme : product complexes,
    191–196
  steady-state enzyme forms and rate
    parameters, 170–173
  substrate analog inhibition, 185–190,
    192
  kinetics of, 126–129
  mechanisms of, 12–13
  misbehaving, 77–82
  kinetic and mechanistic ambiguity, 82
  slope factor not equal to one, 78–81

v_{bkg} greater than zero, 81–82
$v_{control}$ not equal to velocity determined
at zero inhibitor concentration,
77–78
mixed, 90, 91, 93, 99, 102, 103, 104
noncompetitive, 90, 92, 99
phenomenon of, 74–75
potency of, 75
quantitative steps in, 76–77
simplicity of, 75
time course for, 122
transition state analog, 268–272
uncompetitive, 90, 92, 93, 99, 100, 101,
102, 103
Inhibition constant, for ordered mechanism,
176–177
Inhibitor binding site, steric activation and,
201
Inhibitors. *See also* Inhibition; Irreversible
inhibition; Slow-binding inhibition;
Tight-binding inhibition
binding site, 93
classical, 79
competitive, 108
estimating potency of, 116
noncompetitive, 109
partial, 75, 82
poor solubility of, 78, 80
as therapeutic agents, 8–9
tight-binding, 79
time-dependent, 79–80
uncompetitive, 109
velocity and concentration of, 74, 75
Initial reaction velocity (v_u)
analysis of data for enzyme inhibition,
96–108
emergence of mechanism from data,
105–108
global fit, 100–104
method of replots, 96–100
thiazole inhibitors of monomeric Eg51-
367, 103–105
with two independent variables, 96–108
calculating, 38
dependence on inhibitor concentration, 74,
75
for substrate depletion, 113, 114
dependence on substrate concentration,
19–21, 203, 207

mechanistic inadequacy of plots of, *vs.* pH,
220
for ping-pong mechanisms, 156
steady-state kinetic parameters and, 49
steady-state velocity and, 35
Inosine, phosphorylysis of, 246
Insulin-degrading enzyme (IDE), allosteric
modulation of, 214–215
Intermediate polymeric species, 65–66
Intermediates
mechanism for enzyme proceeding
through two reversibly formed, 50,
51–53
multiple, 51–53
Introduction to Systems Philosophy (Laszlo),
211n3
Invertase
early studies of, 5–6, 8
hydrolysis of α-glucosides, 5, 263
hydrolysis of sucrose and, 5, 20–21
Ionic strength, 33
Irreversible inhibition, 115, 137–139
inactivation of HLE by MeoSuc-Ala-Ala-
Pro-Val-CH$_2$Cl, 138–139
Isocratic elution, 29
Isomers, of enzymes, 65
Isotope effects. *See also* α-D (deuterium)
isotope effects; β-D (deuterium)
isotope effects; Kinetic isotopic
effects on enzyme-catalyzed
reactions; Solvent isotope effects
(SIE)
solvent, 240–243
Isotopic substitution of substrate and solvent,
239

Jacob, F., 199
Journal of Biological Chemistry, 4

Kastle, Joseph, 5
k_c, 9, 50, 51–53
pH dependence of, 220–221
k_c/K_m, 49, 51–53
pH dependence of, 220
$(k_c/K_m)_{obs}$, 225–227
$k_{c,obs}$, 225–227
Kekulé, Friedrich, 4
α-Ketoglutarate-dependent oxidation of
NADH, 133

$K_{i,app}$, 76–78, 79, 98
  kinetic dissection of, 89–90
  in tight-binding inhibition, 119, 120–121
  time-dependent, 131
K_i (dissociation constant), 123, 125
  estimation of, for tight-binding inhibitors,
    116–119
  meaning of, for simple one-substrate
    reactions, 93–96
    K_i values and nonequilibrium
      mechanisms of substrate turnover,
      95–96
    multiple, sequentially formed E : I
      species, 93–95
Kinases, 65, 111
  allosteric modulators of, 200
Kinesin spindle protein, 104–105
Kinetic ambiguity, 82–83
Kinetic constants, for one-substrate reaction,
  35
Kinetic isotopic effects on enzyme-catalyzed
  reactions, 9, 10, 239–247
  See also α-D (deuterium) isotope effects
  inhibitor as transition analog, 269–270
  interpretation of, 245–247
  solvent deuterium isotope effects, 243–247
  substrate isotope effects, 240–243
Kinetic mechanism, 11, 12, 13
  of inhibition (*See under* Inhibition)
  for two-substrate reactions, determining,
    156–163 (*See also* Ordered
    mechanism; Ping-pong mechanisms;
    Random mechanism)
    analysis of two-substrate reactions with
      non-Michaelian kinetics, 160–163
    cdk5/p25-catalyzed phosphorylation of
      tau, 158–160
    replots for two-substrate reactions,
      157–158
Kinetic partitioning of enzyme : substrate
  complex, 48–49
Kinetics
  of allosteric modulation, 202–208
  chemical, 5–7
  of enzyme : substrate complex, 46–50
  of inhibition, 126–129
  saturation, 6
Kinetics and Mechanism (Moore & Pearson),
  252

Kinetics-based probes, 219–247
  kinetic isotope effects on enzyme-catalyzed
    reactions, 239–247
  pH dependence of enzymatic reactions,
    220–228
  temperature dependence of enzymatic
    reactions, 229–234
  viscosity dependence of enzymatic
    reactions, 235–239
K_m, 9, 35–36
k_{obs}, 125–126
  dependence on pH, 224
  pH-independent values of, 230
k_{off}, 125
  independent determination of, 134–137
k_{on}, 134
Kuhne, Willy, 4
KXD. *See* αN-Boc-Lys-NH-CH$_2$-CH$_2$-NH-
  dansyl (KXD)

Lag phase, 36–37
Langmuir binding isotherm, 76
Laszlo, Ervin, 211n3
Lavoisier, Antoine, 2
LDDN (Laboratory for Drug Discovery in
  Neurodegeneration), 83, 84, 214
LDN-1487, 214, 215
LDN-27219, 132, 133
LDN-91946, inhibition of ubiquitin
  C-terminal hydrolase-L1 by, 83–84,
  100, 101
Leonard, K., 110
Levene, Phoebus A. T., 4
Linear Eyring plots, 231–232
Linear-pH presentation plots, 223
Lineweaver-Burk plots, 9
Lipase, 6
Lock-and-key model, 5–6, 263–264
Loew, Oscar, 7–8
Log-pH presentation plots, 223
Lowry, Oliver, 9
Lysozyme, X-ray structure of, 10

Macromolecular substrates, enzymes that
  catalyze reactions of, 111
Macroscopic constants, 9
Malachite green molybdate, 28
D/L-Mandelate esters, 6
D-Mandelic acid, 6

Measurement
  direct, 28
  of isotope effects, 242–243
  of product formation, 128
  of solvent isotope effects, 243–244
  of substrate depletion, 128
Mechanisms of action, 11–13. *See also*
    Inhibition; Kinetic mechanism
  emergence from data, 13–14, 105–108
  of inhibition, 12–13
    by substrate depletion, 112
  minimal, 21–22, 24–25
  rapid equilibrium, 22–24
  of substrate turnover, 11–12, 13
Mechanistic ambiguity, 82–83
MEK1, 111
Menten, Maud, 6, 21
MeOSuc, 272
MeoSuc-Ala-Ala-Pro-Val-CH$_2$Cl, 138–139
MeOSuc-Ala-Pro, 272
MeOSuc-Lys(Z)-Ala-Pro, 272
MeOSuc-Pro, 272
MeOSuc-Val-Pro-Val-CF$_3$, 134, 135
Merck, 104, 122
MES buffer, 33–34, 221
7-Methoxycoumarin-3-acetyl-Gly-Gly-
    Phe-Leu-Arg-Lys-Val-Gln-
    Lys(2,4-dinitrophenyl)-NH$_2$, 214
N-Methylsulfonyl-(D)Phe-Gly-Arg-pNA,
    213
Methyltransferases, 111
MgATP, 215–216
Michaelis, Leonor, 6, 21
Michaelis complexes, 11, 51
  in nonprocessive reactions, 66–70
  thermodynamics of formation of, 47
Michaelis-Menten equation, 23, 26, 38–39,
    96–97, 131
  replots and, 158
Michaelis-Menten kinetics, 20, 53
  deviations from, 53–65
    cooperativity, 58–62
    cooperativity or conformational
      transitions, 64–65
    negative cooperativity, 54, 64
    positive cooperativity, 54, 62–63
    substrate concentration-dependent
      conformational transitions, 63–64
    substrate inhibition, 54–58

in presence of inhibitor, 90
two-substrate reactions and, 160, 162
Michaelis-Menten plots, 38
Microscopic constants, 9
Microscopic rate constants, ordered
    mechanism with, 151
Mixed inhibition, 90, 91, 93, 99, 102, 103, 104
Molecular theory, 4
Monad, J., 199
Monomeric enzymes, substrate concentration-
    dependent conformational transitions
    and, 64
MRSA, 111
Multiple, sequentially formed
    enzyme:inhibitor species, 93–95

NADH, α-ketoglutarate-dependent oxidation
    of, 133
NADH-dependent reduction reactions, 28
Negative cooperativity, 54, 59, 63, 64, 160
Neuber, Carl, 3
Neurodegenerative diseases, tissue
    transglutaminase and, 84
Neurofibrillary tangles, 158–159
Nevirapine, 200
Nicotinamide, 212
p-Nitroacetanilide, hydrolysis of, 266–268
p-Nitroanilide, 28, 213, 225, 266, 272
p-Nitroaniline, 28, 135, 268
NMC. *See N,N*-Dimethylated casein (NMC);
    N-methyl casein (NMC)
N-methyl casein (NMC), TGase-catalyzed
    transamidation of, 283–286
NMR crystallographic studies, 65
N,N-Dimethylated casein (NMC), 132
Noncompetitive inhibition, 90, 92, 99
Noncompetitive inhibitors, 109
Nonlinear Eyring plots, 232–233
Nonlinear least squares analysis, 9, 55–56, 99
Non-Michaelian kinetics, two-substrate
    reactions with, 160–163
Nonnucleoside reverse transcriptase inhibitor,
    200
Nonprocessive reaction types, 65–70
Northrop, Dexter, 9
Nucleases, 111

One-step, irreversible, bimolecular reactions,
    253–254

One-step, irreversible, unimolecular
    reactions, 252–253
One-step, reversible reactions, 254–257
One-substrate, two-intermediate reactions,
    inhibition of, 108–110
One-substrate enzymatic reactions
    balancing steady state for, 276, 277
    kinetic constants for, 35
    kinetic mechanism of inhibition of, 89–114
        basic mechanisms and rate expressions,
            90–93
        depletion of substrate, 111–114
        drug discovery and, 89–90
        inhibitor binding site and, 93
        initial velocity date analysis, 96–108
        meaning of K_i for, 93–96
        one-substrate, two-intermediate
            reactions, 108–110
    meaning of K_i for, 93–96
    with two steady-state intermediates, 281,
        282–283
"On the Vital Activity of Enzymes"
    (Kastle), 5
Ordered mechanism, 142, 143, 144–145
    free energy diagrams for, 145, 165, 166
    rapid equilibrium, 150–151
    secondary plots for, 159, 165–167
    steady-state, 151–153, 168
    substrate analog inhibition of, 185–188
    two-substrate reaction, 173, 175–177
        connecting rate parameters to enzyme
            forms, 175
        quantitative analysis, 175–177
Organic chemistry, interplay with
    biochemistry, 2–4
Ostwald, Wilhelm, 7
Oxaloacetate decarboxylase, 10
Oxocarbenium-like transition state, 269

p25, 159
Parallel pathways, 109–110
Parkinson's disease, ubiquitin C-terminal
    hydrolase-L1 and, 83
Partial inhibitors, 75, 82
Pattern recognition, secondary plots and, 163
Pauling, Linus, 10, 264
Payen, Anselme, 3
PBPs. *See* Penicillin-binding proteins
    (PBPs)

PBPx-catalyzed hydrolysis of
    Bx-(D)Ala-^{S}Gly, 227–228
Penicillin-binding proteins (PBPs), 227
Penicillin-resistant *Streptococcus pneumonia*,
    111
Peptide-7-amido-4-methylcoumarins, 28
Peptide-based chloromethane ketones, 138
Peptide phosphoramidates, 270–271
Peptides, binding to Factor VIIa exosite, 213
Peptide trifluoromethyl ketones, 271–272
pH
    effect of temperature on buffer, 231
    optimal, 33
    in temperature-dependence studies of
        enzymatic reactions, 229–230
pH dependence of α-deuterium isotope effect,
    247
pH dependence of enzymatic reactions,
    220–228
    analysis of pH dependencies of steady-
        state kinetic parameters, 223–225
    choice of parameters to plotted in pH
        dependencies, 220
    examples, 225–228
    experimental design, 221–222
        choice of buffers, 221–222
        data collection, 222
        separating effects of pH on catalysis
            from those causing irreversible
            changes in enzyme structure, 221
    fine structure in, 227–228
    mechanistic inadequacy of plots of initial
        velocity *vs.* pH, 220
    pH dependencies reflect reactant state
        ionizations, 220–221
Phenyl esters, acylation and, 49
Phenylethyl chlorides, solvolysis of, 240
Phosphatases, 109, 111
Phosphates, 33, 65
Phosphorimidates, 270–271
Phosphorylation, cdk5/p25-catalyzed, 158–
    160, 161, 179–181
Phosphorylysis of inosine, 246
pH-rate profiles, 223–224
Ping-pong mechanisms, 142, 143, 153–156
    conceptual understanding of, 155–156
    derivation of rate equation for, 153–155
    free energy diagrams for, 156, 157, 167
    replots for, 167–168

secondary plot for, 159
substrate analog inhibition of, 189–190
two-substrate reaction, 173, 181–184
  balancing steady state for, 283–286
  connecting rate parameters to enzyme
    forms, 184
  rate equations and quantitative analysis,
    181–184
PIPES, 221
Planck constant, 43, 229
Plots
  primary, 96, 99, 158, 160, 161
  secondary (*See* Secondary plots)
PNPase. *See* Purine nucleoside phosphorylase
  (PNPase)
Polanyi, Michael, 264*n*1
Polymerases, 111
Polymeric substrates, 65–66
Positive cooperativity, 54, 59, 62–63, 105,
  160
Potency of inhibition, 75, 76
Primary deuterium isotope effects, 10, 243
Primary kinetic isotope effect, 240–241
Primary plots, 96, 99
  for two-substrate reactions, 158, 160, 161
Procepia, 123
Processive reaction types, 65–66, 67, 71
Product formation, measuring, 128
Progress curves, 38
  for irreversible inhibition, 137, 138
  for slow-binding inhibition, 128, 131–134
Prolyl *cis-trans* isomerase, inhibition by
  FK-506, 118–119
Proscar, 123
Proteases, 28, 65, 111
Protein dynamics, catalysis and, 11
Proton bridges, 243
Proton inventory, 244–245, 246
Prout, William, 3
Pseudo-equilibrium constant, 265
Pseudo-first order rate constants, 31, 45, 123,
  144, 254
Purine nucleoside phosphorylase (PNPase),
  246, 269
Purines, inhibition of xanthine oxidase by, 73

Quantitative study of enyzme-catalyzed
  reactions, 4–9
Quastel, J. H., 74

Random mechanism, 142–143, 146–147
  free energy diagrams for, 146, 163, 164
  rapid equilibrium, 148–150, 168
  secondary plots for, 159, 163–164
  substrate analog inhibition of, 188–189
  two-substrate reaction, 173, 177–181
    inhibition of kinase activity of cdk5/p25
      kinase, 179–181
    quantitative analysis, 177–179
Rapid equilibrium assumption, 25–26,
  143
  derivation of rate equation for single-
    substrate reaction, 21–24
  for ordered mechanism, 150–151
  for random mechanism, 148–150
Rapid equilibrium mechanisms, 22–24, 54,
  63, 105, 142, 143
Rapid equilibrium ordered mechanisms, 276,
  278–279
Rapid equilibrium-ordered reactions,
  balancing steady state for,
  276–279
Rapid equilibrium random mechanisms, 143,
  168, 180, 193, 279, 280
Rapid equilibrium random reactions,
  balancing steady state for, 279–282
Rate constants, 6, 9, 12, 26, 41–42, 44,
  45–46, 220, 225
Rate equations
  for ordered mechanism, 175
    of two-substrate reaction, 174
  for ping-pong mechanisms, 153–155,
    181–184
  for random mechanism, 178
  for sequential mechanisms, 147–153
  for single-substrate reaction
    assuming rapid equilibrium, 21–24
    assuming steady state, 24–26, 27
Rate expressions, 22–23, 25
  for competitive inhibition, 94
  for enzymatic reaction, 265
  for mixed inhibition, 91
  for nonprocessive reactions, 67
  for one-step, irreversible, bimolecular
    reactions, 254
  for ordered mechanisms, 151
  for random mechanisms, 177
  for single-substrate reaction with
    intermediate species, 108

Rate laws
  for inhibition of two-substrate enzymatic
    reactions, 169
  for one-substrate reaction, 171
  to predict extent of inhibition, 112–113
Rate parameters
  connecting to enzyme forms in ordered
    mechanism, 175
  connecting to enzyme forms in ping-pong
    mechanisms, 184
  connecting to enzyme forms in random
    mechanism, 177
  steady-state enzyme forms and, 170–173
Reactant state ionizations, pH dependencies
  reflect, 220–221
Reaction rate, temperature and, 42
Reactive centers, substrates with multiple,
  65–71
5α-Reductase, inhibition by finasteride,
  122–123
Relative velocity, dependence on solvent
  viscosity, 236
Relaxation time, 36
Replots, 96–100, 106, 108. *See also* Primary
  plots; Secondary plots
  for inhibition of SIRT1 by TFA-peptide,
    194, 195
  for ping-pong mechanisms, 167–168
  for two-substrate reactions, 158
Resveratrol, 212
Ribonucleotide reductase, modulation by
  deoxyribonucleotides, 215
Robertson, James, 141

SAR. *See under* Structure-activity
  relationship (SAR)
Saturation kinetics, 6
  thermodynamics of, 46–48
Schowen, Richard, 9
Schramm, V. L., 215–216, 270
Science (journal), 5
Secondary deuterium isotope effects, 243
Secondary kinetic isotope effect, 240,
  241–242
Secondary plots, 96, 98, 99, 100
  conceptual understanding of shape of,
    163–168
  for ordered mechanisms, 165–167
  pH-rate profiles, 223–224

  for ping-pong mechanisms, 167–168
  for random mechanisms, 163–164
  for two-substrate reactions, 158, 159, 160
Second-order rate constants, 20, 24, 45,
  60–61, 123, 138, 254
Second-order reaction, 254
Segel, Irwin, 141
Semi-continuous HPLC analysis, 29, 30
Sequential mechanisms, 143. *See also*
  Ordered mechanism; Random
  mechanism
  conceptual understanding of, 143–147
  derivation of rate equations for, 147–153
  in which inhibitor binds to
    enzyme:product complexes, 191–196
Serine, 109
Serine protease irreversible inhibitors, 138
Serine proteases, 10, 225, 234
  acylation of, 243
  mechanism for reactions catalyzed by,
    48–49
Ser/Thr kinase, 159
Shapes, of pH-rate profiles, 223–224
SIE. *See* Solvent isotope effects (SIE)
Sigmoidal kinetics, 162
Sigmoidal shape of pH-rate profiles, 223
Simplicity of inhibition, 75, 76
Simulations, 259–260
Single-substrate enzymatic reactions, kinetics
  of, 19–40
  assay development, 32–35
    assay buffer solution, 33–34
    choice of assay method, 34
    implementation of assay method, 34–35
    selection of substrate, 32–33
  data analysis, 37–40
  data collection, 35–37
  dependence of initial velocity on substrate
    concentration and requirement for
    E:S complex, 19–21
  derivation of rate equation
    with rapid equilibrium assumption,
      21–24
    using steady-state assumption, 24–26
  deviations from Michaelis-Menten kinetics,
    53–65
    cooperativity, 58–62
    cooperativity or conformational
      transitions, 64–65

negative cooperativity, 63
positive cooperativity, 62–63
substrate concentration-dependent
conformational transitions, 63–64
substrate inhibition, 54–58
enzyme assay methods, 26–32
continuous assays, 27–28
coupled enzyme assays, 29–32
discontinuous assays, 28–29
experimental design, 35
free energy diagrams, 44–46
kinetic consequences of E : S complex,
46–50
kinetics of enzymatic action on substrates
with multiple reactive centers, 65–71
reactions with more than one intermediary
complex, 51–53
transition state theory, 41–44
SIRT1
allosteric activation of, 212–213
inhibition of by substrate analogs, 193–196
SIRT1 activating compounds (STACs), 212–213
Sirtuin-catalyzed deacetylation, 193, 194
Sirtuins, 193, 212
The Skeptical Chymist (Boyle), 2
Slope factor, 76
not equal to one, 78–81
Slow-binding inhibition, 115, 122–137
experimental approaches, 131–137
independent determination of k_{off}, 134–137
progress curves, 131–134
titration curves and time-dependent $K_{i,app}$
values, 131
inhibition of 5α-reductase by finasteride,
122–123
kinetics of, 126–129
mechanisms of, 129–131
progress curve for, 128
time course for, 122
time-dependent, 123–126
S_N1 reaction, 269
S_N2 reaction, 269
$[S]_o$, 35
Solvent, isotopic substitution of, 239
Solvent deuterium isotope effects, 243, 247
Solvent isotope effects (SIE), 10, 243. *See
also* Kinetic isotopic effects on
enzyme-catalyzed reactions
measurement of, 243–244

Solvent viscosity. *See* Viscosity dependence
of enzymatic reactions
Spectrophotometric assay, 127
STACs. *See* SIRT1 activating compounds
(STACs)
Statement of rate dependence, 23
State-to-state transitions, 44
Steady state, balanced. *See* Balanced steady
state
Steady-state approximation, 259–260
Steady-state assumption
derivation of rate equations for single-
substrate reaction, 24–26, 27
for ordered mechanism, 151–153, 168
Steady-state derivation of rate law for
competitive inhibition, 97
Steady-state enzyme forms, rate parameters
and, 170–173
Steady-state kinetic parameters, 14, 49–50
estimating, 38–39
initial velocity and, 49
pH dependencies on, 223–225
Steady-state ordered mechanisms, 151–153,
168
Steady-state velocity, initial velocity and, 35
Steric activation, 201
Steric inhibition, 201
Streptococcus pneumoniae, 227
Stromelysin-catalyzed hydrolysis of
substance P, 29, 30
Stromolysin, 28
Structure-activity relationship (SAR), for
allosteric modulators, 212
Structure-activity relationship (SAR)
programs, 170, 196–197
Structure-activity relationship (SAR) studies
of enzyme inhibition, 73
Structure of enzymes, 10–11
Substance P, stromelysin-catalyzed hydrolysis
of, 28, 29, 30
Substrate
dependence of allosteric modulation on
structural features of, 212–216
inhibition by depletion of, 111–114
isotopic substitution of, 239
measuring depletion of, 128
with multiple reactive centers, 65–71
selection of, 32–33
Substrate A, inhibition by analog of, 185–187

Substrate analogs, 185–191
  inhibition mechanisms for, 192
  inhibition of SIRT1 by, 193–196
  ordered mechanisms and, 185–188
  ping-pong mechanisms and, 189–190
  random mechanisms and, 188–189
  for sequential mechanisms in which
    inhibitor bind to enzyme: product
    complexes, 192
Substrate B, inhibition by analog of,
    187–188
Substrate binding, 24
Substrate complexes, 6
Substrate concentration-dependent
    conformational transitions, 63–64, 65
Substrate concentrations
  dependence of initial velocity on, 19–21,
    203, 207
  for high-throughput screens, 275–286
    balancing steady state for
      nonequilibrium reactions involving
      second steady-state intermediate, 281,
      282–283
    balancing steady state for one-substrate
      reactions, 276, 277, 281
    balancing steady state for two-substrate,
      ping-pong reactions, 283–286
    balancing steady state for two-substrate,
      rapid equilibrium ordered reactions,
      276–279
    balancing steady state for two-substrate,
      rapid equilibrium random reactions,
      279–282
  in nonprocessive reactions, 67–68
  in sequential mechanisms, 144
Substrate inhibition, 39, 54–58, 160
  mechanism for, 55
Substrate transformation, 24
Substrate turnover
  K_i values and nonequilibrium mechanisms
    of, 95–96
  mechanism of, 11–12, 13
Suc-Ala-Ala-Pro-Phe-pNA, 225, 234
Suc-Ala-Phe-pNA, 225–226, 232, 234
Succinic dehydrogenase, inhibition of, 74
Suc-Phe-pNA, 225, 234
Sucrose, hydrolysis of, 5, 20–21, 251–252
Sulfonamide antibiotics, 8

Tacrine, 110
tau, cdk5/p5-catalyzed phosphorylation of,
    158–160, 161, 179–181
T-cell-related disease, 270
Temperature dependence of enzymatic
    reactions, 42, 229–234
  experimental design and, 230–231
  interpretation of activation parameters,
    233–234
    acylation of α-chymotrypsin, 234
    pH and, 229–230
    steady-state kinetic parameters, 231–233
      linear Eyring plots, 231–232
      nonlinear Eyring plots, 232–233
    transition state rate theory, 229
Testosterone, 122
Tetrahydroaminoacridine, 110
TFA-peptide. *See* Trifluoracetyl-Lys-
    containing peptide (TFA-peptide)
TGase. *See* Tissue transglutaminase (TGase)
The Theory of Rate Processes (Eyring et al.),
    42
Thermodynamics of saturation kinetics,
    46–48
Thermolysin, inhibition of by
    phosphorimidates, 270–271
Thiazole inhibitors, of monomeric Eg51-367,
    103–105, 106
Thrombin, 213
Thurlow, S., 8, 73
Tight-binding inhibition, 115, 116–121
  different $[E]_o$ for different purposes, 121
  estimation of K_i values for, 116–118
  inhibition of prolyl *cis-trans* isomerase
    activity of FKBP by FK-506, 118–119
  tight-binding inhibitor as active-site titrant,
    119–121
Tight-binding inhibitors, 79
Time dependence, of enzyme inhibition, 122,
    123–126
Time-dependent inhibitors, 79–80
Tissue transglutaminase (TGase)
  balancing steady state for, 283–286
  inhibition by GTP, 84–86
  inhibition by LDN-27219, 133
  kinetic mechanism of, 162–163
Tissue transglutaminase (TGase) assay,
    283–286

Titration curves, in slow-binding inhibition, 131–134
Titration curve simulations, 207
Transglutaminases, 65
Transition state, 42n1
   shape of proton inventory and, 245
   virtual, 245–246
Transition state analog inhibition, 268–272
Transition state analog inhibitors, 10
Transition state theory, 10, 41–44, 229, 264
Transmission coefficient, 229
Traube, Moritz, 3–4
Trifluoracetyl-Lys-containing peptide (TFA-peptide), 194–195
Trifluoromethyl ketones, inhibition of human leukocyte elastase by, 271–272
Triphosphate, 214
Tris buffer, 33, 231
Tris ionization, 231
Two-step, irreversible reactions, 257–259
Two-step reaction, with reversible first step, 259–261
Two-substrate enzymatic reactions, balancing steady state for
   ping-pong mechanisms, 283–286
   rapid equilibrium ordered mechanism, 276–279
Two-substrate enzymatic reactions, kinetic mechanism of inhibition of, 169–197
   conceptual understanding of, 170–173
   drug discovery and, 170
   elucidation of, 173–175
   ordered mechanisms, 175–177, 185–188
   ping-pong mechanisms, 181–184, 189–190
   random mechanisms, 177–181, 188–189
   SAR programs, 196–197
   sequential reactions in which inhibitor binds to enzyme:product complexes, 191–196
   steady-state enzyme forms and rate parameters, 170–173
   substrate analog inhibition, 185–190
Two-substrate enzymatic reactions, kinetics of, 141–168
   basic mechanisms, 142–143
   conceptual understanding of sequential mechanisms, 143–147

derivation of rate equations for sequential mechanisms, 147–153
determining kinetic mechanism for, 156–163
   analysis of two-substrate reactions with non-Michaelian kinetics, 160–163
   cdk5/p25-catalyzed phosphorylation of tau, 158–160
   method of replots for two-substrate reactions, 158
drug discovery and, 141–142
ordered mechanism, 142, 143, 144–145
   rapid equilibrium assumption, 150–151
   secondary plots for, 165–167
   steady-state assumption, 151–153
ping-pong mechanisms, 142, 143, 153–156
   conceptual understanding of, 155–156
   derivation of rate equation for, 153–155
   secondary plots for, 167–168
random mechanism, 142–143, 146–147
   rapid equilibrium assumption, 148–150
   secondary plots for, 163–164
rapid equilibrium random vs. steady-state ordered, 168
shapes of secondary plots, 163–168

Ub-AMC, 83
Ubiquitin C-terminal aldehyde, titration of UCH-L1 by, 120–121
Ubiquitin C-terminal hydrolase-L1 (UCH-L1)
   inhibition by LDN-91946, 83–84, 100, 101
   titration by ubiquitin C-terminal aldehyde, 120–121
UCH-L1. See Ubiquitin C-terminal hydrolase-L1 (UCH-L1)
Uncompetitive inhibition, 90, 92, 93, 99, 100, 101, 102, 103
Uncompetitive inhibitors, 109
Urea, synthesis of, 2–3

Vancomycin, 111–112
v_{bkg}, 76–77, 78
   greater than zero, 81–82
$v_{control}$, 76
   not equal to velocity determined at zero inhibitor concentration, 77–78
Vertical free energy of binding, 266, 268
Virtual transition state, 9, 245–246
Viscogen concentration, 238, 239

Viscosity dependence of enzymatic reactions, 235–239
  data interpretation, 238–239
  experimental design, 238
  interpretational framework, 235–238
Visual inspection, 101, 103–104
$(V/K)_x$, 172, 173
$(V_{max}/K_m)_{obs}$, 14
$(V_{max})_{obs}$, 14
von Bertalanffy, Ludwig, 211n3
V_x, 172, 173

Well position within the plate, 77
West Nile protease, 57, 58
Wilhelmy, Ludwig, 5, 20, 251–252
Wöhler, Friedrich, 2–3
Wolfenden, Richard, 10, 264, 268
Woods, D.D., 8
Wooldridge, W. R., 74
Woolf, Barnet, 7

Xanthine oxidase, 8, 73
x-axis, of free energy diagrams, 44–45
XES, 208, 212
(XES)′, 200, 208
  mode and extent of allosteric modulation and, 211–212
$(X:E:S)′$, mode and extent of allosteric modulation and, 211–212
X-ray crystallographic studies, 65
X-ray structures of enzymes, 10

y-axis, of free energy diagrams, 45

Z-Gln-Gly, 163
Z-Glu(γ-AMC)-Gly, TGase-catalyzed hydrolysis of, 85
Z-Gly-Leu-R, 270
Zwitterionic buffer salts, 33, 221
Zymogen, 213

Printed in the United States
By Bookmasters